Microbial Biotechnology

Microbial Biotechnology

Editor

Ratul Saikia

(Scientist)

Associate Editors

Rajib L Bezbaruah

(Scientist)

&

Tarun Ch. Bora

(Scientist)

Biotechnology Division
North East Institute of Science and Technology (CSIR)
Formerly Regional Research Laboratory
Jorhat - 785 006, Assam

New India Publishing Agency
Pitam Pura, New Delhi- 110 088

Published by
Sumit Pal Jain *for*
New India Publishing Agency
101, Vikas Surya Plaza, CU Block, L.S.C. Mkt.,
Pitam Pura, New Delhi- 110 088, (India)
Phone: 011-27341717, Fax: 011-27341616
E-mail: newindiapublishingagency@gmail.com
Web: www.bookfactoryindia.com

ISBN: 978-89422-80-6

Typeset at: Laxmi Art Creations, Delhi and Printed at: Jai Bharat Printing Press, Delhi

PREFACE

Biotechnology is one of the major branches of biological science of the twenty-first century. Microbial Biotechnology is wide-ranging, multi-disciplinary activities which include recombinant DNA techniques, cloning and the application of microbes to the production of goods from bread to antibiotics. Microbial research and industry is developing rapidly all over the world. Varying in application of concept, practice, scale, style and substance, microbial biotechnology is amongst the latest globalizing frontiers of the corporate world. This branch is increasingly being regarded as a core subject in colleges and universities also. In this book various topics on microbial biotechnology are uniquely combined to provide a complete overview of the subject. The application of bioinformatics in the study of microbial genomics is also included. The book is likely to be useful for the postgraduate student and researchers in developing an insight into the various aspects of the microbial biotechnology as a whole.

August 21, 2007 **Editors**
Jorhat

CONTENTS

CHAPTER **1**

Bioinformatics Tool for PCR Primer Designing in Microbial Genomics

Dinesh Kumar[1*] and Dilip K Arora[2]

[1]Department of Animal Sciences & Centre for Integrated Animal Genomics, Iowa State University, Ames, Iowa, USA 50011-3150
[2]National Bureau of Agriculturally Important Microorganisms, Mau 275 101, India

ABSTRACT

Bioinformatics is a critical tool in designing the molecular biology experiments. The xeroxing of specific part of genome needs correct primer designing. Primer designing has central role in molecular biology right from cloning, sequencing, gene expression profiling by real time PCR, genotyping by RFLP, SNP analysis by real time PCR and auto sequencer. In this chapter we are describing how to proceed for experimental designing right from data mining from GENEBANK to designing of primers for various common molecular genetic analysis work. The write up also describes *in silico* evaluation of primers, various off line commercial and non-commercial software, on line free software and trouble shootings in primer designing, The different strategy described in context of primer designing would be of immense use of researchers.

Key words: bioinformatics tool, PCR primer, microbial genomics, universal primers, Q-PCR assay.

1. INTRODUCTION

One of the critical skillful factors in successful PCR is designing of the two oligonucleotide primers. The increasing use of information from the internet and the sequences held in gene databases are practical starting points when designing primers and reaction conditions for the PCR. A number of software packages such as Oligo, Primer etc. have allowed the process of primer design to be less troublesome. It is also possible to include more than one set of primers in a PCR. It is essential that care is taken in the design of

*Email: dinesh@iastate.edu, dineshkumarbhu@gmail.com

primers for PCR. Several parameters including the length of the primer, %GC content and the 3' sequence need to be optimized for successful PCR. Certain of these parameters can be easily by hand optimized while others are best done with commercial softwares. A poorly designed primer can result in little or no product due to non-specific amplification and/or primer-dimer formation, which can become competitive enough to suppress product formation.

2. GOOD PRIMER'S CHARACTERISTICS

A good primer has following characteristics-i. a melting temperature (Tm) in the range of 52^0 C to 65^0 C; ii. absence of dimerization capability; iii. absence of significant hairpin formation (>3 bp); iv. lack of secondary priming sites; v. low specific binding at the 3' end (i.e. lower GC content to avoid mispriming).

3. ON LINE FREE PRIMER DESIGNING

A number of web-based resources are available for PCR and primer designing. Though most are freely available, they are of variable quality and not well maintained (reviewed by Abd-Elsalam, 2003). This often results in missing links and so sites that may have been useful previously may not be functional at a later date. There are a number of criteria that need to be established in the design of primers and a number of these are listed (Tables 1 & 2).

TABLE 1. Online primer design sites

Tool Name	Description	Website
CODEHOP	Consensus degenerate hybrid Oligonucleotide primers; degenerate PCR primer design; will accept unaligned sequences	http://blocks.fhcrc.org/codehop.html
Gene Fisher	Interactive primer design tool for standard or degenerate primers; will accept unaligned sequences	http://bibiserv.techfak.uni-bielefeld.de/genefisher/
DoPrimer	Easily design primers for PCR and DNA sequencing	http://doprimer.interactiva.de/

Contd...

Table 1. Contd...

Primer3	Comprehensive PCR primer and hybridization probe design tool; many options but easy to accept defaults at first	http://www-genome.wi.mit.edu/cgi-bin/primer/primer3_www.cgi http://www.basic.nwu.edu/biotools/Primer3.html http://www.justbio.com/primer/index.php
Primer Selection	Select PCR primers from nucleotide sequence	http://alces.med.umn.edu/rawprimer.html
Web Primer	Allow alternative design of primers for either PCR & sequencing	http://genome-www2.stanford.edu/cgi-bin/SGD/web-primer
PCR Designer	For restriction analysis of sequence mutations	http://cedar.genetics.soton.ac.uk/public_html/primer.html
Primo Pro 3.4	Reduces PCR noise by lowering the probability of random primering	http://www.changbioscience.com/primo/primo.html
Primo Degenerate3.4	Primo degenerate 3.4 designs PCR primers based on a single peptide sequence or multiple alignments of proteins or nucleotides	http://www.changbioscience.com/primo/primod.html
PCR Primer Design	An application that designs primers for PCR or sequencing purposes	http://pga.mgh.harvard.edu/servlet/org.mgh.proteome.Primer
The Primer Generator	The program analyzes the original nucleotide sequence and desired amino acid sequence and designs a primer that either has a new restriction enzyme site or is missing an old one	http://www.med.jhu.edu/medcenter/primer/primer.cgi
EPRIMER3	Picks PCR primers and hybridization oligos (EMBOSS)	http://bioweb.pasteur.fr/seqanal/interfaces/eprimer3.html
PRIMO	Prediction of forward and reverse oligonucleotide primers	http://bioweb.pasteur.fr/seqanal/interfaces/primo.html3 http://atlas.swmed.edu/primo/primo_form.html

Contd...

Table 1. Contd...

PrimerQuest	A primer design tool.	http://www.idtdna.com/biotools/primer_quest/primer_quest.asp
MethPrimer	Design primers for methylation PCRs	http://itsa.ucsf.edu/~urolab/methprimer/index1.html
Raw primer	A tool for selection of PCR primers	http://alces.med.umn.edu/rawprimer.html
MEDUSA	A tool for automatic selection and visual assessment of PCR primer pairs	http://www.cgr.ki.se/cgr/MEDUSA/
The Primer Prim'er Project	Software suite that completely automates the PCR primer design process	http://www-nmr.cabm.rutgers.edu/bioinformatics/Primer_Primer_Project/Primer.html
Oligonucleotides for the PCR	Seek oligonucleotides on both sides of an area	http://www.citi2.fr/bio2/Oligo2lib.html
GAP Genome	- Wide Automated Primer finder servers	http://promoter.ics.uci.edu/Primers/

TABLE 2. PCR oligonucleotide resources

Tool name	Description	Website
Oligonucléotides pour la PCR	Calculation of melting point of a oligonucleotide	http://www.citi2.fr/bio2/OligoTM.html
Oligonucleotide properties calculator	Prediction of melting temperature	http://www.basic.nwu.edu/biotools/oligocalc.html http://www.microbiology.adelaide.edu.au/learn/oligcalc.htm
Oligonucleotide analyzer	Generates Tm, free energy, molecular weight and hairpin and dimer formation structures	http://www.rnature.com/oligonucleotide.html
Oligo Tm Determination	Prediction of Tm	http://alces.med.umn.edu/rawtm.html
Poland	Prediction of melting temperatures of primers	http://www.biophys.uni-duesseldorf.de/local/POLAND/poland.html
PROLIGO	Oligos parameter calculation	http://www.gensetoligos.com/Calculation/calculation.html

3.1 Software for Primer Design

The use of software in biological applications has given a new dimension to the field of bioinformatics. Many different programs for the designing of primers are now available. Freeware software is available on the internet and many universities have established servers where a user can log on and carry out free analyses of proteins and nucleic acid sequences (Singh and Kumar, 2001). There are a number of simple stand-alone programs as well as complex integrated networked versions of the commercial software available. These software packages may be for complete DNA and protein analysis, secondary structure predictions, primer design, molecular modeling, development of cloning strategies, plasmid drawing or restriction enzyme analyses. Companies engaged in biosoftware development include: Alkami Biosystems, Molecular Biology Insights, PREMIER Biosoft International, IntelliGenetics Inc., Hitachi Inc., DNA Star, Advanced American Biotechnology and Imaging. Some scientists have also developed algorithms and computer programs for various purposes of primer design (Rychlik and Rhoades, 1989; Lowe *et al.*, 1990; Lucas *et al.*, 1991; O'Hara and Venezia, 1991; Tamura *et al.*, 1991; Makarova *et al.*, 1992; Osborne, 1992; Li *et al.*, 1997; Plasterer, 1997; Sze *et al.*, 1998; Gorelenkov *et al.*, 2001). Many programs aiding in the design of primers exist (Table 3).

TABLE 3. PCR primers design software for personal computer

Software name	Description	Website
Primer Select	Analyzes a template DNA sequence and chooses primer pairs for PCR and primers for DNA sequencing.	www.dnastar.com
DNASIS Max	DNASIS Max is a fully integrated program that includes a wide range of standard sequence analysis features.	http://www.medprobe.com/no/dnasis.html
Primer Premier 5	Primer design for Windows and Power Macintosh.	http://www.premierbiosoft.com/primerdesign/primerdesign.html
Primer Premier	Comprehensive primer design for Windows and Power Macintosh	http://www.premierbiosoft.com/

Contd...

Table 3. Contd...

NetPrimer	Comprehensive analysis of individual primers and primer pairs	http://www.premierbiosoft.com/NetPrimer.html
Array Designer 2	For fast, effective design of specific oligos or PCR primerpairs for microarrays	http://www.premierbiosoft.com/dnamicroarray/dnamicroarray.html
Beacon Designer 3.0	Design molecular beacons and TaqMan probes for robustamplification and fluorescence in real time PCR	http://www.premierbiosoft.com/molecular_beacons/taqman_molecular_beacons.html
GenomePRIDE 1.0	Primer design for DNA-arrays/chips.	http://pride.molgen.mpg.de/genomepride.html
Fast PCR	Software for Microsoft Windows has specific, ready-to-use templates for many PCR and sequencing applications: standard and long PCR, inverse PCR, degenerate PCR directly on amino acid sequence, multiplex PCR	http://www.biocenter.helsinki.fi/bi/bare-1_html/manual.htm
OLIGO 6	Primer Analysis Software for Mac and Windows	. http://www.oligo.net/
Primer Designer 4	Will find optimal primers in target regions of DNA or protein molecules, amplify features in a molecule, or create products of a specified length	http://www.scied.com/ses_pd5.htm
GPRIME	Software for primer design.	http://life.anu.edu.au/molecular/software/gprime.htm
Sarani Gold	Genome Oligo Designer is software for automatic large-scale design of optimal oligonucleotide probes for microarray experiments	http://mail.strandgenomics.com/products/sarani/
PCR Help	Primer and template design and analysis	http://www.techne.com/CatMol/pcrhelp.htm

Contd...

Table 3. Contd...

Genorama chip Design Software	Genorama Chip Design software is complete set of programs required for genotyping chip design. The programs can also be bought separately	http://www.asperbio.com/Chip_desin_soft.htm
Primer Designer	The Primer Designer features a powerful, yet extremely simple, real-time interface to allow the rapid identification of theoretical ideal primers for your PCR reactions	http://genamics.com/expression/primer.htm
Primer Premier	Automatic design tools for PCR, sequencing or hybridization probes, degenerate primer design, Nested/Multiplex primer design, restriction enzyme analysis and more.	http://www.biotechniques.com/freesamples/itembtn21.html
PrimerDesign	DOS-program to choose primer for PCR or oligonucleotide probes	http://www.chemie.unimarburg.de/%7Ebecker/pdhome.html

3.2 Common Recommendations for Primer Designing

DNA template and oligonucleotide primers must be considered in greater detail (Linz *et al.*, 1990) and efficacy and sensitivity of PCR mainly depend on the efficiency of primers (He *et al.*, 1994). The ability for an oligonucleotide to serve as a primer for PCR is dependent on several factors including: a) the kinetics of association and dissociation of primer-template duplexes at the annealing and extension temperatures; b) duplex stability of mismatched nucleotides and their location; and c) the efficiency with which the polymerase can recognize and extend a mismatched duplex. The primers which are unique for the target sequence to be amplified should fulfill certain criteria such as primer length, GC%, annealing and melting temperature, 5' end stability, 3'end specificity etc (Dieffenbach *et al.*, 1995). Most of the reviews on PCR optimization (Erlich *et al.*, 1991; Dieffenbach *et al.*, 1995; Roux, 1995) consider different parameters of PCR but generally do not discuss basic concepts of PCR primer design. May be the most critical parameter for successful PCR is the design of primers. All things being equal, a poorly designed primer can result in a PCR reaction that will not work. The primer sequence determines several things such as the length of the product, its melting temperature and ultimately the yield. A badly designed primer can result in little or no product due to non-specific amplification and/or primer-dimer formation, which can become competitive enough to suppress product formation. This application note is provided to give rules that should be taken into account when designing primers for PCR. More comprehensive coverage of this subject can be found elsewhere (Dieffenbach *et al.*, 1995).

The sequences of the primers use for PCR amplification can have a major effect on the specificity and sensitivity of the reaction. When choosing two PCR amplification primers, the following guidelines should be considered:

i. Primer length

Since both specificity and the temperature and time of annealing are at least partly dependent on primer length, this parameter is critical for successful PCR (Wu *et al.*, 1991). For broad-spectrum studies, primers of typically 18-30 nucleotides in length are the best. Primers should be at least 18 nucleotides in length to minimize the chances of encountering problems with a secondary hybridization site on the vector or insert. Primers with long runs of a single base should generally be avoided. It is especially important to avoid 4 or more G's or C's in a row.

ii. Melting Temperature (Tm)

The optimal melting temperatures for primers in the range 52-58°C, generally, primer give better result with lower melting temperatures. Primers with melting temperatures above 65°C should also be avoided because of potential for secondary annealing. It is then advisable to do the sequencing reaction with annealing and extension at 60°C. A good working approximation of this value (generally valid for oligos in the 18–30 base range) can be calculated using the formula of Wallace *et al.* (1979), Tm = 2(A+T) + 4(G+C). Using improved nearest-neighbor thermodynamic values given by SantaLucia *et al.* (1996), an estimate of melting temperature can be obtained for oligonucleotide analysis.

iii. GC Content (Tm and Ta are interrelated)

GC% is an important characteristic of DNA and provides information about the strength of annealing. Primers should have GC content between 45 and 60 percent (Dieffenbach *et al.*, 1995). For primers with a G/C content of less than 50%, may be necessary to extend the primer sequence beyond 18 bases to keep the melting temperature above the recommended lower limit of 50°C. GC content, melting temperature and annealing temperature are strictly dependent on one another (Rychlik *et al.*, 1990).

iv. 3′-End Sequence

It is well established that the 3' terminal position in PCR primers is essential for the control of mis-priming (Kwok *et al.*, 1990). Primers should be "stickier" on their 5' ends than on their 3' ends. A "sticky" 3' end as indicated by a high G/C content could potentially anneal at multiple sites on the template DNA. A "G" or "C" is desirable at the 3' end but the first part of this rule should apply. This GC clamp reduces spurious secondary bands (Sheffield *et al.*, 1989).

v. Dimers and false priming cause misleading results

Primers should not contain complementary (palindromes) within themselves; that is, they should not form hairpins. If this state exists, a primer will fold back on itself and result in an unproductive priming event that decreases the overall signal obtained (Breslauer *et al.*, 1986). Hairpins that form below 50°C generally are not such a problem. Primers should not contain sequences of nucleotides that would allow one primer molecule to anneal to itself or to the other primer used in PCR reactions (primer dimer formation).

vi. Specificity

As mentioned above, primer specificity is at least partly dependent on primer length. It is evident that there are many more unique 24 base oligos, however, primers must be chosen so that they have a unique sequence within the template DNA that is to be amplified. A primer designed with a highly repetitive sequence will result in a smear when amplifying genomic DNA. However, the same primer may give a single band if a single clone from a genomic library is amplified.

vii. Degenerate primers

Degeneracy in primer sequence should also be taken into consideration. Degenerate primers based on the amino acid sequence of conserved regions were also used to search for members of a gene family (Wilks *et al.*, 1989). Computer programs have also been developed specifically for degenerate primer design (Chen and Zhu, 1997).

viii. Complementary primer sequences

Primers need to be designed with absolutely no intra-primer homology beyond 3 base pairs. If a primer has such a region of self-homology, "snap back" can occur. Another related danger is inter-primer homology- partial homology in the middle regions of two primers can interfere with hybridization.

If the homology should occur at the 3' end of either primer, primer dimer formation will occur.

ix. Other recommendations

The concentration of primer in amplification reaction should be between 0.1 and 0.5 μM. If possible, a computer search should be conducted against the vector and insert DNA sequences to verify that the primer and especially the 8-10 bases of its 3' end are unique. Inosine should not be included in sequencing primers. They either do not work or give poor cycle sequencing results. The design of PCR and DNA sequencing primers follows very similar guidelines. Even though primer characteristics can be visually inspected for the presence of the elements listed above, a number of computer programs that have been developed and several guidelines are available for primer selection.

x. Some common default thermodynamic values/acceptable limits for primers.

1. Hairpin maximum dG 3′ end = 2 -kcal/mol
2. Hairpin maximum dG 3′ internal = 3 -kcal/mol
3. 3′ end maximum stable dG = 10 -kcal/mol
4. Self dimer max dG 3′ end = 5 -kcal/mol
5. Self dimer max dG 3′ internal = 6 -kcal/mol
6. Cross dimer maximum dg 3′ end = 5 -kcal/mol
7. Cross dimer maximum dg 3′ end = 5 -kcal/mol
8. Multiplexing maxium dG(kcal/mol) = 6 -kcal /mol

3.3 Trouble Shooting in Primer Designing

i. Try relaxing various parameters, including the self-complementarity parameters and max and min oligo melting temperatures.

ii. For very A-T-rich regions we might have to increase maximum primer size or decrease minimum melting temperature. It is usually unwise to reduce the minimum primer size if template is complex (e.g. a mammalian genome), since small primers are more likely to be non-specific. Make sure that there are adequate stretches of non-Ns in the regions in which we wish to pick primers. If necessary we can also allow an N in our primer and use an oligo mixture containing all four bases at that position.

iii. If there is presence of secondary structure in the template DNA, in such cases substitution of 7-deaza-2′-deoxyGTP for dGTP has been found to be very useful.

iv. The secondary structures of primers can be checked on line by programme mfold at http://scitools.idtdna.com/mfold/mfold.asp and also by off line (Example Beacon Designer 4). The advantage of this programme is the secondary structure, hairpin, cross dimmers, repeat and runs etc can be checked at different conditions of PCR thus it is very helpful in picking of right pair of primers.

v. Use of PCR adjuvant(s) to get to stabilize the secondary structures, magnesium requirements etc.

vi. Use hot start PCR.

4. IMPORTANT STRATEGIES IN PRIMER DESIGNING

Some important strategies should be taken for designing a primer-

4.1 Designing of Multiple Primer for Multiplex PCR

This enables primer pairs to be added in the same tube to do the PCR, thus good for amplifying multiple sites. The multiplex PCR reduces the cost drastically and limitation of inadequate amount of DNA sample. The limitations of this strategy are designing difficulty. The skills in such design are- melting temperatures should be similar; there should not be any cross dimmer with in permissible limit of delta G. The auto-sequencer and fluorescent dyes has made this strategy more advantageous and popular.

4.2 Designing of Universal Primers

Primers can be designed to amplify only one product in most of the PCR reaction; however, primers can also be designed to amplify multiple products. Such primers are called "universal primers". For example, designing of primers to amplify all HPV genes. The strategies of such designing are - i) Align groups of sequences which we want to amplify; ii) Find the most conservative regions at 5′ end and at 3′ end; iii) Design forward primer at the 5′ conservative region; iv) Design reverse primer at the 3′ conservative regions; v) Matching forward and reverse primers to find the best pair; vi) Ensure uniqueness in all template sequences; vii) Ensure uniqueness in possible contaminant sources.

4.3 Designing of Semi-Universal Primers

Primers can be designed to amplify only a subset of template sequences from a large group of similar sequences. For example, designing a primer to amplify HPV type 1 and type 6 genes, but not other types. The strategy of such designing are- i) Align all types of HPV genes; ii) Identify a subset of genes that are more similar to each other than to other subsets. In this case, type 1 and type 6; iii) Find the 5′ and 3′ regions that are conserved between type 1 and type 6, but are variable in other types; iv) Design forward primers from the 5′ region and reverse primers from the 3′ region; v) Matching forward and reverse primers to find the best pair; vi) Ensure uniqueness in all template sequences; vii) Ensure uniqueness in possible contaminant sources.

4.4 Design a Primer When No Template Sequence Available

If we do not have DNA sequence but we have protein from a given species we can still amplify the gene coding of that protein. Here we have to do protein sequencing to get the amino acid sequence, which can be used in mRNA codon usage table to get the potential theoretical DNA sequence coding that particular protein in question. In this process we find certain amino acids which are coded by more than one codon (example-valine coded by 6 different codons, called degeneracy or ambiguity of codons). Still the primer can be designed by avoiding such regions of degeneracy or ambiguity in mRNA codon. The designed primer on guessed mRNA template can be used to amplify the DNA and by sequencing PCR product original sequence of DNA can be obtained in cases where hitherto we do not had DNA sequence to design primers. Subsequently based on the obtained DNA sequence, walking primer and sequencing of subsequent PCR product and their overlapping liner alignment can be used to further sequence the adjoining region of the genes.

4.5 Design a Primer for RFLP

For designing a primer for RFLP, one should first evaluate the template DNA sequence by available RE cutter or taxon specific cutter programme like CLEAVER (Jarman, 2006) to delineate the RFLP designing region. RFLP primer designing may be done by identifying target sequences (recognition site of restriction enzyme) and indicating them as target and primers to be flanked over it. The *in silico* PCR product evaluation/simulation or calculation should be used to optimized the undigested product length, fragments after restriction digestion so that they can be visualized optimally in ordinary gels reducing the cost of RFLP screening.

4.6 Design a Primer for Real Time Q-PCR Assay

For relative gene expression studies one can design a primer for both house keeping as well as target genes by tagging two colours to two different set of primers and the multiplex PCR can be done in single PCR reaction using cDNA. The critical factors are size of amplicons which must be <140 and usually 100bp gives more accurate results.

4.7 Design a Primer for TaqMan Assay

Real-Time PCR offers rapid, simultaneous amplification and sequence specific detection of 2-4 target genes in a single reaction tube. A real-time PCR probe strategy is called the TaqMan assay capitalizes on the 5′-exonuclease activity of Taq polymerase to cleave a labeled hybridization probe during the extension phase of PCR. In a fluorescent TaqMan assay, the probe is labeled at the 5′ end with a fluorescent reporter molecule, usually a tetramethylrhodamine derivative, which acts as a quencher for the reporter. When the two fluorophores are fixed at opposite ends of the 20-30 nucleotides probe and reporter fluorophore is excited by an outside light source, the normal fluorescence of the reporter is absorbed by the nearby quencher, and no reporter fluorescence is detected. Taq polymerase digests the bound probe during extension from one of the primers, freeing the reporter from the quencher, and then reporter fluorescence can be detected and measured. Thus, TaqMan assay demands design of three nucleic acid oligomers, two PCR primers and an internal oligonucleotide probe located.

4.8 Design a Primer for SNP Genotyping (SNaPshot) by Auto Sequencer

SNapShot is generally used where there are no PCR-RFLP tests are possible in the given sequence. In this method the primers are designed based on the DNA sequence just prior to SNPs and different length of primers is designed by adding poly A tails to screen more than one SNP in one shot. The Snap Shot will add one corresponding base to the primer depending upon the SNP base. The Heterozygous individuals will have two peaks of two different bases. The 5′ tailing of primers with different length allows even 10 SNPs to be screened together in one go drastically increasing the throughput SNP screening.

4.9 Design a Primer for Microsatellite or STR

In a species (example fungi and yeast) where, no STR primers are reported, we can develop STR by making genomic library of the fragments cut by combination of tetra and hexa cutter followed by running the agarose gel, cutting 1 kb fragment and cloning it. The clones are hybridized with repeat probes labeled with radioactivity or dig labeling. The clones having repeat sequence are identified and sequenced and the primer designing in the flanking region of iterative repeat gives primer pair for genotyping STR loci generating data on STR alleles.

4.10 Design a Primer for Allele Mining

For SSCP gels one should design the primers of smaller size < 400 base pairs. The walking strategy along with overlapping fragments gives best combination of total gene sequence along with allele mining. The designed primers can be end labeled with fluorescent dyes and can be run at 20-60 degree Celsius in automated capillary based DNA sequencer with CAP polymer to get the gel variants. This technique is better and efficient with respect to SSCP gel run manually in plate mode followed by silver staining.

5. CONCLUSION

The xeroxing or PCR of DNA can easily be done if we design the primers accurately. The concept of primer designing and strategy helps in getting the right primer to be designed for the targeted xeroxing of the specific part of the given genome. *In silico* designing and its evaluation in dry lab before wet lab experiments reduces the cost of PCR drastically and lot of time can be saved besides easy for working.

REFERENCES

Abd-Elsalam, A. and Kamel (2003) Bioinformatics tools and guideline for PCR primer design. *Afr. J. Biotechnol.* 2:91-95

Breslauer, K.J, Ronald, F., Blocker, H. and Marky, L.A.(1986). Predicting DNA duplex stability from the base sequence. *Proc. Natl. Acad. Sci.* 83:3746-3750.

Chen, H. and Zhu, G. (1997). Computer program for calculating the melting temperature of degenerate oligonucleotides used in PCR or hybridization. *Biotechniques* 22:1158-1160.

Dieffenbach, C.W., Lowe, T.M.J. and Dveksler, G.S. (1995). General Concepts for PCR Primer Design. In: PCR Primer, A Laboratory Manual, Dieffenbach, C.W. and, Dveksler GS Ed., (1995) Cold Spring Harbor Labpratory Press, New York, 133-155.

Erlich, H.A., Gelfand, D. and Sninsky, J.J. (1991). Recent advances in the polymerase chain reaction. *Science* 252:1643-1651.

Gorelenkov, V., Antipov, A., Lejnine, S., Darasella, N., and Yuryev, A. (2001). Set of novel tools for PCR primer design. *Biotechniques* 31:1326-1330.

He, Q., Marjamaki, M., Soini, H., Mertsola, J. and Viljanen, M.K. (1994). Primers are decisive for sensitivity of PCR. *Biotechniques.* 17:82-87.

Jarman, S. N. (2006) Cleaver: Software for identifying taxon specific restriction endonuclease recognition sites. *Bioinformatics,* 22: 2160-2161.

Kwok, S., Kellogg, D.E., McKinney, N., Spasic, D., Goda, L., Levenson, C. and Sninsky, J. (1990). Effects of primer-template mismatches on the polymerase chain reaction: Human Immunodeficiency Virus 1 model studies. *Nucleic Acids Res.* 18:999-1005.

Li, PKC., Kupfer, C.J., Davies, D., Burbee, G., Evans, A. and Garner, H.R. (1997). PRIMO: A primer design program that applies base quality statistics for automated large-scale DNA sequencing. *Genomics* 40:476-485.

Linz, U., Delling, U. and Rubsamem-Waigmann (1990). Systematic studies on parameters influencing the performance of the polymerase chain reaction. *J. Clin. Chem. Clin. Biochem.* 28:5-13.

Lowe, T., Sharefkin, J., Yang, S.Q. and Dieffenbach, C.W. (1990). A computer program for selection of oligonucleotide primers for polymerase chain reaction. *Nucleic Acids Res.* 18:1757-1761.

Lucas, K., Busch, M., Mossinger, S. and Thompson, J.A. (1991). An improved microcomputer program for finding gene or gene family-specific Abd-Elsalam 95oligonucleotides suitable as primers for polymerase chain reactionsor as probes. *Comput. Appl. Biosci.* 7:525-529.

Makarova, K.S., Mazin, A.V., Wolf, Y.I. and Soloviev, V.V. (1992). DIROM: An experimental design interactive system for directed mutagenesis and nucleic acid engineering. *Comput. Appl. Biosci.* 8: 425-431.

O'Hara, P.J. and Venezia, D. (1991). PRIMEGEN: a tool for designing primers from multiple alignments. *CABIOS* 7:533-534.

Osborne, BI. (1992). Hyper PCR: A Macintosh Hypercard program for determination of optimal PCR annealing temperature. *Comput. Appl. Biosci.* 8:33-34.

Plasterer, T.N. (1997). PRIMERSELECT: Primer and probe design. *Methods Mol. Biol.* 70:291-302.

Roux, KH. (1995). Optimization and troubleshooting in PCR in: PCR methods and applications. Cold Spring Harbor laboratory. Pp. 5185-5194.

Rychlik, W. and Rhoads, R.E. (1989). A computer program for choosing optimal oligonucleotides for filter hybridization, sequencing and *in vitro* amplification of DNA. *Nucleic Acids Res.* 17:8543-8551.

Rychlik, W., Spencer, W.J. and Rhoads, R.E. (1990). Optimization of annealing temperature for DNA amplification in vitro. *Nucleic Acids Res.* 18:6409-6412.

SantaLucia, J. Jr., Allawi, HAT. and Seneviratne, P.A.(1996). Improved Nearest-Neighbor parameters for predicting DNA duplex stability. *Biochemistry* 35: 3555-3562.

Sheffield, V.C., Cox, D.R., Lerman, L.S. and Myers, R.M. (1989). Attachment of a40 base pair G+C rich sequence (GC-clamp) to genomic DNA fragments by polymerase chain reaction results in improved detection of single base changes. *Proc. Natl. Acad. Sci.* 86:232-236.

Singh, V.K. and Kumar, A. (2001). PCR Primer Design. *Mol. Biol. Today,* 2:27-32.

Sze, S.H., Raytberg, M.A., Gelfand, M.S., Mironov, A.A., Astakhova, T.V. and Pevzner, P.A. (1998). Algorithms and software for support of gene identification experiments. *Bioinformatics* 14:14-19.

Tamura, T., Holbrook, S.R. and Kim, S.H. (1991). A Macintosh computer program for designing DNA sequences that for specific peptides and proteins. *Biotechniques* 10:782-784.

Wallace, R.B., Shaffer, J., Murphy, RF., Bonner, J., Hirose, T. and Itakura, K. (1979). Hybridization of synthetic oligodeoxyribonucleotides to phi chi174 DNA: the effect of single base pair mismatch. *Nucleic Acids Res.* 6:3543-3557.

Wilks, A.F., Kurban, R.R., Hovens, C.M. and Ralph, S.J. (1989). The application ofthe polymerase chain reaction to cloning members of the protein tyrosine kinase family. *Gene* 85: 67-74.

Wu, DYW., Ugozzoli, L., Pal, BK., Qian, J. and Wallace, RB. (1991). The effect of temperature and oligonucleotide primer length on specificity and efficiency of amplification by the polymerase chain reaction. DNA. *Cell Biol.* 10:233-238.

□□□

CHAPTER 2

Detection of Non-Ribosomally Assembled Toxic Cyanopeptides

Maria Estela Silva-Stenico[1*], Augusto Etchegaray[2]
Omar Teschke[3], Marli Fátima Fiore[1]

[1]Universidade de São Paulo, CENA, Lab. Biologia Celulare Molecular, C.P. 96, 13400-970, Piracicaba-SP, Brazil
[2]Faculdade de Química, PUC-Campinas, Caixa Postal 317, 13012-970 Campinas SP, Brazil,
[3]Laboratório de Nanoestruturas e Interfaces, Instituto de Física UNICAMP 13083-970, Campinas, SP, Brazil

ABSTRACT

A number of secondary metabolites are produced by cyanobacteria involving the non-ribosomal peptide biosynthetic pathway. Some of these compounds have potential biotechnological applications, whereas others (the cyanotoxins) are undesired due to their toxicity to animals. The occurrence of toxic cyanobacterial blooms in freshwater bodies that are used as source for human consumption, recreation and irrigation are frequent nowadays due to increasing artificial eutrophication of these environments. It is important to monitor toxin-producing strains in order to prevent the adverse effects caused by their toxins to human and animal health. Thus, the development of rapid and sensitive methods for the detection of cyanotoxins is important as they could be usefui to monitor water supplies and contribute for watershed monitoring strategies. Current methods include cell counting, microscopic identification and toxin analysis by biochemical and analytical techniques. However, in recent years, molecular biology techniques, such as polymerase chain reaction (PCR), have been used to amplify specific regions within biosynthetic genes for rapid and sensitive identification of toxic cyanobacteria. The aim of this chapter is to provide an update of cyanotoxin biosynthesis and discuss the advances and limitations of molecular biology regarding its application as indicative of cyanotoxin production.

Key words: cyanobacterial blooms, cyanotoxins, nonribosomal peptide synthetase, polyketide synthase

*Email: estela@cena.usp.br

1. INTRODUCTION

During the last decades a number of intoxications originated from toxic cyanobacterial blooms have been registered. A major concern was brought by the death of dozens of patients at a hemodialysis center in Caruaru, PE, Brazil. These occurrences contributed to a substantial increase in research on cyanobacteria and their toxins. As a result, a number of publications are available with reference to toxicity, chemical structures, occurrence and analytical methods.

Detection of cyanotoxins in water used for human consumption was considered by the World Health Organization (WHO, 1998) one of the priority areas, which fixed a maximum concentration in drinking water of 1.0 µg/L for microcystin, a toxin produced by several cyanobacteria genera. In Brazil, the same value is accepted for drinking water as established through a Regulation set by the Ministry of Health. Tolerable values for other cyanotoxins are still being discussed.

Cyanobacterial blooms are common in water bodies that receive effluents rich in nitrogen and phosphorous, which is well-known in highly populated areas. Therefore, it is urged that agencies concerned with the administration of water for human consumption be able to predict the formation of blooms and to monitor their development. However, monitoring is complex. If morphologic analysis is carried through optical microscopy, for example, it will not differentiate toxic from non-toxic blooms (Komárek, 1991; Watanabe, 1996), thus requiring the direct analysis for the presence of toxins. Direct methods of toxin detection include: (i) bioassays (high levels of toxins detected by lethal dose determination in tests with mice); (ii) immuno-enzymatic assays (based on enzyme linked immunosorbent assay - ELISA); (iii) biochemical (inhibition of enzymatic activity); (iv) analytical (high performance liquid chromatography (HPLC) with mass-spectrometry detection). In recent years, detection methods based on DNA amplification have been used due to specificity, sensitivity and speed. However, initial attempts that tried to correlate the production of toxins with the presence of specific cyanobacterial genes were not successful. Initial experiments using molecular methods included aloenzyme polymorphism (Kato *et al.*, 1991), composition of nucleotide bases (Fahrenkrug *et al.*, 1992), homologues of the *rpo*D gene (Sakamoto *et al.*, 1993), DNA-DNA hybridizations (Wilmotte, 1994), analysis of intergenic spacer region of phycocyanin genes (Neilan *et al.*, 1995), RAPD (Neilan, 1995) repetitive DNA sequences (Asayama *et a1.*, 1996), intergenic spacer region of rRNA 16S-23S genes (Neilan *et a1.*, 1997), *rbc*L gene (Rudi *et a1*, 1997), rRNA 16S gene (Neilan *et a1.*, 1997; Rudi *et a1.*, 1997). In 1997, however, Dittman *et al.* cloned a part of the genes involved in the biosynthesis of the toxin microcystin in *Microcystis aeruginosa* PCC7806 (Dittmann *et al.*, 1997), leading the exploration of biosynthetic genes for the detection of toxic strains or toxic blooms. Ever since, gene clusters of microcystin biosynthesis produced by *Microcystis, Planktothrix* and *Anabaena* (Christiansen *et al.*, 2003; Nishizawa *et al.*, 2000; Rouhiainen *et al.*, 2004; Tillett *et al.*, 2000) and nodularin (Moffitt and Neilan, 2004) were described and the sequences of genes involved in the biosynthesis of cylindrospermopsin and saxitoxin were identified (Pomati and Neilan, 2004; Schembri *et al.*, 2001). The design of specific primers for the detection of those genes through polymerase chain reaction (PCR) is a complex task. The enzymes, non-ribosomal peptides synthetases (NRPSs) and polyketide synthases (PKSs), involved in cyanotoxin biosynthesis, possess highly conserved sequences, reflecting in their structure a multifunctional character. In other words, the catalysis of several partial reactions is elaborated by a single enzyme. The high conservation degree in the primary structure of those enzymes is also evidenced at the DNA level (Etchegaray *et al.*, 2004a).

The current review will discuss recent studies regarding the involvement of multifunctional enzymes NRPS and PKS in the biosynthesis of cyanotoxins, the organization of genes coding for those enzyme families in cyanobacteria and limitations on the use of molecular biology methods for cyanotoxin detection.

2. CYANOTOXIN BIOSYNTHESIS

Cyanobacteria constitute a group of organisms belonging to the *Bacteria* domain that is capable to produce a number of secondary metabolites. More than 800 bioactive compounds belonging to several classes of substances, including enzyme inhibitors, photosynthesis inhibitors, antimicrobials, antimalaric, imunossupressors and antitumorals were already isolated from cyanobacteria (Burja *et al.*, 2001; Etchegaray *et al.*, 2004b). The toxins produced by cyanobacteria are also synthesized by the secondary metabolism (Carmichael, 1992). A number of cyanotoxins are non-ribosomal peptides, which are allelochemicals elaborated by a mechanism parallel to protein synthesis. The non-ribosomal synthesis of peptides is a biosynthetic pathway used by microorganisms in the production of antibiotics and toxins (Marahiel *et al.*, 1997; von Döhren *et al.*, 1997). The non-ribosomal synthesis uses more than 300 substrates, which consist of non-proteinogenic aminoacids, hydroxyacids and polyketides especially elaborated to be incorporated in the structure of these peptides. Therefore, the structure of cyanotoxins is normally composed of mixed functions that can include a peptide moiety bound to keto functions, insaturations and possibly cyclic, heterocyclic and aromatic functions. As a result, the toxic secondary metabolites produced by cyanobacteria present hepatotoxic, neurotoxic, cytotoxic and dermatotoxic activities (Sivonen and Jones, 1999).

According to their chemical structures cyanotoxins are classified as cyclic peptides (microcystins and nodularins), alkaloids [anatoxin-a, anatoxin-a(S)], saxitoxin, cylindrospermopsin, aplysiatoxins and lyngbyatoxin-a and lipopolysaccharides. The most studied toxins are hepatotoxins and neurotoxins (Fig. 1).

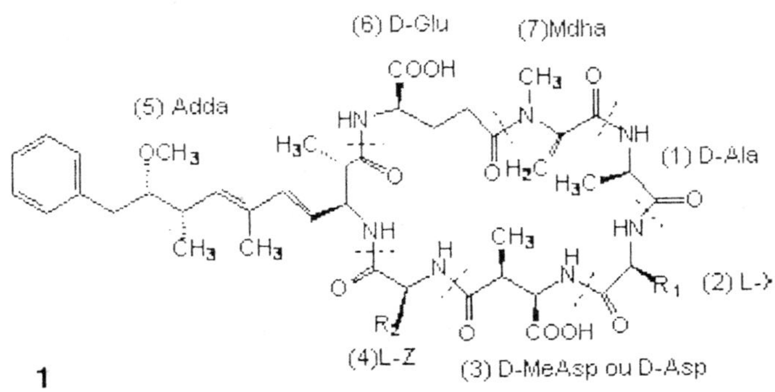

1

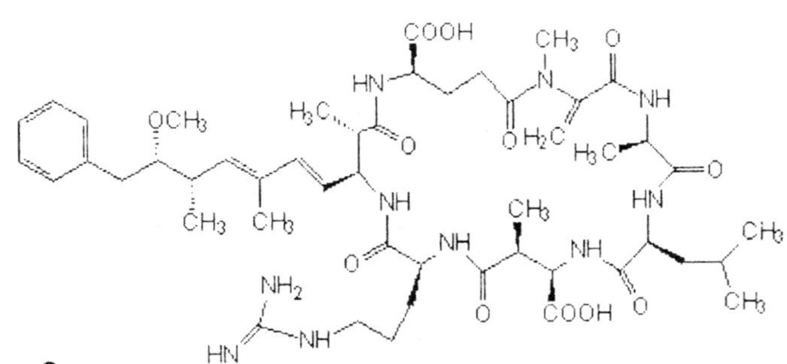

2

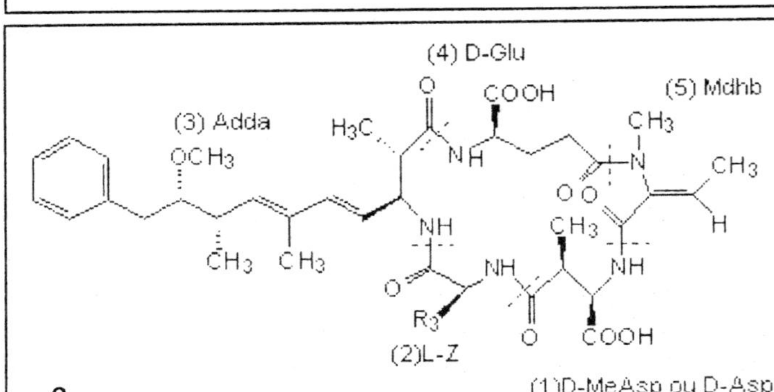

3

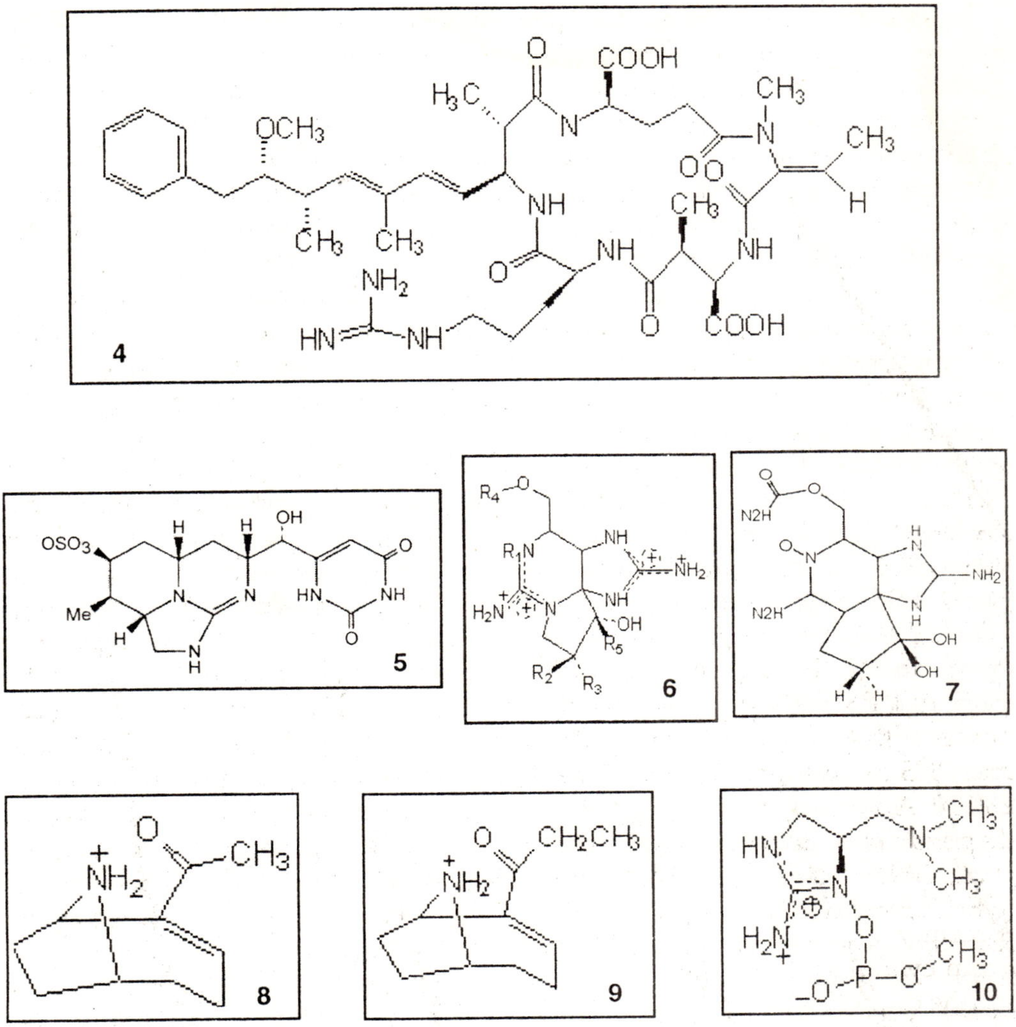

FIG. 1. Toxins (cyclic peptides and alkaloids) produced by cyanobacteria, which have been identified to date. Hepatotoxins correspond to structures (1-5) and neurotoxins structures (6-10). The general structure of microcystin corresponds to cyclic-(D-Ala1-X2-D-MeAsp3-Z4-Adda5-D-Glu6-Mdha) (1). Structure (2) is microcystin-LR the most abundant and toxin isoform. Structures (3 and 4) are isoforms of nodularin that has the generic structure cyclo-(D-MeAsp1-Z2-Adda3-D-Glu4-Mdhb5). Cylindrospermopsin corresponds to structure (5); saxitoxin is represented by the generic structure (6), whereas saxitoxin-X corresponds to structure (7); Structures (8, 9 and 10) are respectively anatoxin-a (8); homoanatoxin-a (9) and anatoxin-a(S) (10). Abbreviations: D-MeAsp (D-erithro-*beta*-methyl aspartic acid); Adda, (2S, 3S, 8S, 9S)-3-amino-9-metoxy-2,6,8-trimethyl-10-phenyl-deca-4,6-dienoic acid; Mdha (N-methyl-dehydroalanine); Mdhb (N-methyl-dehydrobutirine) (Sivonen and Jones, 1999).

The structure of the enzymes, non-ribosomal peptide synthetases (NRPSs) involved in cyanotoxin biosynthesis and the assembly of a number of other bioactive peptides is characterized by a modular arrangement, finding in a corresponding way, blocks of repetitive sequences in the encoded genes (Etchegaray *et al.*, 2004a; Marahiel *et al.*, 1997; von Döhren *et al.*, 1997). Another enzyme family that takes part in the assembly of non-ribosomal peptides corresponds to the polyketide synthases (PKSs), which may also present a modular structure. In bacteria, modular PKSs are termed type I PKSs, which are known to be involved in the biosynthesis of macrocyclic compounds and are here considered because of their involvement in a number of cyanobacterial metabolites, including the cyanotoxins (Fig. 1). Aromatic structures, however, are usually synthesized by non-modular PKSs (Hutchinson and Fuji, 1995; Du *et al.*, 2001). In cyanobacteria, biosynthetic gene clusters of microcystins (Tillett *et al.*, 2000) and nodularin (Moffitt and Neilan, 2004), and part of genes involved in the cylindrospermopsin and saxitoxin production (Pomati and Neilan, 2004; Schembri *et al.*, 2001) were identified and they are discussed below, although several NRPS and PKS genes have been described before in other bacteria and fungi (Du *et al.*, 2001; Hutchinson and Fuji, 1995; Moffitt and Neilan, 2000). In microcystins biosynthesis, the presence of NRPSs, PKSs and hybrid synthetases (NRPS-PKS and PKS-NRPS) was found. Hybrid synthetases are also modular; however they have mixed modules in which the enzymatic activities of NRPS and PKS are present within adjacent modules of the same protein (Du *et al.*, 2001). Analysis of these genes involved in microcystin biosynthesis demonstrates that all of the identified synthetases are modular (Rouhiainen *et al.*, 2004; Tillett *et al.*, 2000). However, not all the NRPSs, PKSs and the hybrid synthetases of cyanobacteria are involved in cyanotoxin biosynthesis. Recently, genes involved in the biosynthesis of other peptides were also isolated from cyanobacteria (Table 2). Most of the non-ribosomal peptides produced by the cyanobacteria present within their structure characteristics that suggest the involvement of both NRPS and PKS activities (Etchegaray *et al.*, 2004b). More to the point that the occurrence of genes of NRPSs in cyanobacteria be a common characteristic (Christiansen *et al.*, 2001) a single strain can produce several non-ribosomal peptides simultaneously, possibly for the co-expression of NRPSs genes in different *loci* (Welker *et al.*, 2004a).

NRPSs work as an industrial assembly line, in other words, the amino acids that constitute the structure of the peptide are incorporated sequentially, through a series of partial and repetitive enzymatic reactions. Modules in NRPSs and PKSs are involved in the stages of elongation of the structure of the metabolite, i.e., in the catalysis of partial reactions. Therefore, for the incorporation of one amino acid in a non-ribosomal peptide, the enzyme uses a module composed of approximately 1,000 amino acids. Each module is composed of small regions involved in specific enzymatic reactions (Fig. 2). There is a correlation between the sequence of modules and the biosynthetic steps. The first module of a NRPS is involved in the activation and incorporation of the N-terminal amino acid and the other modules in the incorporation of the subsequent amino acids, obeying colinearity between the incorporated amino acid and the module involved in its addition. The minimum steps of peptide biosynthesis assembled by NRPSs are adenylation, thioesterification and condensation (Marahiel *et al.*, 1997; von Döhren *et al.*, 1997). The regions involved in these partial reactions are located within domains, i.e., adenylation domain (A domain), thioesterification domain (called Peptidyl Carrier Protein or PCP) - in analogy to the biosynthesis of fatty acids, where the corresponding acyl carrier protein (ACP) domain is the site of thioesterification. Finally a condensation domain (C domain) involved in peptide bond formation (Fig. 2). Several other domains have been described, although they are not essential for the incorporation of amino acids. Among these are the N-methylation and epimerization domains, where the epimerization domain constitutes a variation of the condensation domain.

The N-methyl transferase domain (NMT or M) is involved in methylation of the amino group of the corresponding amino acid (Marahiel *et al.*, 1997). The epimerization domain is involved in conversion of the L-amino acids to its D-isomers. Concluding the enzymatic synthesis of peptides, the product is liberated of the enzyme through a thioesterase activity, usually catalyzed by the thioesterase domain (T). The T domain is generally inserted in the last module of NRPSs (Fig. 2). The M domain is usually inserted between the adenylation domain and the PCP domains (Marahiel *et al.*, 1997). The C domain also presents some variations that are associated with transference of intermediate peptides, formed by other synthetases of the same operon (C domain - Fig. 2), or in the transfer of fatty- or aromatic acids (Etchegaray *et al.*, 2004a). Examples of biosynthetic systems composed of peptide synthetases are also shown in Figure 2. There are systems composed of a single NRPS, which is normally found in fungi such as the enzyme ACV synthetase of *Penicillium chrysogenum* and there are bacterial systems in which more than one synthetase is involved in the biosynthesis of a nonribosomal peptide (von Döhren *et al.*, 1997).

As the partial reactions described above are repetitive, it is obvious that certain regions of the primary structure of the synthetases (within the modules of NRPSs) are highly conserved. Thus, when these sequences are aligned a high degree of sequence similarity is found (Etchegaray *et al.*, 2004a).

Therefore, the amino acid sequences of domains (A, PCP and C) of different modules and synthetases, also present high degree of similarity, although they are involved in the incorporation of different amino acids within the structure of a novel antibiotic or toxin. In addition, the DNA sequences coding for modular peptide synthetases are also highly conserved. Following this rational, using degenerate primers in a PCR reaction, similar DNA sequences can be amplified in several non-ribosomal peptide producing organisms such as cyanobacteria, *Bacillus* sp., actinomycetes and fungi. In fact, the NRPSs constitute a family of enzymes that are codified by very similar genes. Thus, in order to amplify the DNA of a particular NRPS system, it is important to obtain previous information of its genes so that specific primers can be designed (no degenerated) to avoid amplification of similar or additional NRPS genes (Fig. 3).

To be active, NRPSs must be converted from an apo to a holo-enzyme form by a specific post-translational reaction. Post-translational modification is carried by the transfer of the cofactor 4'-phosphopanthetein (PAN) to a serine residue in the conserved motif (LGGDSL) of PCP domain. PAN is covalently attached to the OH group of serine through a phosphodiester bound (segment T of Fig. 2). The SH moiety of the cofactor is involved in formation of active intermediates with substrates in the form of thioesters, an important stage of nonribosomal peptide biosyntheses in analogy to fatty acid and polyketide biosynthesis. Enzymes involved in the transfer of PAN were already described, the phosphopanthotenyl transferases (Walsh *et al.*, 1997), which are normally present in NRPS operons. Thus the biosynthesis of nonribosomal peptides is very complex. The expression of peptide synthetases requests the concomitant expression of other enzymes, some involved in the synthesis of specific substrates, and also of enzymes as the phosphopanthotenyl transferases, required for the production of holo-enzyme, therefore, for enzymatic activity. In addition, under certain environmental conditions, the biosynthesis can be induced or repressed (Kaebernick *et al.*, 2000). Accordingly, it can be concluded that the presence of NRPS genes in a particular microorganism is a condition but not a confirmation of the production of nonribosomal peptides. Even the production of specific messenger RNA may have a vague meaning, because the involved proteins require post-translational modifications.

The interest in NRPS and PKS biosynthesis is not only devoted for identification of genes involved in the production of toxins, since several antibiotics and other compounds with therapeutic activity produced by fungi and actinomycetes in industrial scale are elaborated with the participation of NRPSs and PKSs. Among them are the penicillins, vancomycins and the

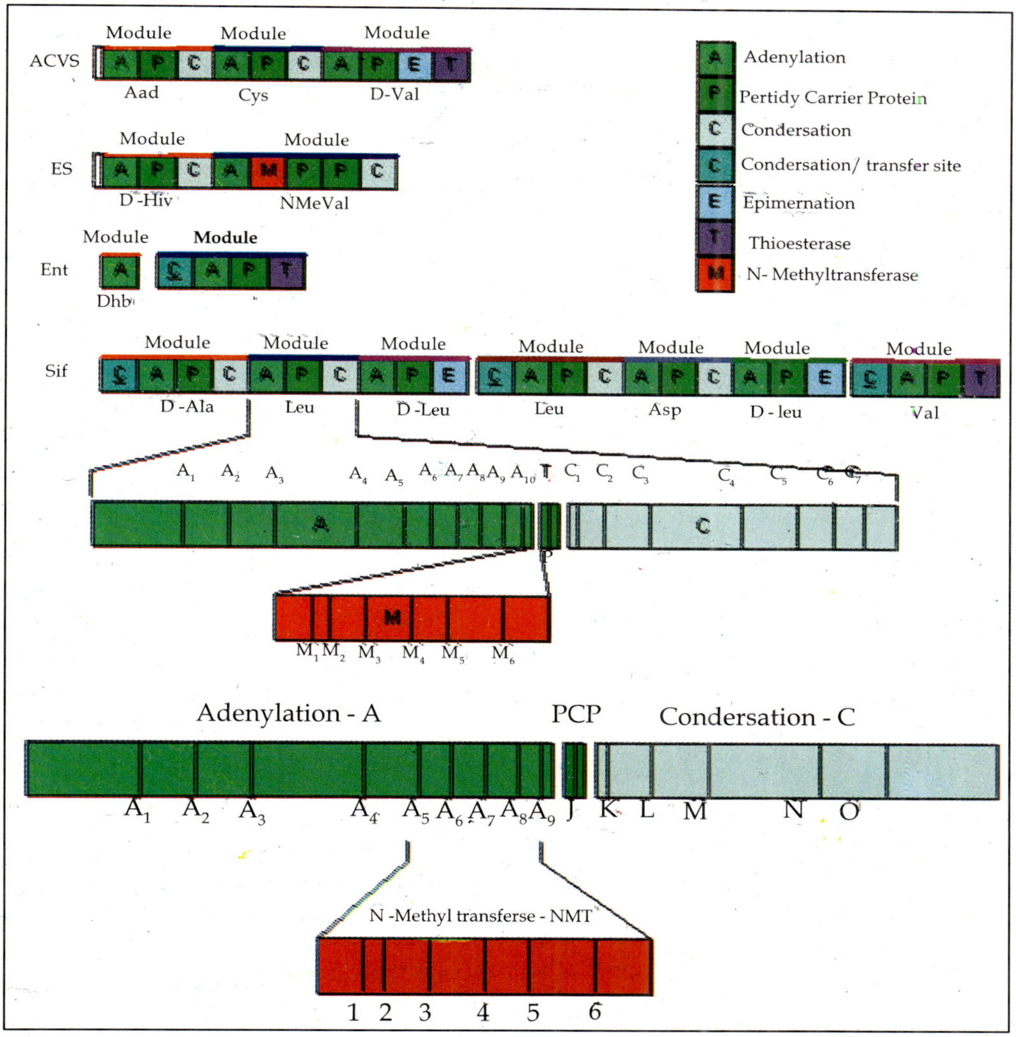

FIG. 2. Modular organization of NRPSs. In general, the minimal module architecture is composed by domains involved in partial reactions such as adenylation (A), thioesterification (PCP or P) and condensation (C) partial reactions. Other domains include condensation/transfer site (C̲), epimerization (E), thioesterase (T) and N-methyl-transferase (M). Conserved sequences of the adenylation, PCP and condensation domains are indicated in A_1 to A_{10} (adenylation domain). As indicated, when present, the N-methyltransferase domain is inserted between the adenylation and thioesterification domains. Sites of conserved sequences within the M domain are indicated by numbers 1 to 6. The nonribosomal peptide synthetases represented here are, ACVS (γ-(L-α-aminoadipyl)-L-cysteinyl-D-valine synthetase), ES (enniatin synthetase), Ent (enterobactin synthetase), Srf (surfactin synthetase), where modules and the aminoacids that are incorporated in the peptide structure by the enzymatic activity of each module are indicated (von Döhren *et al.*, 1997).

immunosupressor cyclosporin A. Although the mentioned antibiotics have varied applications, one of the largest problems faced nowadays is related to the resistance of the microorganisms. To reduce this problem, several strategies have been used, since the search for new drugs to the manipulation of biosynthetic genes to produce new structures (Du *et al.*, 2001).

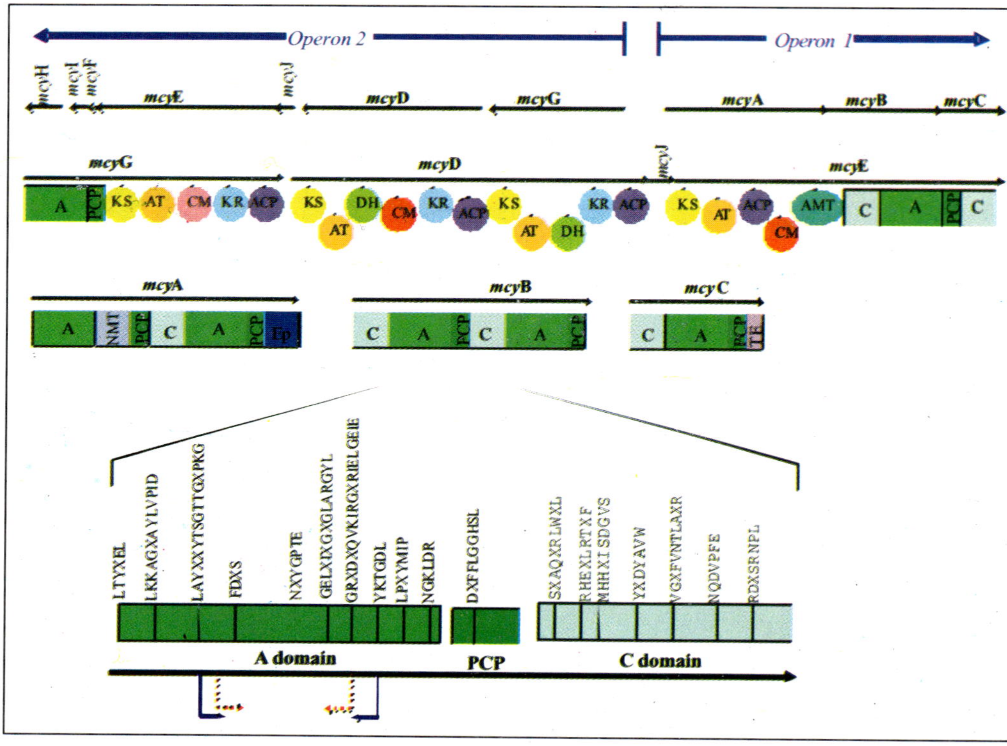

FIG. 3. Grouping of genes involved in microcystin biosynthesis of *Anabaena* sp. strain 90. Genes are organized in two major operons transcribed in opposed senses extending a total size of 55.4 kb (Rouhiainen *et al.*, 2004). Strategies used in the amplification and detection of these genes may involve the amplification with degenerated primers based in conserved sequences (full arrows), allowing the amplification of NRPSs genes of several microorganisms, or, alternatively, amplification with specific primers (hatched arrows) designed from non-conserved regions of the nucleotide sequence. In the total, three NRPSs genes (*mcy*A, *mcy*B and *mcy*C), a modular PKS (*mcy*D) and two genes of hybrid synthetases, respectively NRPS-PKS (*mcy*G) and PKS-NRPS (*mcy*E) are involved in the biosynthesis of microcystins. The several enzymatic activities of the modular synthetases involved are indicated. Details show the adenylation, PCP and condensation domains of the first module of *mcy*B, where one can see motifs of highly conserved sequences. Abbreviations: (A) adenylation domain, (PCP) Peptidyl Carrier Protein, (C) condensation domain, (Ep) epimerase, (KS) keto synthase, (AT) acyl transferase, (CM) carboxy methylase, (KR) keto reductase, (ACP) acyl carrier protein, (DH) dehydratase, (AMT) amino transferase.

In one of the alternatives the aspect of substrate specificity of peptide synthetases was approached. Thus the adenylation domain of the first module of gramicidin A synthetase was cloned in *Escherichia coli* and the protein produced in wide scale to obtain the crystal structure (Finking and Marahiel, 2004). This important step in the characterization of peptide synthetases made possible the identification of residues involved in specificity. As a result, today, the specificity amino acids of novel systems can be identified from the primary structure of adenylation domains through sequence alignment. The residues that check the specificity are located within the conserved motifs A_3 and A_6 of the adenylation domain (Fig. 2). The amino acids that check specificity are nine, which can be analyzed using a database to predict the theoretical specificity (Challis *et al.*, 2000). This information was also used after the sequence of the adenylation domains of each NRPS module of the synthetases involved in microcystin biosynthesis. Surprisingly, a variation was detected among the amino acids that check specificity exactly for modules involved in the incorporation of variable aminoacids (Fig. 3). The activation and incorporation of different amino acids is, therefore, dependent on the amino acid residues that align the binding pocket of the adenylation domain of the individual modules (Stachelhaus *et al.*, 1999; Challis *et al.*, 2000; Schwarzer *et al.*, 2003). It is believed that the wide distribution and variation in the microcystin biosynthesis can be related with its biological function (Mikalsen *et al.*, 2003; Welker *et al.*, 2004b) which is still not established. However, it is unlikely that the diversity of microcystin isoforms has a restrictive effect in the function of this toxin within cyanobacterial cells. It seems more probable that diversity is crucial for several strains in their natural habitats. Besides, the processes that generate genetic diversity, particularly in the amino acids that have specificity, can play an important role in their biological effect.

3. ORGANIZATION OF GENES OF CYANOTOXIN BIOSYNTHESIS

Among the hepatotoxins, microcystins have received more attention since they are found with more frequently in water bodies. Microcystins are synthesized mainly by three genera of planktonic cyanobacteria, such as *Microcystis*, *Anabaena* and *Planktothrix*, and more rarely in *Anabaenopsis*, *Hapalosiphon* and *Nostoc* (Sivonen and Jones, 1999). There are more than 70 variants of microcystins (Codd *et al.*, 2005). All isoforms contain the general cyclic structure D-Ala-L-X-D-MeAsp-L-Z-Adda-D-Glu-Mdha, where Adda is the 3-amino-9-methoxy-2,6,8-trimethyl-10-phenyl-4,6-decadienoic acid, D-MeAsp is 3-methyl-aspartic acid, Mdha is N-methyldehydroalanine and X and Z are variable L-amino acids that contribute for the different isoforms (Fig. 1) (Carmichael *et al.*, 1988; Honkanen *et al.*, 1990; Rinehart *et al.*, 1994).

In those amino acids the variable X is commonly leucine (L), arginine (R) or tyrosine (Y) and Z is usually arginine (R), alanine (A) or methionine (M). For instance, microcystin-LR contains leucine and arginine, while microcystin-YA has tyrosine and alanine (Carmichael, 1992). Other amino acids variations include different degrees of methylation of Mdha and/or MeAsp and configuration of the Adda chain (Rinehart *et al.*, 1994). Demethylation of the MeAsp amino acid, for instance, is termed as [D-Asp] microcystin-LR. In addition, there are many optical isomers of the microcystins.

Although several microcystins can be produced during a single bloom, microcystin-LR is the most common occurrence and one of the most toxic variants (Carmichael, 1992; Fawell *et al.*, 1993). For this reason it is the best characterized among all cyanotoxins. Molecular studies aiming to identify the biosynthetic genes of cyanotoxin were initiated with the *Microcystis aeruginosa* species, a microcystin-LR producer. In 1996, Meissner *et al.* using the degenerated primers, coreI and coreII (Borchert *et al.*, 1992), designed to amplify by PCR sequences of NRPS adenylation domain began the search of homologous sequences in toxic and non-toxic strains of *M. aeruginosa* (Meissner *et al.*, 1996). This preliminary work resulted in the amplification of a DNA fragment of approximately 200 bp that showed significant similarity with regions corresponding to genes of NRPSs. Soon after, those authors constructed a genomic library of the toxic strain *M. aeruginosa* HUB524 and they isolated and sequenced regions adjacent to the 200 bp fragment. In this way, it was possible to isolate a fragment of DNA of 2982 bp, denominated *mapep*1 that showed high identity with a complete NRPS module, which hybridized exclusively with DNA of hepatotoxins-producing strains. A specific primer set was designed to amplify this NRPS module and it was tested in toxic and non-toxic strains of *Microcystis*. The results showed that only toxic strains contained the DNA fragment of interest, suggesting the involvement of this coding region of peptide synthetases in the production of microcystins (Dittmann *et al.*, 1996). Those authors also sequenced more regions in both extremities of the *mapep*1 fragment and they observed that *M. aeruginosa* HUB524 presented more than one module of NRPS. To prove the involvement of that region in the production of microcystins, a mutant of the *M. aeruginosa* PCC7806 was produced by the insertion of a chloramphenicol resistance cassette within the *mapep*1 sequence, which inactivated the NRPS gene in this cyanobacterium (Dittmann *et al.*, 1997). This mutant was unable to produce microcystins, proving the involvement of the NRPS in the biosynthesis of these toxins. The two NRPS genes partially characterized in this study were denominated *mcy*A and *mcy*B. The antibiotic resistance cassette in the mutant strain was introduced in the adenylation domain of *mcy*B. As no microcystin variant was produced by the mutant, and the wild strain was capable to synthesize several isoforms, it was suggested that the NRPS module did not have high substrate specificity, being possibly responsible for the introduction of variable amino acids. Soon after, a primer set (FAA/RAA) was developed targeting a less conserved region of the adenylation domain of the *mcy*B (Fig. 3), which was specific enough to amplify fragments of microcystin synthetase from several microcystin-producing cyanobacteria (Neilan *et al.*, 1999). The DNA fragment of 758 bp obtained with FAA/RAA primer was used to screen in a lambda Zap library of *M. aeruginosa* PCC7806 one clone containing a fragment of 7 kb of the gene cluster denominated *mcy* (Tillett *et al.*, 2000). The remaining sequence of *mcy* was obtained by the authors using walking PCR approach. In total, 63.6 kb of the *mcy* cluster and adjacent regions were isolated from *M. aeruginosa* PCC7806. In this way, the *mcy* cluster of 55 kb located in the chromosome was entirely identified. This cluster contains 10 genes (*mcy*A-*mcy*J) (Fig. 3), where genes *mcy*A-G and J encode modules of PKS/NRPS (involved in polyketide/polypeptide chain elongation) and additional accessory domains (catalyzing modification of polyketide/ polypeptide chains), while *mcy*H encode a putative protein homologue to a ABC transporter, whose function may be the extracellular or subcellular transport of microcystins. Japanese researchers also described a gene cluster involved in microcystin-LR production in *M. aeruginosa* K-139 (Nishizawa *et al.*, 1999; 2000). The DNA sequence of this gene cluster was identical to that described by Tillet *et al.* (2000). In 2003, the gene cluster involved in microcystin-RR production of *Planktothrix agardhii* CYA126/8 was published (Christiansen *et al.*, 2003). More recently, it was identified the gene cluster responsible by microcystin-LR biosynthesis in the filamentous heterocytous cyanobacterium *Anabaena* 90 (Rouhiainen *et al.*, 2004) (Fig. 4). The comparison of gene clusters responsible for microcystins production in the three different cyanobacterial genera (*Microcystis, Planktothrix* and *Anabaena*) showed that most of genes was in a different order, not all genes were present in the three organisms and the identity among genes was low (Figure 4). Therefore, there is a need to characterize microcystins gene cluster for each organism (Rouhiainen *et al.*, 2004).

Other hepatotoxin, nodularin, only found in strains of the genus *Nodularia* (Sivonen and Jones, 1999) possesses similar chemical structure of the microcystins (Annila *et al.*, 1996) (Fig. 1). However, the structure of nodularin (Adda-D-Glu-Mdhb-D-Asp-Arg), a cyclic pentapeptide, is smaller than the cyclic heptapeptide microcystin, which is the largest known cyanotoxin. Nodularin differs from microcystin by lacking the aminoacid D-Ala and X and having Mdhb (*N*-methyl-dehydrobutyrine) in place of Mdha. Neilan *et al.* (1999) were the first to detect DNA sequences involved in nodularin production. The PCR products were obtained using genomic DNA of the toxic *Nodularia spumigena* PCC73104 and degenerated primers (MRF2/MTR), specific for NRPS, and also the set primer FAA/RAA, designed to target a *mcy*B region of *Microcystis*. The authors concluded that *N. spumigena* PCC7310 contained NRPS genes, what was expected since nodularin possesses chemical structure similar to the microcystins. Putative sequences of genes encoding NRPSs and PKSs of nodularin were obtained later, which showed high homology with sequences of *mcy*C and *mcy*D genes, respectively, of microcystins (Moffitt and Neilan, 2001). In 2004, the complete gene cluster required for nodularin synthesis in *N. spumigena* NSOR10 was finally sequenced and was called *nda* (Moffitt and Neilan, 2004). Two modules that catalyze the incorporation of D-Ala (present in *mcyA* gene) and X (present in *mcy*B gene) in microcystins were deleted in nodularin. Molecular phylogenetic study suggests that nodularin synthetase genes were derived from microcystin synthetase genes and that the nodularin may be the more extreme structural variant of microcystins (Rantala *et al.*, 2004).

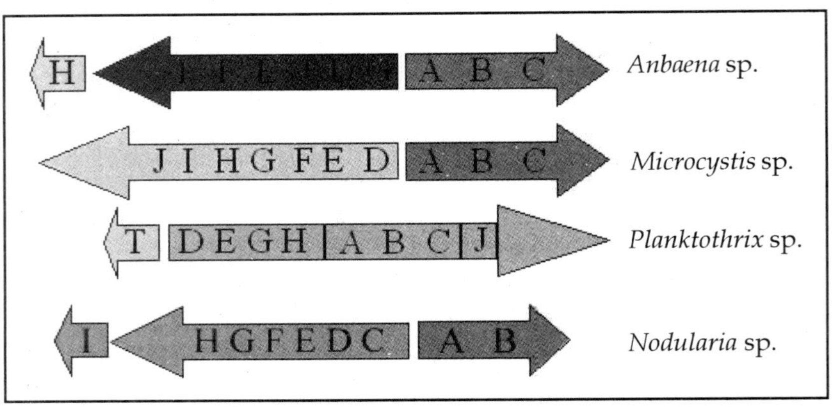

FIG. 4. Organization of the operons involved in microcystin and nodularin biosynthesis in the cyanobacteria *Anabaena, Microcystis, Planktothrix* and *Nodularia*. The illustration shows the order of the genes and the direction of its transcription (modified from Rouhiainen *et al.*, 2004).

The third known hepatotoxin is cylindrospermopsin ($C_{15}H_{21}N5O_7S$), belonging to a new class of alkaloids that possesses a guanidine function repeated three times, combined with the structure hydroxymethyluracyl (Ohtani *et al.*, 1992) (Fig. 1). Cylindrospermopsin has been found in some strains of cyanobacteria of the genera *Cylindrospermopsis* (Ohtani *et al.*, 1992), *Umezakia* (Harada *et al.*, 1994), *Anabaena* (Schembri *et al.*, 2001), *Aphanizomenon* (Banker and Carmeli, 1997) and *Raphidiopsis* (Li *et al.*, 2001).

Biochemical studies and structural analysis suggest a NRPS-PKS hybrid system for cylindrospermopsin synthesis in *C. raciborkii* (Burgoyne *et al.*, 2000). Recently, putative DNA sequences encoding NRPS and PKS were identified in two *C. raciborskii* strains (AWT205 and CYP020B) and the direct involvement of these enzymes in cylindrospermopsin production was demonstrated (Schembri *et al.*, 2001). The authors also found the putative NRPS and PKS sequences in *Anabaena bergii* ANA283A, cylindrospermopsin-producing strain. Four *Aphanizomenon ovalisporum* strains were identified as cylindrospermopsin producers by molecular technique (Fergusson and Saint, 2003). However, further studies are necessary for the knowledge of complete genes cluster involved in cylindrospermopsin biosynthesis.

Three different neurotoxins were found in cyanobacteria: anatoxin-a ($C_{10}H_{15}NO$), anatoxin-a(S) ($C_7H_{17}N_4O_4P$) and saxitoxin ($C_{10}H_{17}N_7O_4$) (Fig. 1). Anatoxin-a is produced by strains of the genera *Anabaena, Oscillatoria* and *Aphanizomenon* while anatoxin-a(S) was only found in *Anabaena*, and saxitoxin was observed in *Aphanizomenon, Anabaena, Lyngbya, Cylindrospermopsis* and *Planktothrix* (Lagos *et al.*, 1999; Sivonen and Jones, 1999). The molecular bases of neurotoxins biosynthesis are not yet known, although they are studied for more than 30 years. Recently, genes potentially involved in the biosynthesis and regulation of saxitoxin in *A. circinalis* strains 131C and 3448 were identified (Pomati and Neilan, 2004). The authors developed a technique called PCR-based positive hybridization (PPH), specifically to obtain genes associated with saxitoxin. This technique was based partly on the same bases of subtractive suppressive hybridization. During the study, a specific putative transposase of the toxic strain *A. circinalis* 131C was isolated, suggesting that the molecular base of saxitoxin production is encoded by a large mobile region of the genome of *A. circinalis*. Putative sequences coding for acetyltransferase (clone 183) and of the chaperonine GroEL (clone 118), correlated with saxitoxins biosynthesis, were obtained. Those authors also showed the recovery of several hypothetical proteins with high specificity with proteins found in toxic strains, but it is not known any similar biosynthetic pathway to the proposal for saxitoxins, it is difficult to identify these proteins through the search of homologue sequences in database.

4. DETECTION OF GENES INVOLVED IN CYANOTOXIN PRODUCTION

The development of techniques for detection of toxic cyanobacterial blooms using DNA amplification started after the work of Dittmann *et al.* (1997), where part of the biosynthetic genes of microcystin produced by *M. aeruginosa* PCC7806 were cloned and sequenced. Knowledge of those sequences has facilitated the design of specific primers for PCR amplification. Regions coding for the adenylation domain of the *mcy*B gene were initially considered the most promising for PCR amplification, once they are responsible for the different isoforms of this toxin (Figure 3). In 1999, Neilan *et al.* and Nishizawa *et al.* reported the first primers sets denominated FAA/RAA and core1/core2R (degenerated), respectively, which target region of adenylation domain of the *mcy*B gene (Neilan *et al.*, 1999; Nishizawa *et al.*, 1999) (Table 1). Although those two primers sets have shown good correlation with toxin production, several specificity problems were observed (Tillett *et al.*, 2001). The authors suggested that the lack of specificity could be due to high degree of similarity in the sequence of *mcy*B adenylation domain with other NRPSs adenylation domains (Dittmann *et al.*, 1997) or to the occurrence of multiple adenylation domains in toxic and non-toxic *Microcystis* sp. (Nishizawa *et al.*, 1999). The adenylation domains are fundamental in the mechanism of the NRPS and NRPS genes were found in a great number of cyanobacteria (Christiansen *et al.*, 2001). In that way, searching for an unusual NRPS domain, Tillett *et al.* (2001) designed a primer set (MSF/MSR) specific to N-methyltransferase (NMT) domain, which is only found in the *mcy*A region (Table 1, Fig. 2 and 3). This primer set was tested in 37 strains of *Microcystis* and although it discriminated toxic and non-toxic strains of most cyanobacteria, two non-producing microcystins strains showed positive results in the PCR amplification. This result may suggest that the primer set MSF/MSR recognized a sequence region in the two strains identical to the region of the microcystin synthetase, which is not involved in the toxin synthesis or that the genes encoding for microcystins biosynthesis were inactive.

New primer sets are being described frequently in the literature mainly due to increase in the knowledge of genes involved in cyanotoxin biosynthesis. A list of the most recent primer sets designed to detect genes involved in microcystin biosynthesis published in the literature is shown in Table 1. In several studies using some of those primer sets, the PCR technique was applied in environmental samples aiming to detect earlier blooms of potentially toxic *Microcystis* (Baker *et al.*, 2002). In our laboratory, potentially toxic blooms of the Billings reservoir (São Paulo, SP) and Salto Grande (Americana, SP) were detected with success using the primer set OmetF/OmetR (Lorenzi, 2004). However, the PCR technique, in spite of identifying potentially toxic cyanobacteria or blooms, is not quantitative. A molecular technique denominated real time polymerase chain reaction (qRT - PCR), which provide quantitative data of the PCR product, was used recently to detect and quantify putative microcystins produced by *Microcystis* in a Germany lake (Kurmayer and Kutzenberger, 2003) and also for *Microcystis* and *Anabaena* in several

TABLE 1. Primers used to detect genes responsible for cyanotoxin production

Cyano toxins	Target Genes	Primers	Sequences	Fragment Size (Pb)	Reference
Microcy stin	*mcyB*	FAA RAA	CTATGTTATTTATACATCAGG CTCAGCTTAACTTGATTATC	758	Neilan *et al.* (1999)
		core1 core2R	TTIAA(A/G)GC(A/G)GG(C/T)GGITATGTGCCGAT(C/T)C C CCTTTTGGCTTICCTGTIGTICC(A/G)GAIGT(G/A)TAIAT	260	Nishizawa *et al.* (1999)
		McyBMafwd1 tox4r	AATCAACGGTTAGTTGCTTATGT CACTAACCCCTATTTTGGATACC	288	Kurmayer *et al.* (2003)
	mcyA	MSF MSR	ATCCAGCAGTTGAGCAAGC TGCAGATAACTCCGCAGTTG	1300	Tillett *et al.* (2001)
		mcyA-Cd 1R mcyA-Cd 1F	AAAAGTGTTTTATTAGCGGCTCAT AAAATTAAAAGCCGTATCAAA	300	Hisbergues *et al.* (2003)
		QmetF QmetR	TTATTCCAAGTTGCTCCCCA GGAAATACTGCACAACCGAG	200	Lorenzi (2004)
		Cho09F Ch09R	TGGCGACTTCAGAAAAAGGGC GCGGGGCAAAAATCACATC	1100	Cho *et al* (2004)
	mcyD	mcyDF mcyDR	GATCCGATTGAATTAGAAAG GTATTCCCCAAGATTGCC	818	Rantala *et al.* (2004)
		mcyDF2 mcyDR2	GGTTCGCCTGGTCAAAGTAA CCTCGCTAAAGAAGGGTTGA	297	Bertasi *et al.* (2003)
	mcyE	mcyE-F2 MicmcyE-R8 AnamcyE-12R	GAAATTTGTGTAGAAGGTGC CAATGGGAGCATAACGAG CAATCTCGGTATAGCGGC	-	Vaitomaa *et al.* (2003)
		mcyE-F2 mcyE-R4	GAAATTTGTGTAGAAGGTGC AATTCTAAAGCCCAAAGACG	812	Rantala *et al.* (2004)

Contd....

Table 1. Contd....

Nodularin	PS	NPF NPR	TATTTTGTGGTGGAGAAGCACTA GGAACTATCTGATAATTAGAC	150	Moffitt and Neilan (2001)
Cylindrospermopsin	PS	M13 M14	GGCAAATTGTGATAGCCACGAGC GATGGAACATCGCTCACTGGTG	597	Schembri *et al.* (2001)
	PKS	M4 M5	GAAGCTCTGGAATCCGGTAA AATCCTTACGGGATCCGGTGC	650	Schembri *et al.* (2001)
	PS	CPSF CPSRC	AGTATATGTTGCGGGACTCG CCCGCCAAGACAGAAGG	500	Burns *et al.* (2004)

Finland lakes (Vaitomaa *et al.*, 2003). In the real time PCR reaction, amplicons are marked with fluorescent probes allowing quantification in each cycle.

Examples of other cyanopeptides of non-ribosomal origin are presented in Table 2.

TABLE 2. Cyanopeptides produced non-ribosomaly involving NRPS and PKS genes.

Cyanobacteria	Cyanopeptide	Reference
Anabaena 90	anabaenopeptilide	Rouhiainen *et al.* (2000)
Nostoc sp. GSV224	nostopeptolide	Hoffmann *et al.* (2003)
Lyngbya majuscula	curacin A	Chang *et al.* (2004)
Cylindrospermopsis raciborskii	cylindrospermopsin	Schembri *et al.* (2001)
Lyngbya majuscula	jamaicamide	Edwards *et al.* (2004)
Nostoc sp ATCC53789	nostocyclopeptide	Becker *et al.* (2004)
Lyngbya majuscula / Schizothrix sp.	Somocystinamide A	Nogle and Gerwick (2002)
Lyngbya majuscula	barbamide	Chang *et al.* (2002)

5. NON-TOXIC CYANOPEPTIDES

Siderophores are a class of peptides that play an important physiological role for the producing microorganisms due to the capacity to chelate iron (III). These biomolecules are produced by fungi and bacteria in conditions of low iron concentrations (Silva-Stenico *et al.*, 2005). Siderophores are connected with binding and transportation of Fe (III). The main classes of siderophores (hydroxamates and catechols) are elaborated by NRPSs, also involved in the biogenesis of other antibiotics and toxins. Some siderophores also present the capacity to bind other metals such as Ga(III), Cr(III), Al(III) (Birch and Bachofen, 1990; Hu and Boyer, 1996). The biosynthesis of siderophores is regulated by iron and also influenced by other environmental factors (Moffitt and Neilan, 2000). The biosynthesis of siderophores is regulated by iron and also influenced by other environmental factors (Moffitt and Neilan, 2000). NRPSs have been implicated in the biosynthesis of cyanobacterial toxins microcystin, nodularin and cylindrospermopsin (Dittmann *et al.*, 1997; Neilan *et al.*, 1999). In some cases they constitute hybrids between NRPSs and PKSs or the sequences (NRPSs and PKSs) are contiguous. The genes of toxin biosynthesis are to some extent different from the genes of siderophore biosynthesis, specifically in size and organization. Besides the cyanotoxin production by cyanobacteria, a wide variety of interesting bioactive secondary metabolites are described for these microorganisms. Examples for that are described in Table 2. The identification of biosynthetic genes allows the selection of producing microorganisms for bioprospection of new metabolites.

In our study using extracts of *M. aeruginosa* NPCD-1 containing bioactive peptides, we used atomic force microscopy (AFM) in order to image the details of the peptide damage to a bacterial structure. Samples were incubated with *Bacillus subtilis* cells and observed on AFM. The morphology changes of *B. subtilis* cells were investigated by observing cell deposited on mica, dried in ambient conditions (21°C, 40% humidity) and scanned in air. Figure 5A and B shows no damage in cells when methanol or cell extract of *M. aeruginosa* NPCD-1 were incubated with *B. subtilis* cells. Figure 5C and D shows the damage in *B. subtilis* cell wall, when extracts from cyanobacteria that contained bioactive compounds were added.

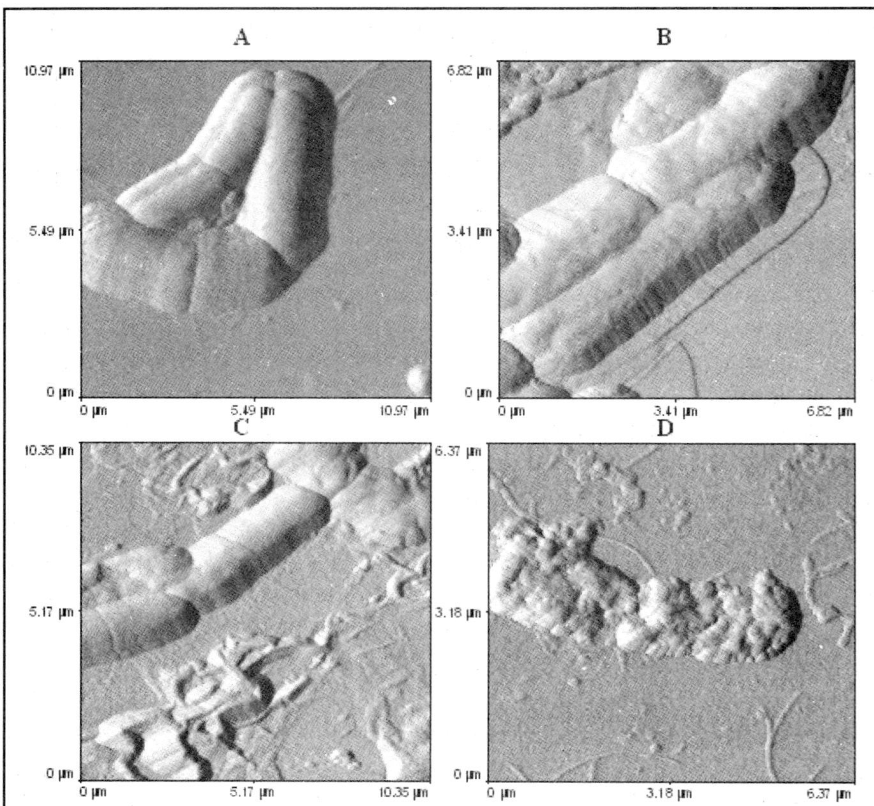

FIG. 5. Topographic images of *B. subtilis* cells obtained by AFM. (A) Methanol (control); B) *M. aeruginosa* NPCD-1 cells extract; C) Culture medium ASM-1 extracted in ethyl acetate (control); D) *M. aeruginosa* NPCD-1 extract in ethyl acetate.

6. LIMITATIONS OF MOLECULAR BIOLOGY TO MONITOR CYANOTOXIN PRODUCTION

Some studies concerning the identification of genes responsible for microcystin biosynthesis suggested that the difference among microcystins-producing strains (toxic) and non-producers (non-toxic) was related primarily with the presence or absence of microcystin synthetase gene cluster (Meissner *et al.*, 1996; Nishizawa *et al.*, 2000). However, for the elucidation of cyanotoxin production in the aquatic environments, it is necessary to obtain both quantitative and qualitative information. Thus it is necessary to know the regulation of those genes; infer the mRNA levels of proteins and to qua ntify toxin production. The main problem concerning with PCR amplification of genes to monitoring cyanotoxin production is that not always the presence of these specific sequences is indicative of toxin production. Several studies showed that in some *Microcystis* (Kaebernick *et al.*, 2001; Mikalsen *et al.*, 2003; Nishizawa *et al.*, 1999; Tillett *et al.*, 2001) and *Planktothrix* (Kurmayer *et al.*, 2004) strains containing microcystin synthetase (*mcy*) genes, the toxin was not produced. It was suggested that the expression of the *mcy* genes might have been repressed in those strains, in particular as a consequence of environmental factors, for example, light intensity (Kaebernick *et al.*, 2000). Other concerning factor was the verification that even with the transcription of the genes *mcy*A and *mcy*B in *Microcystis* strain, the production of microcystin was not detected (Mikalsen *et al.*, 2003). This result raise questions whenever the regulation of microcystins production is not at transcription level, but at translation (or even post-translation). Several authors suggested that the inactivation of microcystin biosynthetic genes may occur due to spontaneous mutations in the *mcy* genes during long cultivation periods (Kaebernick *et al.*, 2001; Tillett *et al.*, 2001). However, the natural occurrence of nine *Planktothrix* strains containing the *mcy* genes, but not producing microcystins, was observed at several lakes in Europe (Kurmayer *et al.*, 2004). A phylogenetic study suggests that the sporadic distribution of microcystin synthetase genes in cyanobacteria is an inheritance of their old ancestral and that the absence of this metabolic pathway in several cyanobacterial genera and species is a result of repetitive deletions of the *mcy* genes during the evolution (Rantala *et al.*, 2004). However, the occurrence of lateral transfer events followed by recombination of *mcy* genes among different *Microcystis* strains has also been described as a possible cause of the genetic variation of the *mcy* gene cluster (Tanabe *et al.*, 2004). The authors hypothesize that plasmids or phages still unknown mediate genetic changes among different *Microcystis* strains. Supporting this idea, a possible transposase was identified in toxic *A. circinalis* (Pomati and Neilan, 2004), which, according to the authors, supports the hypothesis that lateral gene transfer may occur among toxic cyanobacterial strains. According to the studies mentioned above, it can predict that in some strains, microcystin genes are present, but were inactivated, or that their activation is controlled by certain unknown factors.

7. CONCLUSION

The molecular methods have revolutioned the systematic of organisms bringing advances in ecological studies. The amplification of DNA using PCR technique and specific primers targeting regions house keeping genes of cyanobacteria, such as rRNA 16S, *cpcBA*-IGS, *rpoC1*, has been successful used in studies of cyanobacterial diversity in several ecosystems. Those genes have low mutation rate and, therefore, they are quite conserved, since they are involved with the organism's primary metabolism. Nevertheless, they are not capable to distinguish toxic from nontoxic cyanobacteria. The PCR amplification of those genes can be used to identified genera and/or species known as toxins producers. However, the cyanotoxin biosynthesis occurs at the strains level. It fits to point out that a same bloom can contain toxic or non-toxic strains. Therefore, the detection of those genes does not necessarily indicate that there is production of cyanotoxins.

The discovery of *mcy* gene cluster in *Microcystis*, *Planktothrix* and *Anabaena* made possible the elaboration of specific primers for the detection of genes responsible for the synthesis of microcystins by PCR. Initially, it was believed that it was possible to relate the genotype of a microcystin-producing cyanobacterium (presence of the *mcy*) with its phenotype (microcystin synthesis). In this way, all the strains containing *mcy* genes would produce microcystins. However, molecular studies evidenced the differential expression of the genes and the production of those complex molecules in response to environmental factors. Research carried out both in laboratory and in the field showed that in some strains or blooms of *Microcystis* and *Planktothrix* containing the *mcy* genes, microcystins are not detected. There are evidences that the synthases expression, and consequent biosynthesis of the toxins, can be restricted to certain environmental conditions. It should be taken in consideration that the regulation studies of those genes were initiated only recently, therefore, more information are necessary to identify the factors that are interfering in this process. In a near future, the sequencing of toxic cyanobacterial genomes and the functional analysis of identified genes is expected to bring more information. Additional studies concerning the regulation of the toxin-biosynthetic genes, mainly to identify environmental or cellular factors that affect their transcription and expression are necessary to elucidate cyanotoxin production in the natural ecosystems. The discovery of the biological function of these toxins, which is still unknown, will probably contribute to understand the lack of expression of *mcy* genes in some strains. PCR amplification is, therefore, applicable only to detect the potentialities for cyanotoxin production. Therefore, it cannot be always considered as an indicative of the presence of these toxins. Thus, it is concluded that is still premature to use only DNA amplification procedures for routine analysis of water quality, considering the presence or absence of cyanotoxins.

Acknowledgements

The authors are grateful for the financial support provided by the Fundação de Amparo à Pesquisa do Estado de São Paulo (FAPESP 2003/12529-4 and FAPESP 2005/56303-5) and to L.O. Bonugli for technical assistance M.E. Silva-Stenico was the recipient of post-doctoral fellowship from FAPESP (2004/16042-5).

REFERENCES

Annila, A., Lehtimaki, J., Mattila, K., Eriksson, J.E., Sivonen, K., Rantala, T.T. and Drakenberg, T. (1996). Solution of nodularin: an inhibitor of serine/threonine-specific protein phosphatases. *J. Biol. Chem.* 271:16695-16702.

Asayama, M., Kabasawa, M., Takahashi, I., Aida, T. and Shirai, M. (1996). Highly repetitive sequences and characteristics of genomic DNA in unicellular cyanobacterial strains. *FEMS Microb. Let.* 137:175-181.

Baker, J.A., Entsch, B., Neilan, B.A. and Mckay, D.B. (2002). Monitoring changing toxigenicity of a cyanobacterial bloom by molecular methods. *Appl. Environ. Microb.* 68:6070-6076.

Banker, R. and Carmeli, S. (1997). Identification of cylindrospermopsin in *Aphanizomenon ovalisporum* (Cyanophyaceae) isolated from Lake Kinneret, Israel. *J. Phycol.* 33:613-616.

Becker, J.E., Moore, R.E. and Moore, B.S. (2004). Cloning, sequencing and biochemical characterization of the nostocyclopeptide biosynthetic gene cluster: molecular basis for imine macrocyclization. *Gene,* 325:35-42.

Bertasi, B., Bignotti, E., Ferrando, L., D'abrosca, F., Scaratti, L. and Pomati, F. (2003). The standardization of a molecular biology method to verify the presence of *Microcystis aeruginosa. Veterinary Research Communication* 27:277-279.

Birch, L. and Bachofen, R. (1990). Complexing agents from microorganisms. *Experientia,* 46:827-834.

Borchert, S., Patil, S.S. and Marahiel, M.A. (1992). Identification of putative multifunctional peptide synthetase genes using highly conserved oligonucleotide sequences derived from known synthetases. *FEMS Microb. Lett.* 92:175-180.

Burgoyne, D.L., Hemscheidt, T.K., Moore, R.E. and Runnegar, M.T.C. (2000). Biosynthesis of cylindrospermopsin. *J. Org.Chem.* 65:152-156.

Burja, A.M., Banaigs, B., Abou-Mansour, E., Burgess, J.G. and Wright, P.C. (2001). Marine cyanobacteria - a prolific source of natural products. *Tetrahedron* 57:9347-9377.

Burns, B.P., Saker, M.L., Moffitt, M.C. and Neilan, B.A. (2004). Molecular detection of genes responsible for cyanobacterial toxin production in the genera *Microcystis, Nodularia,* and *Cylindrospermopsis. Methods Mol. Biol.* 268:213-22.

Carmichael, W.W. (1992). Cyanobacterial secondary metabolites - the cyanotoxins. *J. Appl. Bacteriol.* 72:445-459.

Carmichael, W.W., Beasly, V., Bunner, D.L., Eloff, J.N., Falconer, I., Gorham, P., Harada, K.-I., Krishnamurty, T., Min-Juan, I., Moore, R.E., Rinehart, K., Runnegar, M., Skulberg, O.M. and Watanabe, M. (1988). Naming cyclic heptapeptide toxins of cyanobacteria (blue-green algae). *Toxicon* 26:971-973.

Challis, G.L., Ravel, J. and Townsend, C.A. (2000). Predictive, structure-based model of amino acid recognition by nonribosomal peptide synthetase adenylation domains. *Chem. and Biol.* 7:211-224.

Chang, Z., Flatt, P., Gerwick, W.H., Nguyen, V.A., Willis, C.L. and Sherman, D.H. (2002). The barbamide biosynthetic gene cluster: a novel marine cyanobacterial system of mixed polyketide synthase (PKS)-non-ribosomal peptide synthetase (NRPS) origin involving an unusual trichloroeucyl starter unit. *Gene* 296:235-247.

Chang, Z., Sitachitta, N., Rossi, J.V., Roberts, M.A., Flatt, P.M., Jia, J., Sherman, D.H., Gerwick, W.H. (2004). Biosynthetic pathway and gene cluster analysis of curacin A, an antitubulin natural product from the tropical marine cyanobacterium *Lyngbya majuscula. J. Nat. Prod.* 67:1356-1367.

Cho, J.E., Bang, S.W. and Han, M.S. (2004). Development of oligonucleotide primers for the detection of harmful Microcystis in water. *Bulletin of Environmental Contamination and Toxicology* 72:655-662.

Christiansen, G., Dittmann, E., Ordorika, L.V., Rippka, R., Herdman, M. and Börner, T. (2001). Nonribosomal peptide synthetase genes occur in most cyanobacterial genera as evidenced by their distribution in axenic strains of the PCC. *Arch. Microb.* 176:452-458.

Christiansen, G., Fastner, J., Erhard, M., Börner, T. and Dittmann, E. (2003). Microcystin biosynthesis in *Planktothrix*: genes, evolution and manipulation. *J. Bacteriol.* 185:564-572.

Codd, G. A., Lindsay, J., Young, F.M., Morrison, L.F. and Metcalf, J.S. (2005). Harmful cyanobacteria: from mass mortalities to management measures. In: Huisman, J., Matthijs, H.C.P. and Visser, P.M. Ed., Harmful cyanobacteria. Springer, Dordrecht, The Netherlands. pp. 1-24.

Dittmann, E., Meissner, K. and Börner, T. (1996). Conserved sequences of peptide synthetase genes in the cyanobacterium *Microcystis aeruginosa. Phycologia* 35:62-67.

Dittmann, E., Neilan, B.A., Erhard, M. von Döhren, H. and Börner, T. (1997). Insertional mutagenesis of a peptide synthetase gene that is responsible for hepatotoxin production in the cyanobacterium *Microcystis aeruginosa* PCC7806. *Molec. Microb.* 26:779-787.

Du, L., Sánchez, C. and Shen, B. (2001). Hybrid peptide-polyketide natural products: biosynthesis and prospects toward engineering novel molecules. *Metab. Engin.* 3:78-95.

Edwards, D.J., Marquez, B.L., Nogle, L.M., McPhail, K., Goeger, D.E., Roberts, M.A. and Gerwick, W.H. (2004). Structure and biosynthesis of the jamaicamides, new mixed polyketide-peptide neurotoxins from the marine cyanobacterium *Lyngbya majuscule. Chem. Biol.* 11:817-833.

Etchegaray, A., Silva-Stenico, M. E., Moon, D. H. and Tsai, S. M. (2004a). *In silico* analysis of nonribosomal peptide synthetase of *Xanthomonas axonopodis* pv. *citri*: identification of putative siderophore and lipopeptide biosynthetic genes. *Microbiol. Res.* 159:425-437.

Etchegaray, E., Rabello, E., Dieckmann, R., Moon, D.H., Fiore, M.F., Von Döhren, H., Tsai, S.M. and Neilan, B.A. (2004b). Algicide production by the filamentous cyanobacterium *Fischerella* sp. CENA19. *J. Appl. Phyc.* 16:237-243.

Fahrenkrug, P.M., Bett, M.B. and Parker, D.L. (1992). Base composition of DNA from selected strains of the cyanobacterial genus *Microcystis. Internat. J. System. Bacteriol.* 42:182-184.

Fawell, J., Hart, J., James, H. and Parr, W. (1993). Blue-green algae and their toxins: analysis, toxicity, treatment, and environmental control. *Water Supply* 11:109-121.

Fergusson, K.M. and Saint, C.P. (2003). Multiplex PCR assay for *Cylindrospermopsis raciborskii* and cylindrospermopsin producing cyanobacteria. *Environ. Toxicol.* 18:120-125.

Finking, R. and Marahiel, M.A. (2004). Biosynthesis of nonribosomal peptides. *Ann. Rev. Microbiol.* 58:453-488.

Harada, K., Ohtani, I., Iwamoto, K., Suzuki, M., Watanabe, M.F., Watanabe, M. and Terao, K. (1994). Isolation of cylindrospermopsin from a cyanobacterium *Umezakia natans* and its screening method. *Toxicon* 32:73-84.

Hisbergues. M., Christiansen, G., Rouhiainen, L., Sivonen, K. and Borner, T. (2003). PCR-based identification of microcystin-producing genotypes of different cyanobacterial genera. *Archives Microbiol.* 180:402-410.

Hoffmann, D., Hevel, J.M., Moore, R.E. and Moore, B.S. (2003). Sequence analysis and biochemical characterization of the nostopeptolide. A biosynthetic gene cluster from *Nostoc* sp. GSV224. *Gene* 311:171-180.

Honkanen, R.E., Zwiller, J., Moore, R.E., Daily, S.L., Khatra, B.S., Dukelow, M. and Boynton, A.L. (1990). Characterization of microcystin-LR, a potent inhibitor of type 1 and type 2A protein phosphatases. *J. Biolog. Chem.* 265:19401-19404.

Hu, X. and Boyer, G.L. (1996). Siderophore-mediated aluminum uptake by *Bacillus megaterium* ATCC 19213. *Appl. Environ. Microbiol.* 62:4044-4048.

Hutchinson, C.R. and Fujii, I. (1995). Polyketide synthase gene manipulation: a structure-function approach in engineering novel antibiotics. *Ann. Rev. Microbiol.* 49:201-238.

Kaebernick, M., Neilan, B.A., Börner, T. and Dittmann, E. (2000). Light and the transcriptional response of the microcystin biosynthesis gene cluster. *Appl. Environ. Microbiol.* 66:3387-3392.

Kaebernick, M., Rohrlack, T., Christoffersen, K. and Neilan, B.A. (2001). A spontaneous mutant of microcystin biosynthesis: genetic characterization and effect on *Daphnia. Environ. Microbiol.* 3:669-679.

Kato, T., Watanabe, M.F. and Watanabe, M. (1991). Allozyme divergence in *Microcystis* (Cyanophyceae) and its taxonomic inference. *Arch. Hydrobiol.* 92:129-140.

Komárek, J. (1991). A review of water-bloom forming *Microcystis* species, with regard to populations from Japan. *Algolog. Studies* 64:115-127.

Kurmayer, R. and Kutzenberger, T. (2003). Application of real-time PCR for quantification of microcystin genotypes in a population of the toxic cyanobacterium *Mycrocystis* sp. *Appl. Environ. Microbiol.* 69:6723-6730.

Kurmayer, R., Christiansen, G. and Chorus, I. (2003). The abundance of microcystin-producing genotypes correlates positively with colony size in *Microcystis* sp. and determines its microcystin net production in Lake Wannsee. *Appl. Environ. Microbiol.* 69:787-795.

Kurmayer, R., Christiansen, G., Fastner, J. and Börner, T. (2004). Abundance of active and inactive microcystin genotypes in population of the toxic cyanobacterium *Planktothrix* spp. *Environ. Microbiol.* 6:831-841.

Lagos, N., Onodera, H. Zagatto, P. A., Andrinolo, D., Azevedo, S.M.F.O. and Oshima, Y. (1999). The first evidence of paralytic shellfish toxins in the freshwater cyanobacterium *Cylindrospermopsis raciborskii*, isolated from Brazil. *Toxicon* 37:1359-1373.

Li, R.H., Carmichael, W.W., Brittain, S., Eaglesham, G.K., Shaw, G.R., Liu, Y.D. and Watanabe, M.M. (2001). First report of the cyanotoxins cylindrospermopsin and deoxycylindrospermopsin from *Rhaphidiopsis curvata* (Cyanobacteria). *J. Phycol.* 37:1121-1126.

Lorenzi, A.E. (2004). Abordagens moleculares para detectar cianobactérias e seus genótipos produtores de microcistinas presentes nas represas Billings e Guarapiranga, São

Paulo, Brasil. Piracicaba. M.Sc. thesis, Programa de Pós-Graduação em Energia Nuclear na Agricultura, Universidade de São Paulo, pp 92.

Marahiel, M.A., Stachelhaus, T. and Mootz, H.D. (1997). Modular peptide synthetases in nonribosomal peptide synthesis. *Chem. Rev.* 97:2651-2673.

Meissner, K., Dittmann, E. and Börner, T. (1996). Toxic and non-toxic strains of the cyanobacterium *Microcystis aeruginosa* contain sequences homologous to peptide synthethase genes. *FEMS Microbiol. Lett.* 133:295-3003.

Mikalsen, B., Boison, G., Skulberg, O.M., Fastner, J., Davies W., Gabrielsen, T.M., Rudi, K. and Jakobsen, K.S. (2003). Natural variation in the microcystin synthetase operon *mcyABC* and impact on microcystin production in *Microcystis* strains. *J. Bacteriol.* 185:2774-2785.

Moffitt, M.C. and Neilan B.A. (2000). The expansion of mechanistic and organismic diversity associated with non-ribosomal peptides. *FEMS Microbiol. Lett.* 191:159-167.

Moffitt, M.C. and Neilan B.A. (2001). On the presence of peptide synthetase and polyketide synthase genes in the cyanobacterial genus *Nodularia*. *FEMS Microbiol. Lett.* 196:207-214.

Moffitt, M.C. and Neilan, B.A. (2004). Characterization of the nodularin synthetase gene cluster and proposed theory of the evolution of cyanobacterial hepatotoxins. *Appl. Environ. Microbiol.* 70:6353-6362.

Neilan, B.A. (1995). Identification and phylogenetic analysis of toxigenic cyanobacteria by multipelx randomly amplified polymorphic DNA PCR. *Appl. Environ. Microbiol.* 61:2286-2291.

Neilan, B.A., Jacobs, D. and Goodman, A.E. (1995). Genetic diversity and phylogeny of toxic cyanobacteria determined by DNA polymorphisms within the phycocyanin locus. *Appl. Environm. Microbiol.* 61:3875-3883.

Neilan, B.A., Jacobs, D., Del Dot, T., Blackall, L.L., Hawkins, P.R., Cox, P.T. and Goodman, A.E. (1997). rRNA sequences and evolutionary relationships among toxic and nontoxic cyanobacteria of the genus *Microcystis*. *Internat. J. System. Bacteriol.* 47:693-697.

Neilan, B.A., Dittmann, E., Rouhiainen, L., Bass, R.A., Schaub, V., Sivonen, K. and Börner, T. (1999). Nonribosomal peptide synthesis and toxigenicity of cyanobacteria. *J. Bacteriol.* 181:4089-4097.

Nishizawa, T., Asayama, M., Fujii, K., Harada, K. and Shirai, M. (1999). Genetic analysis of the peptide synthetase genes for a cyclic heptapeptide microcystin in *Microcystis* spp. *J. Biochem.* 126:520-529.

Nishizawa, T., Ueda, A., Asayama, M., Fujii, K., Harada, K. and Shirai, M. (2000). Polyketide synthase gene coupled to the peptide synthetase module involved in the biosynthesis of the cyclic heptapeptide microcystin. *J. Biochem.* 127:520-529.

Nogle, L.M. and Gerwick, W.H. (2002). Somocystinamide A, a novel cytotoxic disulfide dimer from a Fijian marine cyanobacterial mixed assemblage. *Org. Lett.* 4:1095-1098.

Ohtani, I., Moore, R.E. and Runnegar, M.T.C. (1992). Cylindrospermopsin: a potent hepatotoxin from the blue-green alga *Cylindrospermopsis raciborskii*. *J. Amer. Chem. Soc.* 114:7942-7944.

Pomati, F. and Neilan, B.A. (2004). PCR-based positive hybridization to detect genomic diversity associated with bacterial secondary metabolism. *Nuc. Ac. Res.* 32:1-9.

Rantala, A., Fewer, D.P., Hisbergues, M., Rouhiainen, L., Vaitomaa, J., Börner, T. and Sivonen, K. (2004). Phylogenetic evidence for the early evolution of microcystin synthesis. *Proc. Natl. Acad. Sci.* U.S.A. 101:568-573.

Rinehart, K.L., Namikoshi, M. and Choi, B.M. (1994). Structure and biosynthesis of toxin from blue-green algae (Cyanobacteria). *J. Appl. Phycol.* 6:159-176.

Rouhiainen, L., Paulin, L., Suomalainen, S., Hyytiäinen, H., Buikema, W., Haselkorn, R. and Sivonen, K. (2000). Genes encoding synthetases of cyclic depsipeptides anabaenopeptilides, in *Anabaena* strain 90. *Molec. Microbiol.* 37:156-167.

Rouhiainen, L., Vakkilainen, T., Siemer, B.L., Buikema, W., Haselkorn, R. and Sivonen, K. (2004). Genes coding for hepatotoxic heptapeptides (Microcystins) in the cyanobacterium *Anabaena* strain 90. *Appl. Environ. Microbiol.* 70:686-692.

Rudi, K., Skulberg, O.M., Larsen, F. and Jakobsen, K.S. (1997). Strain characterization and classification of oxyphotobacteria in clone cultures on the basis of 16S rRNA sequences from the variable regions V6, V7, and V8. *Appl. Environ. Microbiol.* 63:2593-2599.

Sakamoto, T., Shirai, M., Asayama, M., Aida, T., Sato, A., Tanaka, K., Takahashi, H. and Nakano, M. (1993). Characteristics of DNA and multiple *rpoD* homologs of *Microcystis* (*Synechocystis*). *Int. J. Syst. Bacteriol.* 43:844-847.

Schwarzer, D., Finking, R. and Marahiel, M.A. (2003). Nonribosomal peptides: from genes to products. *Nat. Prod. Rep.* 20:275-287.

Schembri, M.A., Neilan, B.A. and Saint, C.P. (2001). Identification of genes implicated in toxin production in the cyanobacterium *Cylindrospermopsis raciborskii*. *Environm. Toxicol.* 16:472-482.

Silva-Stenico, M.E., Pacheco, F.T.H., Rodrigues, J.L.M. and Tsai, S.M. (2005). Growth and siderophore production of *Xylella fastidiosa* under iron-limited condition. *Microbiol. Res.* 160:429-436.

Sivonen, K. and Jones, G. (1999). Cyanobacterial toxin. In: Ed. Chorus, I. and Bartram, J. Toxic cyanobacterial in water: A guide to their public health consequences, monitoring, and management. London: Spoon. pp. 41-111.

Stachelhaus, T., Mootz, H.D. and Marahiel, M.A. (1999). The specificity-conferring code of adenylation domains in nonribosomal peptide synthetases. *Chem. Biol.* 6:493-505.

Tanabe, Y., Kaya, K. and Watanabe, M.M. (2004). Evidence for recombination in the microcystin synthetase (*mcy*) genes of toxic cyanobacteria *Microcystis* spp. *J. Molec. Evol.* 58:633-641.

Tillett, D., Dittman, E., Erhard, M., von Döhren, H., Börner, T. and Neilan, B.A. (2000). Structural organization of microcystin biosynthesis in *Microcystis aeruginosa* PCC7806: an integrated peptide-polyketide synthetase system. *Chem. Biol.* 7:753-764.

Tillett, D., Parker, D.L. and Neilan, B.A. (2001). Detection of toxigenicity by a probe for the microcystin synthetase A gene (*mcyA*) of the cyanobacterial genus *Microcystis*, comparison of toxicities with 16S rRNA and phycocyanin operon (phycocyanin intergenic spacer) phylogenies. *Appl. Environ. Microbiol.* 67:2810-2818.

Vaitomaa, J., Rantala, A., Halinen, K., Rouhiainen, L., Tallberg, P., Mokelke, L. and Sivonen, K. (2003). Quantitative real-time PCR for determination of microcystin synthethase E copy numbers for *Microcystis* and *Anabaena* in lakes. *Appl. Environ. Microbiol.* 69:7289-7297.

von Döhren, H., Keller, U., Vater, J. and Zocher, R. (1997). Multifunctional peptide synthetases. *Chem. Rev.* 97:2675-2705.

Walsh, C.T., Gehring, A.M., Weinreb, P.H., Quadri, L.E. and Flugel, R.S. (1997). Post-translational modification of polyketide and nonribosomal peptide synthases. *Curr. Opin. Chem. Biol.* 1:309-315.

Watanabe, M. (1996). Isolation, cultivation, and classification of bloom-forming *Microcystis* from Japan. In: Watanabe, M.F., Harada, K.-I, Carmichael, W.W. and Fujiki, H. (Ed.) Toxic *Microcystis*. Boca Raton: CRC Press, pp. 13-34.

Welker, M., Christiansen, G. and von Döhren, H. (2004a). Diversity of coexisting *Planktothrix* (Cyanobacteria) chemotypes deduced by mass spectral analysis of microcystins and other oligopeptides. *Arch. Microbiol.* 182:288-298.

Welker, M., Brunke, M., Preussel, K., Lippert, I. and von Döhren, H. (2004b). Diversity of distribution of *Microcystis* (Cyanobacteria) oligopeptide chemotypes from natural communities studied by single-colony mass spectrometry. *Microbiology* 150:1785-1796.

WHO (1998). Guidelines for Drinking-Water Quality. 2nd ed, Addendum to Vol. 2: Health criteria and other supporting information. Geneva: World Health Organization.

Wilmotte, A.(1994). Molecular evolution and taxonomy of the cyanobacteria. In: Bryant, D.A. (Ed.) The molecular biology of the cyanobacteria. The Netherlands: Kluwer, pp. 1-25.

❑❑❑

CHAPTER **3**

Fungal Biotransformations : A Powerful Tool in Drug Metabolism Studies

Mônica Tallarico Pupo[2], Keyller Bastos Borges[1], Warley de Souza Borges[2] and Pierina Sueli Bonato[1*]

[1]Departamento de Física e Química, Faculdade de Ciências Farmacêuticas de Ribeirão Preto, Universidade de São Paulo, Ribeirão Preto, Brasil
[2]Departamento de Ciências Farmacêuticas, Faculdade de Ciências Farmacêuticas de Ribeirão Preto, Universidade de São Paulo, Ribeirão Preto, Brasil

ABSTRACT

Biotransformation is an essential step in the elimination of drugs and reduction of their toxicity. The use of fungi in biotransformation reactions can provide information about the metabolism of new drugs and facilitate the attainment of necessary amount of metabolites for toxicological and pharmacological evaluation. Biotransformation studies require the right choice of the microorganisms and also of the culture conditions (pH, substrate, oxygen, salts, temperature, etc). The optimization of these procedures is essential to achieve good results. The fundamental aspects of drug biotransformation by fungi are discussed here as well as some applications showing that many fungi are used in biotransformation studies, mainly *Cunninghamella* species. Several species mimic the mammalian metabolism producing the same metabolites observed in humans.

Key words: fungal biotransformation, drug metabolism, stereoselective biotransformation.

1. INTRODUCTION

Before new chemical entities (NCE) have become ready to be marketed as a medicine, the three main phases that govern the activity of the drug, i. e., the *pharmaceutical phase*, the *pharmacokinetic phase* and the *pharmacodynamic phase* must be studied (Table 1). Thus, the knowledge of drug proprieties and other relevant aspects are required for that the new drug presents acceptable

*E mail: psbonato@fcfrp.usp.br

absorption, distribution, metabolism, excretion and toxicity (ADMET) properties (Li, 2001). Drugs have the therapeutically utility measured in function of the beneficial action that they exert on biological system. Drug pharmacological action depends on the administrated amount capable to reach, in suitable concentration, the therapeutical target (receptor, enzyme, and transport group, cellular or *in vivo* model). Thus concentration is influenced, among other things, by *in vivo* metabolism. Drug metabolism is a biochemical modification usually promoted by specialized enzymatic systems, to transform the drugs in to more hydrophilic species, so that they can be excreted *via* urine or faeces. Usually, the formation of water-soluble compounds not only enhances drug elimination, but also transforms drugs and xenobiotics into substances relatively nontoxic or inactive metabolites (Drayer, 1976). However, many drugs are biotransformed to pharmacologically active metabolites and many toxic side effects (e.g. tissue necrosis, carcinogenicity, teratogenicity) can be attributed to the formation of these reactive toxic metabolites (Drayer, 1982). In other cases, the administrated drug requires previous metabolism to express a pharmacological effect; this drug is called a prodrug.

Drug metabolism reactions are usually classified into two categories: Phase I (*Functionalization*) and Phase II (*Conjugation*) reactions. Phase I reactions refer to a set of reactions that result in more functionalized compounds. These reactions include, hydrolysis, reduction and oxidation, and they either introduce new functional groups (e.g. aromatic or aliphatic hydroxylation) or modify a previously existent polar functional group (e. g. -OH, -SH, $-NH_2$, -COOH, -COOR) in the molecule. Some examples of Phase I reactions are reduction of ketones and aldehydes to alcohols, oxidation of alcohols to acids, hydrolysis of ester and amines to COOH, NH_2 and OH groups, reduction of azo and nitro compounds to NH_2 moieties, *S*-, *N*- and *O*- dealkylation to SH, NH_2 and OH groups. Usually Phase I metabolized drugs are not hydrophilic enough to be excreted, and they can still retain biological activity. Then, these formed metabolites can, undergo subsequent enzymatic Phase II reactions, called conjugation, yielding more hydrophilic compounds that are easily excreted *via* urine or bile (Low, 1998).

Phase II reactions involve the conjugation of hydrophilic moieties on a suitable chemical group of the molecule. In most cases, Phase I reactions precede Phase II reactions because the functional group that will be added in Phase II requires a modification of the molecule, but Phase II reactions can occur in the original molecule if the compound has an appropriate site. The attached endogenous groups are small and polar compounds such as glucuronic acid, sulfate, glutathione and amino acids (e.g. glycine, taurine and glutamine), all of them increase the hydrophilicity of the molecule. Other conjugation reactions, such as *N*– acetylation of amines and *S*–, *N*– and *O*–methylation, can terminate or attenuate the biologic activity and generally result in more lipophilic products (Low, 1998).

TABLE 1. The three main phases for studies biological activity of a drug

Phase	Process	Research objectives
Pharmaceutical phase	Selection of the administration route	Optimization of the distribution
	Suitable pharmaceutical formulation	Optimization of the absorption
	Liberation and dissolution of drug	Eliminate unwanted organoleptic properties
Pharmacokinetic phase	Pharmaceutical availability of the drug in the organism: absorption, distribution, metabolism (biotransformation), excretion and reabsorption.	Optimization of the bioavailability: - Administered dose x Concentration at the site of action in function of the time
Pharmacodynamic phase	Interaction of drugs and metabolites with:	Optimization of therapeutics effects:
	- Targets of therapeutics activity (receptors, etc.)	- To obtain maximal activity
	- Targets of toxic effects	- To obtain maximal selectivity
	Type and intensity of the biological response	- To obtain minimal toxicity

*Adapted from Wermuth (2003).

2. DRUG METABOLISM MODELS

The cytochrome P450 (CYP450) is a super family of membrane bound, heme-containing enzymes that is the main system responsible for endogenous substrates and xenobiotics metabolism in humans. The current nomenclature for the several CYP isoforms was created according to conventions of molecular biology and it is based on three classifications: family, represented by an Arabic numeral; sub-family, represented by a capital letter; and an individual gene, represented by another Arabic numeral. In humans, about 90% of drug metabolism can be attributed to seven main isoenzymes: CYP1A2, CYP2B6, CYP2C9, CYP2C19, CYP2D6, CYP2E1 and CYP3A (Nebert *et al.*, 1991; Nelson *et al.*, 1996). Expression and enzymatic activity are regulated by genetic factors and external factors such as: alcohol, sex, diseases, tobacco and drug-drug interactions (Bauer *et al.*, 1994; Waltersack and Klotz, 1996; Porter and Coon, 1991). In addition, some studies have demonstrated that potential relationships exist between enzymatic activity and diseases, such as cancer and Parkinson's disease (Fleming *et al.*, 1992; Puchetti *et al.*, 1994; Romkes-Sparkes *et al.*, 1994; Rostami-Hodjegan *et al.*, 1998).

In vivo models could be used for drug metabolism studies. In this case, therapeutic doses of the drug are administered to healthy individuals, patients or animals and the plasmatic or urinary concentrations of the drug and its metabolites are evaluated. The studies of drug metabolism have also been carried out using probe substrate drugs to estimate the *in vivo* enzymatic activity. The antipyrine was widely used as a substrate because it is extensively metabolized by all enzymes. However, due to the multiplicity of CYP450 enzymes and substrate selectivity, the use of only one enzymatic marker does not demonstrate correctly the evaluation of all responsible enzymes for the metabolism. Thus, it is necessary to use a specific enzymatic marker whose metabolism is predominantly or exclusively mediated just by one cytochrome P450 isoform (Frye *et al.*, 1997). Several drugs have been used as specific substrates for a target CYP isoform. Debrisoquine is a well-known example of *in vivo* probe substrate of CYP2D6, recommended by FDA (Tucker *et al.*, 2001; Streetman *et al.*, 2000).

In vitro models have become an attractive alternative for drug metabolism studies mediated by cytochrome P450 enzymes. The search and development of *in vitro* methods for biotransformation studies started due to ethical, scientific and economic reasons. Biotransformation reactions in animals or humans are complex processes and the use of simple models, such as *in vitro* models, has its advantage (Brandon *et al.*, 2003). *In vitro* metabolism models currently available are based on sub-cellular fractions of human and animal liver (microsomes, supersomes, cytosol, S9 fraction, cell lines, transgenic cell lines), tissues, animals, commercially available enzymes, primary hepatocytes, liver slices and perfused liver (Brandon *et al.*, 2003; Tingle and Helsby, 2006).

Another alternative for *in vitro* metabolism studies consists in using biomimetic systems that reproduce the action of cytochrome P450 enzymes. Isolated enzyme systems or intact whole organisms (fungi, bacteria and yeasts) may be used for biotransformations (Aleu and Collado, 2001; Garcia-Granados *et al.*, 2004). These methods allow the study of specific biotransformation reactions under strictly controlled conditions. It is possible to use high concentrations of xenobiotic, that combined with a relatively clean matrix facilitate the analysis of the products. In addition, high amounts of products can be obtained. Currently, biocatalysis has become a promising tool in the development of new technologies for the synthesis of compounds with high aggregate value. Microbial models are useful in predicting, and sometimes necessary for preparing the metabolites of drugs administered to animals and to man (Vézina, 1987). Isolated enzymes are compatible with each other: sequential reactions can be performed by using multienzymatic systems in order to simplify reactions processes, in particular, if the isolation of an unstable intermediate can be omitted (Faber, 1997).

Further the use of microorganisms and isolated enzymes is cheaper compared to metallic catalysts and reduced number of reactional steps is required. Biotransformation reactions usually proceed in a pH range of 5-8, typically around 7, at room temperatures (20-40 °C), preferably at around 30 °C, and atmospheric pressures (0-34 kPa), an advantage over chemical processes which require significant input of energy. These typical conditions often minimize problems of undesired side-reactions such as decomposition, isomerization, racemization and rearrangement that are common in the traditional methodology (Faber, 1997). Biotransformations are generally environmentally safer, since the reactions are carried out in aqueous media, the enzymes are atoxic, and the cells can be completely degraded in the environment (Vézina, 1987). The drug concentration is higher than utilized in animal tissues, leading to increased amounts of the products that can be extracted and identified more easily.

In many cases the biological transformations are enantioselective, allowing the production of chiral compounds from racemic mixtures. Almost all enzymes are made from L-amino acids and thus are chiral catalysts. A prochiral substrate may be transformed into an optically active product and both enantiomers of a racemic substrate may react at different rates, affording a kinetic resolution. If microbial catalysts are used, specifically in the case of sulfoxides, high rates of enantioselectivity can be obtained (in some cases arrives until 99%).

Fungi can perform chemoselective reactions at one single functional group, even in the presence of another reactive chemical group. Moreover, regioselectivity and diastereoselectivity are also observed. Fungi may distinguish between functional groups which are chemically situated in different regions of the same substrate molecule. Some biocatalysts can accomplish reactions at sites that are difficult to access by organic synthesis e.g. the selective functionalization of non-activated positions in organic molecules, such as the hydroxylation of aliphatics chains (Faber, 1997).

Fungi are able to catalyze a broad spectrum of chemical reactions. They also exhibit a high substrate tolerance by accepting a large variety of synthetic substance (Faber, 1997). Their high surface-volume ratio confers rapid growth and high rates of metabolism leading to efficient transformation of the added substrate. Therefore, fungi play an important role in the metabolism studies that must accompany the clinical evaluation of drugs. Furthermore, the maintenance of stock cultures of microorganisms is clearly simpler and cheaper than the maintenance of cell or tissue cultures or laboratory animals.

In spite of a high number of advantages in using microorganisms for drug metabolism studies, this approach presents some disadvantages. Newly introduced microbial processes often require important investment to secure the equipment necessary to handle large volumes of biological reagents under aseptic conditions (Vézina, 1987). Enzymes are provided by nature in only one enantiomeric form, since there is no general way of creating mirror-image enzymes (from D-amino acids) it is impossible to invert the chiral induction of a given enzymatic reaction by choosing the other enantiomer of the biocatalyst, a strategy which is possible if chiral chemical catalysts are involved (Faber, 1997). Microbial biotransformations require narrow operation parameters since elevated temperatures as well as extreme pH lead to deactivation of the protein, as do high salt concentrations. Enzymes display their highest catalytic activity in water, but the majority organic compounds are only poorly soluble in aqueous media. However, this problem can be overcome by the solubilization of the substrate in an appropriate water soluble solvent, such as dimethyl sulfoxide, dimethylformamide, methanol etc. Enzymes are susceptible to inhibition phenomena: many enzymatic reactions are prone to substrate or product inhibition, which causes the enzyme to cease to work at higher substrate and/or product concentrations, a factor that could limit the efficiency of the process. In some cases, many products of biotransformation may be obtained also in few quantities that difficult the chemical analysis.

Whole cells have low tolerance to organic solvents, subproducts and produced metabolites. The large amount of biomass present in the reaction medium causes low overall yields and makes product recovery troublesome, particularly when the product is stored inside the cells and not excreted into the medium (Faber, 1997).

3. FUNGAL BIOTRANSFORMATION OF DRUGS

Biotransformations have been utilized by man for a long time. One of the first application of biotransformation was made in 1858 by Pasteur, using the fungus *Penicillium glaucum*. He obtained the ammonium L-tartrate from ammonium DL-tartrate, due to a selective degradation of the D-enantiomer (Hanson, 1995). However, only in 20th century, biotransformation studies have received more attention, since the development of microbial hydroxylation of bioactive steroids or intermediary products for corticosteroids synthesis. The a-hydroxylation of progesterone using *Rhizopus arrhizus* was described in 1952, in a single microbial step (Peterson *et al.*, 1952). This reaction was very important for the economic synthesis of adrenocortical hormones (corticosterone, cortisone and hydrocortisone) and afforded vast possibilities for the preparation of bioactive derivatives (prednisone, prednisolone and triamcinolone) (Vézina, 1987).

The 11b-hydroxylation is also crucial for the biological activity of steroids. This reaction was first reported in 1953 using *Cunninghamella blakesleeana* and *Curvularia lunata*. Cortexolone was hydroxylated at 11b position producting hydrocortisone in 60-70% yield by *C. lunata*, higher yields were obtained when the substrates were acetylated at positions 17a and 21 (Vézina, 1987).

Although fungal biotransformations have been extensively explored in the literature, the main purpose of these studies has been the production of additional bioactive compounds. The use of fungal biotransformation to study drug metabolism has been less explored and will be covered here.

Fungi produce enzymes that usually are secreted to the extracellular medium, which facilitates their recovery from the fermentation medium. The use of fungi in biotransformation process is also a promising field since drug metabolites and NCE can be generated in industrial scale. In addition, biocatalysis presents high degree of regio- and stereospecificity, versatility and high reaction rate in relatively mild reactions conditions.

The biotransformation experiments should be carefully planned. Some of the relevant aspects that require attention are: i) selection of the target compound and desired reaction (oxidation, reduction, hydroxylation, etc.), ii) evaluation of the microbial enzymatic activity or isolated enzymes, iii) determination of the ideal reaction conditions (temperature, time, pH, solvent, substrate concentration), iv) the choice of the appropriate analytical method (CG, CE, HPLC, RMN, etc), v) elucidation of the absolute configuration of products and vi) the scale-up of the reaction to large scale (industrial or preparative).

3.1. Basic Conditions for Biotransformations

Biotransformation experiments should be based on some previously described procedures in order to increase the efficiency of the process. The kind of chemical reaction of interest is very important. In early days, the selection of the microorganisms was done randomly, but today it can be based on prior observation that certain microorganisms perform the desired reaction with a related compound (Vézina, 1987). Screenings in small scale should be carried out in order to evaluate the microbial enzymatic activity. The reaction conditions, the method of analysis of the reaction (generally chromatographic and/or spectrometric methods), and the procedure for extraction of produced compounds should also be established.

One of the main aspects in biotransformations is the determination of the reaction conditions. A synthetic fermentative medium, with an exact and known chemical composition, allows an easier analysis of the metabolites. However, complex media are often employed in the routine laboratory work (Becker *et al.*, 1990). Those media are composed by complex nutrients, such as plant (malt, potato, soya) and animal extracts (peptone, tryptone), and their exact compositions are not known. Any medium for the cultivation of fungi must provide certain basic nutritional requirements, including a carbon source (glucose is an excellent carbon and energy source for microbial growth), water, nitrogen source, phosphate source, sulfate source, and various mineral nutrients, such as iron and magnesium (Becker *et al.*, 1990). In general, the culture media are easily prepared and are inexpensive. Media preparation usually requires an autoclave treatment for elimination of all microorganisms. Generally, a temperature of 121 °C for 20 minutes is used to heat sterilize the culture media (Becker *et al.*, 1990).

A suspension of vegetative cells or spores from agar plates or slants is used for the fungi inoculation in small flasks containing a suitable medium. After 1-2 days of incubation, the obtained mycelial mass is transferred to bigger flasks containing the substrate soluble in the fermentative medium, and incubated for an additional period of time. This procedure is convenient for the screening and allows the analysis of several conditions at the same time (medium, temperature, pH, carrier solvent, substrate concentration, incubation time). The cells must adjust to the new medium before start growing at a rapid rate. The length of the lag phase depends on a number of factors: age and genotype of the inoculums, temperature, nutrients levels of both media, aeration, and the concentration of toxins that may have been formed in the old medium (Becker *et al.*, 1990). Samples should be taken at time intervals and analyzed by chromatographic methods, to follow the course of transformation. This process usually takes 1-5 days (Vézina, 1987).

The substrate to be transformed is often dissolved in a small amount of suitable water-miscible solvent (acetonitrile, propylene glycol, ethanol, methanol, tert-butanol, acetone, dioxane, dimethyl sulfoxide, dimethylformamide) and can be added to growing cells, resting cells, spores, dried cells or to enzymes isolated from these cells. The transforming agents can be suspended into a medium for use in agitated aerated vessels, or packed in columns for continuous operation; they can be immobilized in solid support for increased stability and prolonged operation (Vézina, 1987). The most employed method is the addition of the substrate during or after the culture growth in agitated aerated vessels, since oxygen is frequently the limiting factor in fungal cultures. Better results can be obtained by the addition of the substrate during the logarithmic or exponential phase, when the fungus has reached the maximum growth. In this stage the microorganisms present fast growth and easy formation of the multienzymatic system. The addition of the substrate in this stage can induce the formation of enzymes which normally are not produced by the microorganism, increasing the efficiency of the process. The recuperation of the products may be done by extraction with appropriate solvents, precipitation, adsorption, and also distillation of the culture media when the compounds are volatile.

3.2. Relevant Examples of Drug Metabolism by Fungi

The biotransformation of drugs by fungi has been revised by Azerad (1999). Some examples cited in this review as well as more recent examples are discussed below. In recent addition, among several fungi used in process of biotransformation, the genus *Cunninghamella* deserves prominence. Figures 1 - 3 show the structures of the drugs discussed in this section.

3.2.1 Cunninghamella species

The filamentous fungus C. elegans has been extensively studied for drug biotransformation. This fungus metabolized chlorpromazine and methdilazine within 72h. The major metabolites obtained from chlorpromazine were chlorpromazine sulfoxide (36%), N-desmethylchlorpromazine (11%), N-desmethyl-7-hydroxychlorpromazine (6%), 7-hydroxychlorpromazine sulfoxide (5%), and chlorpromazine N-oxide (2%), all of which have been found in animal studies. The major metabolites of methdilazine were 3- hydroxymethdilazine (35%), methdilazine sulfoxide (30%), methdilazine N-oxide (4%), phenothiazine (3%) and 2-hydroxymethdilazine (Zhang *et al.*, 1996).

This fungus has also been used in the biotransformation of artemether to five metabolites, 9β-hydroxyartemether, 9α-hydroxyartemether, ring-rearranged 9β-hydroxyartemether, 3α-hydroxydeoxyartemether and 14-hydroxyartemether (Abourashed and Hufford, 1996). Artemisinin, a similar compound, was biotransformed into 7β-hydroxyartemisinin (78.6%), 7β-hydroxy-9α-artemisinin (6.0%), 4α-hydroxy-1-deoxoartemisinin (5.4%), 6β-hydroxyartemisinin (6.5%) (Parshikov *et al.*, 2004).

C. elegans biotransformed approximately 74% of the drug diphenhydramine. The major metabolites observed were diphenhydramine-N-oxide (3.0%), N-desmethyldiphenhydramine (30.0%), N-acetyldidesmethyldiphenhydramine (13.0%) and N-acetyl-N-desmethyldiphenhydramine (12.0%) (Moody *et al.*, 2000).

Moody *et al.* (2001) also used this fungus to biotranform amoxapine into three metabolites: the major mammalian metabolite, 7-hydroxyamoxapine, and two other novel metabolites, N-formyl-7-hydroxyamoxapine and N-formylamoxapine.

Choudhary *et al.* (2005) reported that microbial biotransformation of prednisone (17α-21-dihydroxy-pregna-1,4-diene-3,11,20-trione) by *C. elegans* presents two metabolites, 17α-21-dihydroxy-5α-pregn-1-ene-3,11,20-trione and 17α-20S,21-trihydroxy-5-α-pregn-1-ene-3,11-dione, while the fermentation of prednisone with *Fusarium lini*, *Rhizopus stolonifer* and *Curvularia lunata* formed the 1,4-pregnadiene-17α-20S,21-triol-3,11-dione metabolite.

Another species frequently used in biotransformation of drugs is *C. blakesleeana*. Ibrahim (1997) reported the microbial biotransformation of emetine into O-methylpsychotrine by *C. blakesleeana* MR-198. Further, this fungus was used to oxidize the tert-butyl group of terfenadine to the corresponding alcohol 1-[4-(1,1-dimethyl-2-hydroxyethyl)phenyl]-4-[4-hydroxydiphenylmethyl)-1-piperidinyl]-1-butanol (Schmitz *et al.*, 2000).

More recently, the filamentous fungus *C. blakesleeana* AS 3.153 was used as a microbial model of mammalian metabolism to transform verapamil into 23 metabolites (Liu *et al.*, 2002; Sun *et al.*, 2004).

After 96 hours of incubation with *C. blakesleeana* AS 3.153, approximately 92.5% of pantoprazole was metabolized to six metabolites: pantoprazole sulfone (1.7%), pantoprazole tioether (12.4%), 6-hydroxy-pantoprazole thioether (1.3%), 4'-O-demethyl-pantoprazole thioether (48.1%), pantoprazole thioether-1-N-b-glucoside (20.6%) and a glucoside conjugate of pantoprazole thioether (8.4%) (Xie *et al.*, 2005).

Three phase I metabolites of indomethacin (O-desmethylindomethacin, N-deschlorobenzoylindomethacin and O-desmethyl-N-deschlorobenzoylindomethacin) were formed by *C. blakesleeana* (Zhang *et al.*, 2006), whereas 95% of the hypoglycemic drug tolbutamide was biotransformed into 4'-hydroxytolbutamide in 96 hours (Huang *et al.*, 2006).

3.2.2 Other Species of Fungi Used in Drug Biotransformation

The metabolites 39-O-demethylrapamycin, 27-O-demethylrapamycin, 16-O-demethylrapamycin, 40-O-phosphoric ester of rapamycin were isolated from the culture of *Thamnidium elegans* incubated with rapamycin. Biotransformation of rapamycin using Syncephalastrum racemosus or *Gliocladium deliquescen* yielded the conversion products 24-hydroxyrapamycin, secorapamycin A, secorapamycin B, secorapamycin C, 16-O-demethylrapamycin B. Some derivatives exhibited a stronger immunosuppressive effect than the parent compound (Kuhnt *et al.*, 1997).

The drug podophyllotoxin was biotransformed to picropodophyllin (8%) by *Penicillium citreoviride* and *Penicillium melini* (Guo *et al.*, 1998).

Ciprofloxacin and norfloxacin were biotransformed by *Pestalotiopsis guepini*. The major metabolites of ciprofloxacin were N-acetylciprofloxacin (52.0%), desethylene-N-acetylciprofloxacin (9.2%), N-formylciprofloxacin (4.2%), 7-amino-1-cyclopropil-6-fluoro-4-oxo-1,4-dihydroquinoline-3-carboxylic acid (2.3%). Norfloxacin metabolites included N-acetylnorfloxacin (55.4%), desethylene-N-acetylnorfloxacin (8.8%), N-formylnorfloxacin (3.6%), 7-amino-1-ethyl-6-fluoro-4-oxo-1,4-dihydroquinoline-3-carboxylic acid (2.1%) (Parshikov *et al.*, 2001). The fungus *Mucor ramannianus* metabolized also ciprofloxacin to N-acetylciprofloxacin (Parshikov *et al.*, 1999).

The ability of the fungus *Beauveria bassiana* ATCC 7159 to biotransform the cinoxacin into two metabolites (1-ethyl-1,4-dihydro-3-(hydroxymethyl) [1,3] dioxolo [4,5-g] cinnolin-4-one and 1-ethyl-1,4-dihydro-6,7-dihydroxy-3-(hydroxymethyl)cinnolin-4-one) was reported by Parshikov *et al.* (2002).

The fungus *Aspergillus ochraceus* biotransformed the irbesartan into keto derivative, two monohydroxylated resulting from the oxidation of the spirocyclopentane ring (Alexandre *et al.*, 2004).

Artemisinin was transformed into 7β-hydroxy-9α-artemisinin and 5β-hydroxyartemisinin by fungi Eurotium amstelodami and *Aspergillus niger* (Parshikov *et al.*, 2006). In addition, the authors reported the biotrasformation of artemisinin into 7β-hydroxyartemisinin (88%) and 6β-hydroxyartemisinin (51%) by *Mucor ramannianus* (Parshikov *et al.*, 2005).

Among several fungi studied, the *Cunninghamella echinulata* and *Rhizopus arrhizus* biotransformed the trimegestone in hydroxylated derivatives (Lacroix *et al.*, 1999).

3.3 Fungal Stereoselective Biotransformation of Drugs

Individual enantiomers of racemic drugs often are metabolized at different rates and by different routes. In addition, drug biotransformation process may create a new asymmetric center in the metabolite (i.e. stereoisomeric or enantiomeric products) (Low, 1998). The previous examples reported here demonstrated that some fungi can generate a similar metabolic profile to mammalian. In addition, the metabolism could also be stereoselective (Clark and Hufford, 1991). Thus, stereoselective microbial biotransformation of drugs is a powerful tool to study mammalian metabolism.

The fungi *Rhizopus arrhizus* and *Geotrichum candidum* biotransformed (±)-atenolol to (S)-atenolol in 99% and 91%, respectively. The same fungi biotransformed (±)-propranolol to (S)-propranolol in good optical yield of 88% and 78%, respectively (Damle *et al.*, 2000).

About 91% of the drug mirtazapine was biotransformed by *Cunninghamella elegans* into the seven metabolites: 8-hydroxymirtazapine, N-desmethyl-8-hydroxymirtazapine, N-desmethylmirtazapine, 13-hydroxymirtazapine, mirtazapine N-oxide, 12-hydroxymirtazapine and N-desmethyl-13-hydroxymirtazapine. Circular dichroism spectral analysis of unused mirtazapine indicated that it was slightly enriched with the R(-)-enantiomer. When the fungus was treated with the optically pure forms of the drug, the S(+)-enantiomer produced all the seven metabolites whereas the R(-)-enantiomer produced only 8-hydroxymirtazapine, N-desmethyl-8-hydroxymirtazapine and mirtazapine N-oxide. (Moody *et al.*, 2002).

Freitag *et al.* (1997) demonstrated that *Cunninghamella echinulata* stereoselectively converted the drug rac-mexiletine into hydroxymethyl mexiletine (4% of added drug) and p-hydroxymexiletine (32% of added drug). About 30.1 mg/mL of R + S-hydroxymethylmexiletine and 64.6 mg/mL of R + S-p-hydroxymexiletine were formed.

Thomason *et al.* (1996) described the incubation of ibuprofen in non-growing cultures of *Verticillium lecanii*, and the results indicated that racemic ibuprofen is converted into a 70:30 (S:R) enantiomeric mixture after 6 days of incubation in the absence of any metabolism. This suggests that *V. lecanii* could be an useful model for the investigation of the mechanism of mammalian chiral inversion of ibuprofen.

4. CONCLUSIONS

Today, microbial biotransformation is important for drug mammalian metabolism studies. These in vitro models do not require animals and offer a

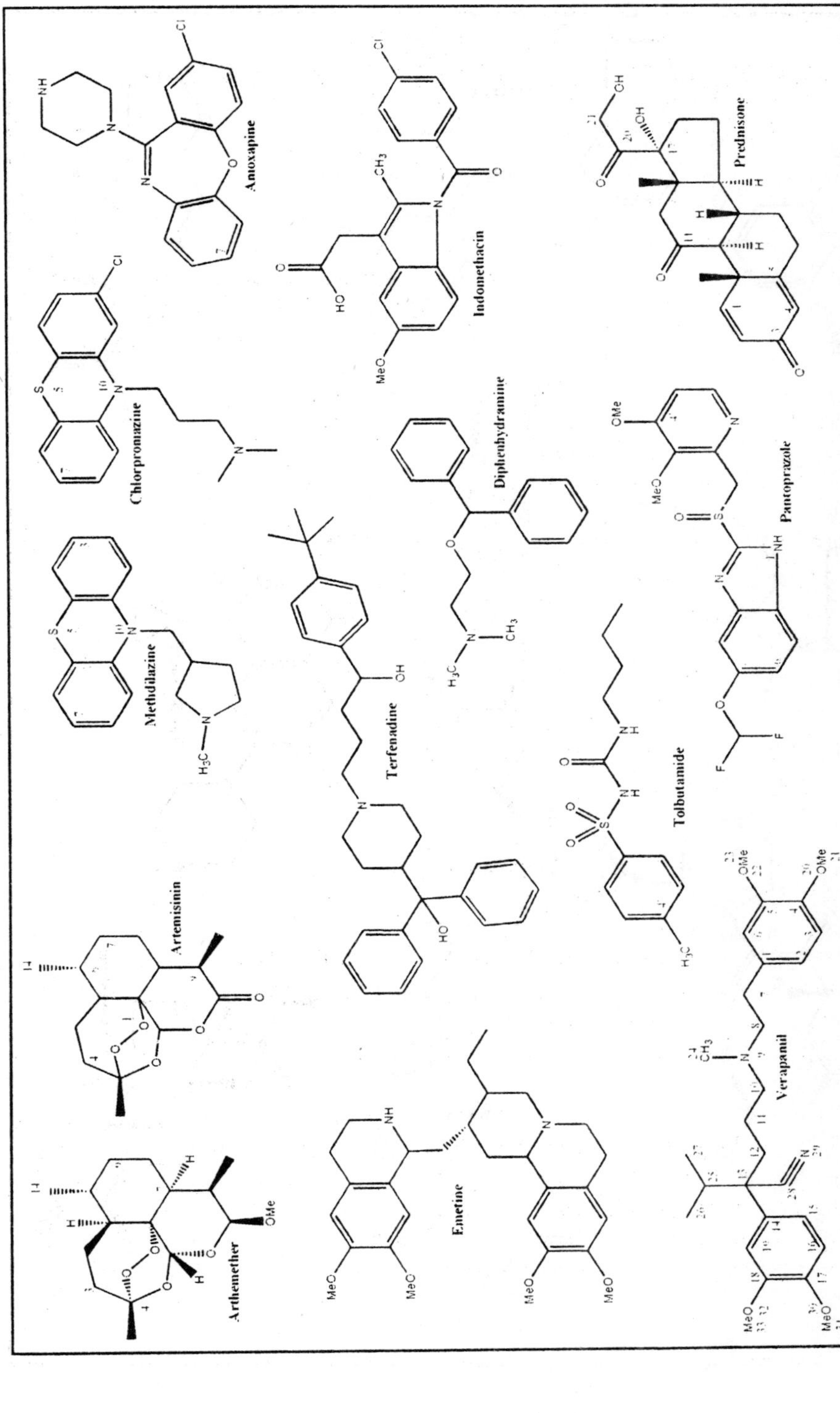

FIG. 1. Chemical structure of drugs used in biotransformation studies by *Cunninghamella* species.

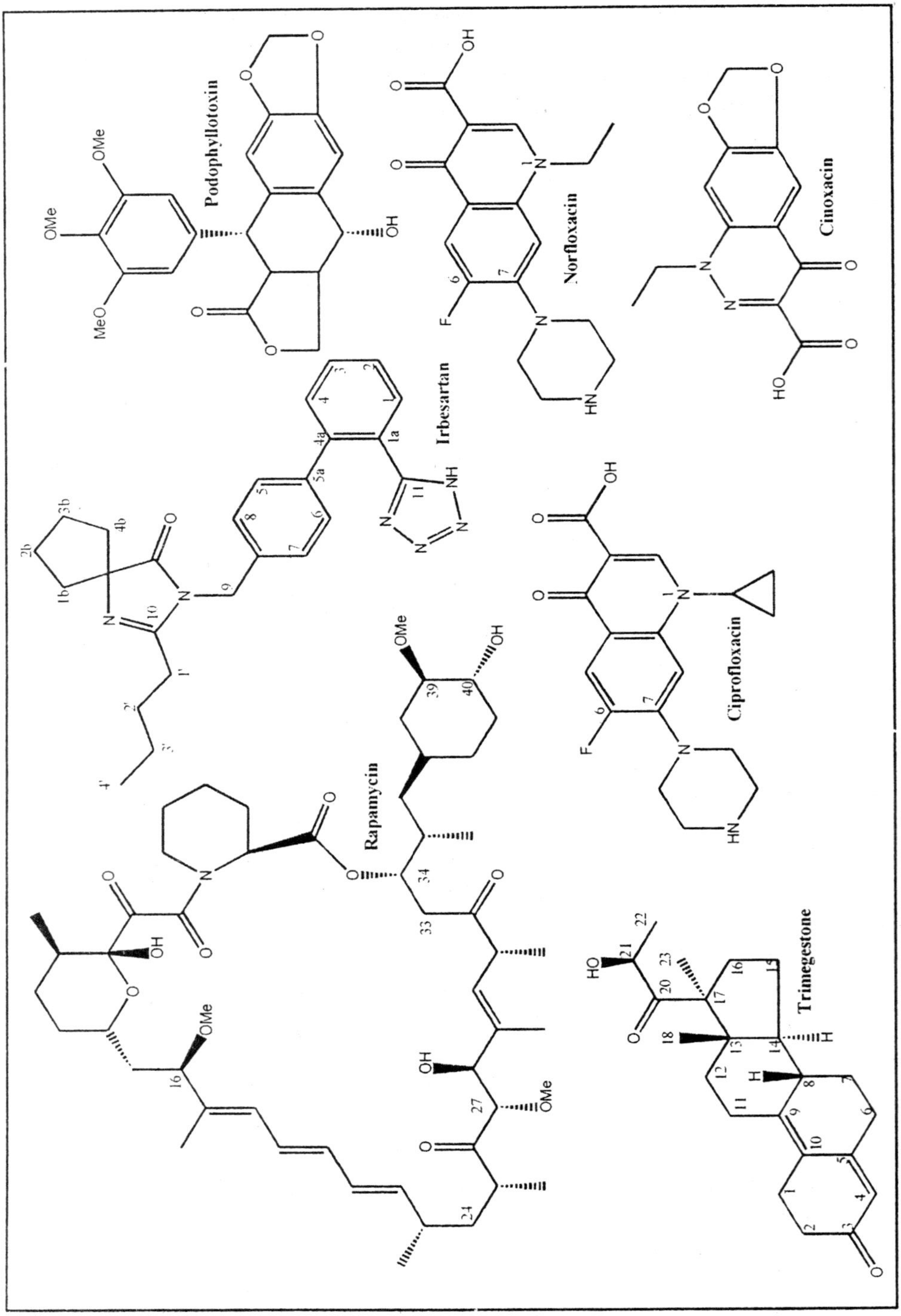

FIG. 2. Chemical structure of drugs used in biotransformation studies by fungi species.

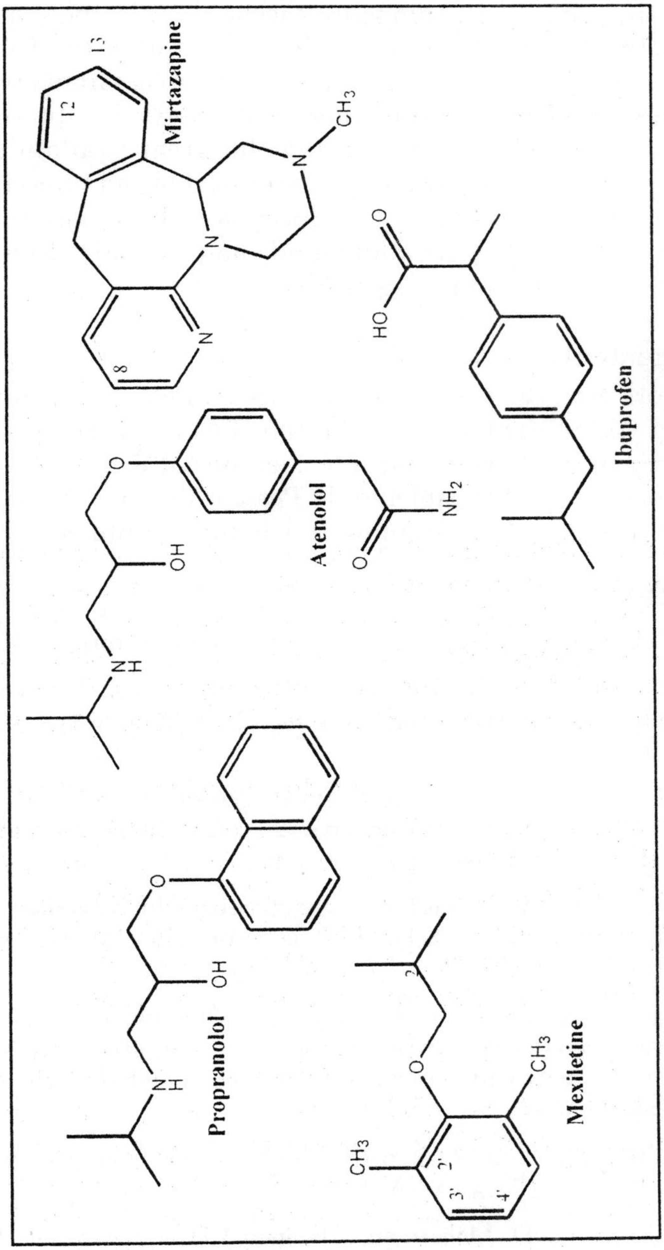

FIG. 3. Chemical structure of drugs used in stereoselective biotransformation studies by fungi species

less complex way to elucidate the human biotransformation pathway of a new drug. In addition, biotransformation reactions are considered as economically and ecologically competitive technology by the synthetic organic chemists, in the search of new production routes for fine chemical, pharmaceutical and agrochemical compounds (García-Granados *et al.*, 2004). However, the use of microorganisms by industry to obtain interesting compounds is still modest, considering the great availability of useful microorganisms and the large scope of reactions that can be accomplished by them. Although stereoselective biotransformation studies are very important to obtain enantiomerically pure compounds and to study drug metabolism, few studies are described in the literature.

Acknowledgements

The authors are grateful to Fundação de Amparo à Pesquisa do Estado de São Paulo (FAPESP), sub-program BIOTA/FAPESP (Rede BIOprospecTA), Conselho Nacional de Desenvolvimento Científico e Tecnológico (CNPq) and to Coodernação de Aperfeiçoamento de Pessoal de Nível Superior (CAPES) for financial support and for granting research fellowships.

REFERENCES

Abourashed, E. A. and Hufford, C. D. (1996). Microbial transformation of artemether. *J. Nat. Prod.* 59:251-253.

Aleu, J. and Collado, I. G. (2001). Biotransformations by *Botrytis* species. *J. Mol. Catal. B-Enzym.* 13:77-93.

Alexandre, V., Ladril, S., Maurs, M. and Azerad, R. (2004). Microbial models of animal drug metabolism Part 5. Microbial preparation of human hydroxylated metabolites of irbesartan. *J. Mol. Cat. B-Enzym.* 29:173-179.

Azerad, R. (1999). Microbial models for drug metabolism. In: Advances in Biochemical Engineering Biotechnology, Scheper, T. (ed). Springer - Verlag, Berlin, Germany, pp.169-218.

Bauer, L. A., O'Sullivan, T., Reiss, W. G., Horn, J. R., Opheim, K. and Strandness, D. E. (1994). Liver blood flow, antipyrine clearance, and antipiryne metabolite formation clearance in patients with chronic active hepatitis and alcoholic cirrhoris. *Br. J. Clin. Pharmacol.* 37:375-381.

Becker, J. M., Caldwell, G. A. and Zachgo, E. A. (1990). Biotechnology: a laboratory course. Academic Press, San Diego, EUA, 232pp.

Brandon, E. F. A., Raap, C. D., Meijerman, I., Beijnen, J. H. and Schellens, J. H. M. (2003). An update on *in vitro* test methods in human hepatic drug biotransformation research: pros and cons. *Toxicol. Appl. Pharmacol.* 186:233-246.

Choudhary, M. I., Siddiqui, Z. A., Musharraf, S. G., Nawaz, S. A. and Rahman, A. U. (2005). Microbial transformation of prednisone. *Nat. Prod. Res.* 19:311-317.

Clark, A. M. and Hufford, C. D. (1991). Use of microorganisms for the study of drug metabolism: an update. *Med. Res. Rev.* 11:473-501.

Damle, S. V., Patil, P. N. and Salunkhe, M. M. (2000). Biotransformations with *Rhizopus arrhizus* and *Geotrichum candidum* for the preparation of (*S*)-Atenolol and (*S*)-Propranolol. *Bioorgan. Med. Chem.* 8:2067-2070.

Drayer, D. E. (1976). Pharmacologically active drug metabolites: therapeutic and toxic activities, plasma and urine data in man, accumulation in renal failure. *Clin. Pharmacokinet.* 1:426-443.

Drayer, D. E. (1982). Pharmacologically active metabolites of drugs and other foreign compounds. Clinical, pharmacological, therapeutic and toxicological considerations. *Drugs* 24:519-542.

Faber, K. (1997). Biotransformations in organic chemistry: a text book. Springer - Verlag, 3 ed. Berlin, Germany, pp.402.

Fleming, C. M., Kaisary, A., Wilkinson, G. R., Smith, P. and Branch, R. A. (1992). The ability to 4-hydroxylate debrisoquine is related to recurrence of bladder cancer. *Pharmacogenetics* 2:128-134.

Freitag, D. G., Foster, R. T., Coutts, R. T., Pickards, M. A. and Pasutto, F. M. (1997). Stereoselective metabolism of *rac*-mexiletine by the fungus *Cunninghamella echinulata* yields the major human metabolites hydroxymethylmexiletine and *p*-hydroxymexiletine. *Drug Metab. Dispos.* 25:685-692.

Frye, R. F., Matzke, G. R., Adedoyin, A., Porter, J. A. and Branch, R. A. (1997). Validation of the five-drug "Pittsburgh cocktail" approach for assessment of selective regulation of drug-metabolizing enzymes. *Clin. Pharmacol. Ther.* 62:365-376.

García-Granados, A., Fernández, A., Gutierrez, M. C., Martínez, A., Quirós, R., Rivas, F. and Arias, J. M. (2004). Biotransformation of ent-13-epi-manoyl oxides difunctionalized at C-3 and C-12 by filamentous fungi. *Phytochemistry* 65:107-115.

Guo, H. Z., Guo, D. A., Fei, X. Y., Cui, Y. J. and Zheng, J. H. (1998). Biotransformation of podophyllotoxin to picropodophyllin by microbes. *J. Asian Nat. Prod. Res.* 1:99-102.

Hanson, J. R. (1995). An introduction to biotransformations in Organic Chemistry. In: J. Man (ed.), Biochemical & Medicinal Chemistry Series, W. H. Freeman, Oxford, England: pp.1-89.

Huang, H., Yang, X., Li, Q., Sun, L. and Zhong, D. (2006). Biotransformation of tolbutamide to 4'-hydroxytolbutamide by the fungus *Cunninghamella blakesleeana*. *Appl. Microbiol. Biotech.* 72:486-491.

Ibrahim, A. R. S. (1997). Microbial transformation of emetine into *O*-methylpsychotrine. *Saudi Pharm. J.* 5:52-55.

Kuhnt, M., Bitsch, F., Ponelle, M., Fehr, T. and Sanglier, J. J. (1997). Microbial conversion of rapamycin. *Enzyme Microb. Tech.* 21:405-412.

Lacroix, I., Biton, J. and Azerad, R. (1999). Microbial models of drug metabolism: microbial transformations of trimegestone® (RU27987), a 3-Keto-Ä$^{4,9(10)}$-19-norsteroid drug. *Bioorg. Med. Chem.* 7:2329-2341.

Li, A. P. (2001). Screening for human ADME/Tox drug properties in drug discovery. *Drug Discov. Today* 6:357-366.

Liu, L., Huang, H. H. and Zhong, D. F. (2002). Capability of *Cunninghamella blakesleeana* to transform verapamil. *Chinese J. Pharmacol. Toxicol.* 16:292-298.

Low, L. K. (1998). Metabolic changes of drugs and related organic compounds. In: Delgado, J. N. and Willian, A. R. (eds.) Textbook of organic medicinal and pharmaceutical chemistry. 10th Edition. Lippincott-Raven Publishers, Philadelphia, USA, pp. 43-122.

Moody, J. D., Heinze, T. M., Hansen, Jr. E. B. and Cerniglia, C. E. (2000). Metabolism of the ethanolamine-type antihistamine diphenhydramine (Benadryl)^{TM} by the fungus *Cunninghamella elegans. Appl. Microbiol. Biotechnol.* 53:310-315.

Moody, J. D., Heinze, T. M. and Cerniglia, C. E. (2001). Fungal transformation of the tricyclic antidepressant amoxapine: identification of *N*-carbomethoxy compounds formed as artifacts by phosgene in chloroform used for the extraction of metabolites. *Biocatal. Biotransfor.* 19:155-161.

Moody, J. D., Freeman, J. P., Fu, P. P. and Cerniglia, C. E. (2002). Biotransformation of mirtazapine by *Cunninghamella elegans. Drug Metab. Dispos.* 30:1274-1279.

Nebert, D. W., Nelson, D. R., Coon, M. J., Estabrook, R. W., Feyereisen, R., Fujii-Kuriyama, Y., Gonzalez, F. J., Guengerich, F. P. and Gunsalus, I. C. (1991). The P450 superfamily: update on new sequences, gene mapping, and recommended nomenclature. *DNA Cell Biol.* 10:1-14.

Nelson, D. R., Koymans, L., Kamataki, T., Stegeman, J. J., Feyereisen, R., Waxman, D. J., Waterman, M. R., Gotoh, O. and Coon, M. J. (1996). P450 superfamily: update on new sequences, gene mapping, accession numbers and nomenclature. *Pharmacogenetics* 6:1-42.

Parshikov, I. A., Freeman, J. P., Lay-Jr, J. O., Beger, R. D., Williams, A. J. and Sutherland, J. B. (1999). Regioselective transformation of ciprofloxacin to *N*-acetylciprofloxacin by the fungus *Mucor ramannianus. FEMS Microbiol. Lett.* 177:131-135.

Parshikov, I. A., Heinze, T. M., Moody, J. D., Freeman, J. P., Willians, A. J. and Sutherland, J. B. (2001). The fungus *Pestalotiopsis quepini* as a model for the biotransformation of ciprofloxacin and norfloxacin. *Appl. Microbiol. Biotech.* 56:474-477.

Parshikov, I. A., Moody, J. D., Heinze, T. M., Freeman, J. P., Willians, A. J. and Sutherland, J. B. (2002). Transformations of cinoxacin by *Beauveria bassiana. FEMS Microbiol. Lett.* 214:133-136.

Parshikov, I. A., Muraleedharan, K. M. and Avery, M. A. (2004). Transformation of artemisinin by *Cunninghamella elegans*. *Appl. Microbiol. Biotechnol.* 64:782–786.

Parshikov, I. A., Miriyala, B., Muraleedharan, K. M., Illendula, A., Avery, M. A. and Williamson, J. S. (2005). Biocatalysis of the antimalarial artemisinin by *Mucor ramannianus* strains. *Pharm. Biol.* 43:579-582.

Parshikov, I. A., Miriyala, B., Muraleedharan, K. M., Avery, M. A. and Williamson, J. S. (2006). Microbial transformation of artemisinin to 5-hydroxyartemisinin by *Eurotium amstelodami* and *Aspergillus niger*. *J. Ind. Microbiol. Biotechnol.* 33:349-352.

Peterson, D. H., Murray, H. C., Eppstein, S. H., Reineke, L. M., Weintraub, A., Meister, P. D. and Leigh, H. M. (1952). Microbial transformations of steroids. I. Introduction of oxygen at Carbon-11 of progesterone. *J. Am. Chem. Soc.* 74:5933-5936.

Porter, T. D. and Coon, M. J. (1991). Cytochrome P450: multiplicity of isoforms, substrates and catalytic and regulatory mechanisms. *J. Biol. Chem.* 266:13469-13472.

Puchetti, V., Faccini, G., Micciolo, R., Ghimenton, F., Bertrand, C. and Zatti, N. (1994). Dextromethorphan test for evaluation of congenital predisposition to lung cancer. *Chest* 105:449-453.

Romkes-Sparkes, M., Mnuskin, A., Chen, H. - D., Persad, R., Fleming, C. and Sibley, G. N. A. (1994). Correlation of polymorphic expression of CYP2D6 mRNA in bladder mucosa and tumor tissue to *in vivo* debrisoquine hydroxylase activity.. *Carcinogenesis* 15:1995-1961.

Rostami-Hodjegan, A., Lennard, M. S., Woods, H. F. and Tucker, G. T. (1998). Metabolism of studies of the CYP2D6 polymorphism in relation to lung cancer and Parkinson's disease. *Pharmacogenetics* 8:227-238.

Schmitz, G., Franke, D., Stevens, S., Takors, R., Weuster-Botz, D. and Wandrey, C. (2000). Regioselective oxidation of terfenadine with *Cunninghamella blakesleeana*. *J. Mol. Cat. B-Enzym.* 10:313-324.

Streetman, D. S., Bertino, J. S. Jr. and Nafziger, A. N. (2000). Phenotyping of drug-metabolizing enzymes in adults: a review of *in vivo* cytochrome P450 phenotyping probes. *Pharmacogenetics* 10:187-216.

Sun, L., Huang, H. H., Liu, L. and Zhong, D. F. (2004). Tranformation of verapamil by *Cunninghamella blakesleeana*. *Appl. Environ. Microb.* 70:2722-2727.

Tingle, M. D. and Helsby, N. A. (2006). Can *in vitro* drug metabolism studies with human tissue replace *in vivo* animal studies? *Environ. Toxicol. Pharmacol.* 21:184-190.

Thomason, M. J., RhysWilliams, W., Hutt, A. J., Lloyd, A. W. and Hanlon, G. W. (1996). Demonstration of the chiral inversion of ibuprofen by *Verticillium lecanii* in non-growing cultures. *Lett. Appl. Microbiol.* 23:417-420.

Tucker, G. T., Houston, J. B. and Huang, S. M. (2001). Optimizing drug development: strategies to assess drug metabolism/transporter interaction potential towards a consensus. *Br. J. Clin. Pharmacol.* 86:215-221.

Vézina, C. (1987). Biotransformation. In: Bu'lock, J. and Kristiansen, B. (eds). Basic biotechnology, Academic Press, San Diego, USA. pp.463-482.

Waltersack, I. and Klotz, U. (1996). Influence of diet and nutritional status on drug metabolism. *Clin. Pharmacokinet.* 31:375-381.

Wermuth, C. G. (2003). Medicinal Chemistry: definition and objectives, the three main phases of drug activity, drug and disease classifications. In: Camille G. Wermuth (Ed.) The Practice of Medicinal Chemistry, 2nd Edition. Elsevier Academic Press London, UK, pp. 29-40.

Xie, Z. Y., Huang, H. H. and Zhong, D. F. (2005). Biotransformation of pantoprazole by the fungus *Cunninghamella blakesleeana*. *Xenobiotica* 35:467-477.

Zhang, D. L., Freeman, J. P., Sutherland, J. B., Walker, A. E., Yang, Y. and Cerniglia, C. E. (1996). Biotransformation of chlorpromazine and methdilazine by *Cunninghamella elegans*. *Appl. Environ. Microbiol.* 62:798-803.

Zhang, P., Lin, L. H., Huang, H. H., Xu, H. Y. and Zhong, D. F. (2006). Biotransformation of indomethacin by the fungus *Cunninghamella blakesleeana*. *Acta Pharmacol. Sin.* 27:1097-1102.

❑❑❑

CHAPTER **4**

Methylotrophy in Bacteria Concept and Significance

G. Selvakumar*, Sehar Nazim and S. Kundu

Vivekananda Institute of Hill Agriculture, (ICAR) Almora-260601, Uttarakhand, India

ABSTRACT

Methylotrophy is a phenomenon by which microorganisms derive energy, and in many cases, cell carbon from reduced molecules that have no C–C bond (also called C1compounds). This gains significance because of its unique mode of nutrition, the ability to utilize more than one metabolic pathway and its vast distribution among various groups of bacteria and yeasts. It is one of the earliest modes of energy metabolism present in microorganisms, through which microorganisms are able to utilize a wide range of compounds, which are considered unsuitable for most forms of life. In the present chapter we focus on various aspects of methylotrophy, and the application of methylotrophic bacteria in a spectrum of biological activities and commercial possibilities.

Key words: methylotrophy, PPFM's, enzymes, bioremediation

1. INTRODUCTION

An integrating theme in biology is the enormous diversity of living organisms on earth. This is especially true of microorganisms and bacteria in particular which have evolved on earth over a period of nearly 3.86 billion years, due to which they excel at using various organic and inorganic substances as sources of nutrients and energy. This vast diversity of potential substrates has led to the evolution of enzymes capable of transforming many unrelated compounds by different metabolic pathways. Mutations in metabolic enzymes have generated catalysts capable of utilizing new substrates providing the growth advantage over other organisms, thus

*Email: gselva74@rediffmail.com

providing an opportunity to exploit new ecological niches. Furthermore, most microorganisms exhibit more than one mode of nutrition with distinct metabolic pathways. This metabolic diversity which has evolved with adaptation in microorganisms confers upon them the distinctive edge for survival on a variety of substrates considered quite unsuitable for any living organism.

The phenomenon of methylotrophy gains significance in this regard due to its unique mode of nutrition, the ability to utilize more than one metabolic pathway and its vast distribution among various groups of bacteria and yeasts. It is one of the earliest modes of energy metabolism and one of the most important natural ways to get rid of green house gases (methane derivatives) in today's world. In this chapter we focus on the concept of methylotrophy, and the application of methylotrophic bacteria in a spectrum of biological activities and commercial possibilities. Though the definition of a methylotroph would ideally include yeasts and other methylotrophic microbes, we would be limiting our discussion to bacteria in particular.

2. METHYLOTROPHY

Methylotrophs are microorganisms that derive energy, and in many cases, cell carbon from reduced molecules that have no C–C bond (also called C1compounds). The phenomenon of utilizing such C1compounds is known as methylotrophy. The organisms involved are primarily aerobic heterotrophs. Though the functional definition of a methanotroph encompasses the organisms that oxidize methane and utilize 1C compounds, methylotrophs differ from methanotrophs due to their inability to oxidize methane, due to absence of methane mono oxygenase, the enzyme which converts methane to methanol (Anthony, 1986).

2.1 Methylotrophy V/s Autotrophy

Methylotrophy is often confused with autotrophy since autotrophs utilize $CO_{2,}$ a 1C source as substrate. The 1C metabolism differ from autotrophism in that the organism involved utilizes a substrate more reduced then CO_2 which is the staple 'C' source for autotrophs. As a matter of fact, methylotrophy does not describe one type of metabolism; it includes under a common name a bunch of different pathways of utilizing C1 compounds. De Marco (2004) has outlined the following possibilities as the reason for the prevailing confusion between methylotrophy and autotrophy:

- Some methylotrophs do grow on C1compounds autotrophically, fixing the CO_2 produced (e.g. *Ralstonia*, *Xanthobacter*, *Paracoccus*, the methylotrophic *Archea,* and methylotrophic clostridia). Thus, for these organisms (exclusively) heterotrophic would be the proper antonym of methylotrophic. Furthermore, other autotrophic microorganisms (lithotrophs or phototrophs) can utilize C1compounds (methanol, formate) as a supplementary source of energy besides their more typical ones

- Some methylotrophs are mixotrophic. *Methylococcus capsulatus* fixes carbon derived from methane mostly at the level of formaldehyde, through the ribulose monophosphate (RuMP) pathway (heterotrophic), but also in part at the level of CO_2 through the Calvin–Benson–Bessham (CBB) cycle (autotrophic) (Taylor *et al.,* 1980; Baxter *et al.,* 2002). Also, the serine cycle for carbon fixation is intrinsically mixotrophic in that it incorporates one molecule of CO_2 for every two of formaldehyde fixed
- Many of the reactions that compose the RuMP pathway are common to the CBB cycle. Some authors (Quayle and Ferenci, 1978) have hypothesized that the CBB cycle actually originated from the RuMP pathway and that organisms like *Methylococcus capsulatus,* in which both pathways work simultaneously, may be 'transition organisms' between the two metabolic types (Taylor *et al.,* 1980)
- Formaldehyde, which is the molecule fixed by the heterotrophic methylotrophs, is obviously not inorganic, but it is not 'far' from CO_2 (just two dehydrogenation steps away), hence it is easy to slip into the mistake of considering it as 'almost inorganic'

2.2 How Old are Methylotrophs?

The answer to this question could be obtained from recent evidences that the genes encoding the enzymes and co-enzymes of the methanogenic sulfate reducing Archea are present in *Methylobacterium extorquens* AM1. This bacterium has nine gene clusters involved in growth on C1compounds, the largest of which is a ~ 30 kb segment containing the gene for NADP-dependent methylene tetrahydrofolate dehydrogenase (*mtdA*) as well as a number of other methylotrophic genes. A 14 kb region of this 30 kb gene cluster was found to contain 13 new open reading frames. The translated products of 12 of these genes were similar to those of genes in methanogens or *Archeoglobus fulgidus,* or both. The products of six of these methanogen like genes (*orfY, orf4, orf7, orf9, orf17 and orf18*) are similar to unidentified gene products in archeal genome sequences. The other six were similar to the translated products for known genes of the tetrahydromethanopterin / methanofuran pathway in methanogens and sulfate reducing *Archeoglobus fulgidus* and to other archeal proteins. The studies of Chistoserdova *et al.* (1998) showed that the methanogenic and sulfate reducing Archea and methylotrophs have in common an energy metabolism based on the interconversion of C1compounds, and most probably, closely related coenzymes and enzymes are used for these catabolic interconversions. The genes encoding these enzymes have either been conserved, because these organisms evolved from a common ancestor, or they were horizontally transferred between more recent ancestors (Chistoserdova *et al.*, 1998).

Recently, the presence of the genes in question has been reported in *Pirellula* sp. strain1, a representative of planctomycetes (Glockner *et al.*, 2003). Planctomycetes, an enigmatic group, within bacteria, reveal features not common in the bacterial domain, such as a cell wall lacking peptidoglycan that is reminiscent of Archea, intricate cell compartmentalization reminiscent of eukaryotes, division by budding similar to yeast and unique ladderane lipids in some autotrophic ammonium oxidizing representatives (Fuerst, 1995; Lindsay *et al.* 2001; Damste *et al.* 2002). Although no physical record of the antiquity of Planctomycetes exists, their ubiquitous nature, the large size of their genomes and the variety of metabolic capabilities encoded, support their position near the root of universal phylogenetic tree. These facts suggest that the origin of methanopterin / methanofuran linked C1pathway in planctomycetes preceded the divergence of Proteobacteria, pointing to the antiquity of this pathway. These evidences suggest that methanogens and methylotrophs would have developed from the same ancestral group, most probably the planctomycetes. It also suggests that it is not necessary that methanogenesis preceded methylotrophy and the *vice versa* is very much possible, however evidence exists for both being present as long ago as 2.78 billion yrs (Brocks *et al.*,2003; Chistoserdova *et al.*, 2006).

2.3 Classification of C1 Compounds Utilizing Bacteria

Whittenbury *et al.* (1970) had classified the one carbon substrate utilizing bacteria into type I (*Methylococcus, Methylomonas, and Methylobacter)* and type II (*Methylosinus* and *Methylocystis)*. The major division between types I and II was based upon differences in intracytoplasmic membrane arrangement and carbon incorporation pathways. Most workers accept this as a basis for development of a formal classification, and most organisms isolated since this work fall more or less into one of this groups. As more detailed knowledge has accumulated, a new genus of facultative methylotrophs *viz.*, *Methylobacterium* has been accommodated in the type II group, which utilizes only reduced C1substrates, and incorporates them via serine pathway (Patel *et al.*, 1978). Recent DNA homology studies however suggests that this classification is inappropriate, for there is little homology between the DNA of *Methylobacterium organophilum* XX and a representative type II species, *Methylosinus trichosporium* OB3b. Attempts are now being made to devise a formal taxonomy and nomenclature of C1utilizing bacteria, although there is a continuing need to accommodate the unexpected, e.g., *Methylococcus capsulatus* which has three pathways for incorporation of C1units *viz.* Ribulose di phosphate, ribulose monophosphate (RMP) and serine pathway (Whittenbury and Dalton, 1981).

2.3.1 Mechanism of C1 assimilation in to bacterial cell

In methanotropic bacteria that have the ability to oxidize methane the enzyme methane monooxygenase incorporates an O_2 molecule into the carbon compound. In the genus *Methylosinus* the electron needed for the oxidation of CH_4 to CH_3OH comes from the cytochrome C. Because of this no ATP is produced in the oxidation of methane to methanol. This results in equal quantity of cell biomass being produced whether methane or methanol is used as growth substrate. The subsequent steps are common to both the methanotrophs and the strict methylotrophs. The substrates generally utilized by methylotrophs are given in Table 1.

TABLE 1. Common substrates used in methylotrophy

Substrate used for growth		**Substrate used for cometabolism**	
Methane	CH_4	Ammonium	NH_4^+
Methanol	CH_3OH	Ethylene	$H_2C{=}CH_2$
Methylamine	CH_3NH_2	Chloromethane	CH_3Cl
Dimethylamine	$(CH_3)_2NH$	Bromomethane	CH_3Br
Trimethylamine	$(CH_3)_3N$	Ethane	CH_3CH_3
Tetramethylamine	$(CH_3)_4N^+$	Propane	$CH_3CH_2CH_3$
Trimethylsulfonium	$(CH_3)_3S^+$		
Formate	$HCOO^-$		
Formamide	$HCONH_2$		
Carbon mono oxide	CO		
Dimethylether	$(CH_3)_2O$		

2.3.2 *Pathways of C1Assimilation*

2.3.2.1 Serine pathway

The serine pathway was first studied by Lawrence and Quayle (1970). Formaldehyde enters the pathway by reacting with tetrahydrofolate to form N^5, N^{10}-methylene tetrahydrofolate. The first enzymatic step catalyzed by trans-hydroxymethylase, involves the condensation of methylene tetrahydrofolate with glycine to form serine. This enzyme present in most bacteria is involved in the synthesis of serine and purines, but in methylotrophic bacteria using the serine pathway, it assumes a more central role in metabolism.

There is little information concerning regulation of the channeling of formaldehyde into either oxidative or assimilative metabolism. Harder and Attwood (1975), proposed that the reactivity of formaldehyde with tetrahydrofolate is such that when all coenzyme is derivatized, only then formaldehyde is available for oxidative energy generation. Control would hence be affected through the relative levels of reduced NADH, ATP and tetrahydrofolate. However, Quayle (1980) pointed out that a tetrahydrofolate mediated mechanism can participate in oxidation of formaldehyde to formate and can bypass such control.

Half of the 2-phosphoglycerate generated by formaldehyde fixation is converted to the net product of the pathway, 3-phosphoglycerate. The remainder is converted to phosphoenolpyruvate (PEP), the accepter molecule for the second important carbon fixation step in this pathway. PEP is carboxylated to form oxaloacetate, which is then reduced to malate. Carbon dioxide is derived mainly from the oxidation of reduced C1 substrates, it has been calculated that approximately 50% of the carbon derived from methanol is assimilated as carbon dioxide by the facultative methylotroph *Pseudomonas* AM1 (Large *et al.*, 1961).

Phosphoenolpyruvate carboxylase plays a key role in methylotrophic metabolism. The studies of Newaz and Hersh (1975) suggested that this enzyme is at the crossroads between carbon assimilation and energy generation. The phosphoenolpyruvate carboxylase of methylotrophic strains examined to date is predominantly of the class 3 types, subject neither to activation nor inhibition. However, studies of methylamine grown *Pseudomonas* strain MA have indicated that carboxylase is activated by NADH, in this organism, and thus becomes a focal point for regulation of metabolism.

The generation and cleavage of malyl CoA is another important step in the serine pathway. No detectable ATP and CoA dependent malate lyase have been found in *M. trichosporium* OB3, but enzyme extracts of this bacteria do catalyze Mg^{++} dependent cleavage of malyl Co-A, which suggests that malate thiokinase, the enzyme responsible for generation of malyl CoA, might be limiting in the first case, and hence any malyl CoA generated would be rapidly hydrolysed by the lyase. Alternatively, an acyl CoA might be involved. Malyl CoA has been detected at high activities in methanol grown *M. organophilum* (O'Connor, 1981).The action of malyl CoA lyase regenerates one accepter molecule of glyoxalate and one molecule of acetyl CoA. Glycine is derived from glyoxalate which enters the cycle along with methylene tetrahydrofolate.

In this way 2C unit acetyl CoA are synthesized from one molecule of formaldehyde and one molecule of CO_2, utilizing two molecules each of NADH and ATP for each acetyl CoA synthesized (Fig. 1).

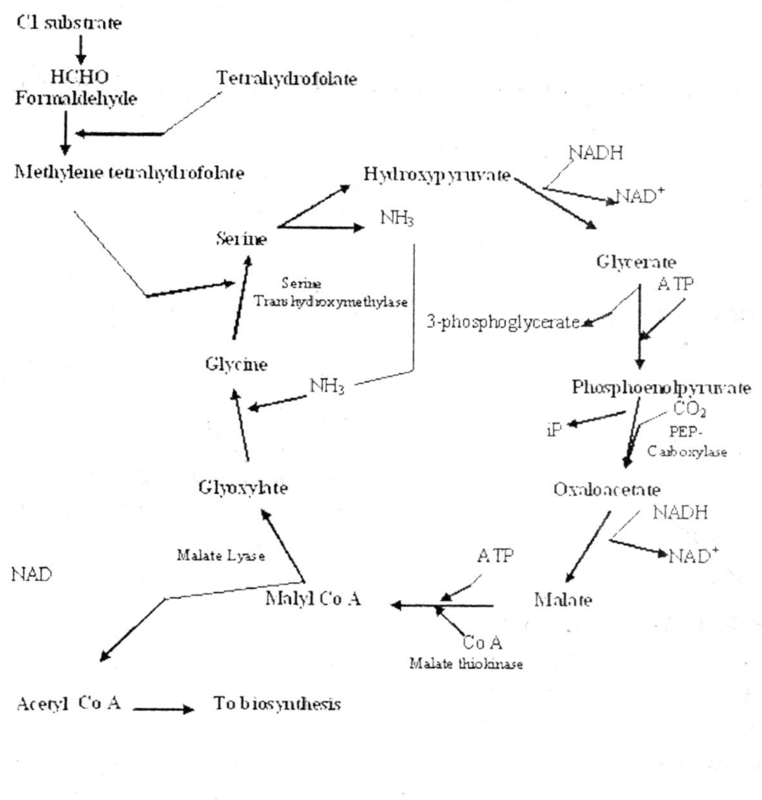

Overall: Formaldehyde + CO_2 + CoA + 2 NADH + $2H^+$ + 2 ATP $\rightarrow$ Acetyl CoA + 2 NAD^+ + 2ADP + 2 iP + 2 H_2O

FIG.1. A schematic representation of serine pathway in methylotrophs

2.3.2.2 Ribulose mono phosphate pathway

It is more efficient than serine pathway since all the carbon atoms are derived from formaldehyde, and since formaldehyde is at the same oxidation level as the cell material no reducing power is needed. The RuMP pathway requires the addition of one molecule of ATP for each molecule of glyceraldehyde -3- PO_4 synthesized. Due to the low energy requirement of the RuMP pathway the cell yield per mole of substrate is higher in the RuMP pathway. The enzyme hexulose phosphate synthase and hexulose-6-PO4 isomerase are unique to this pathway. The RuMP pathway is also called Quayle pathway after its discoverer John Quayle (1961) (Fig. 2).

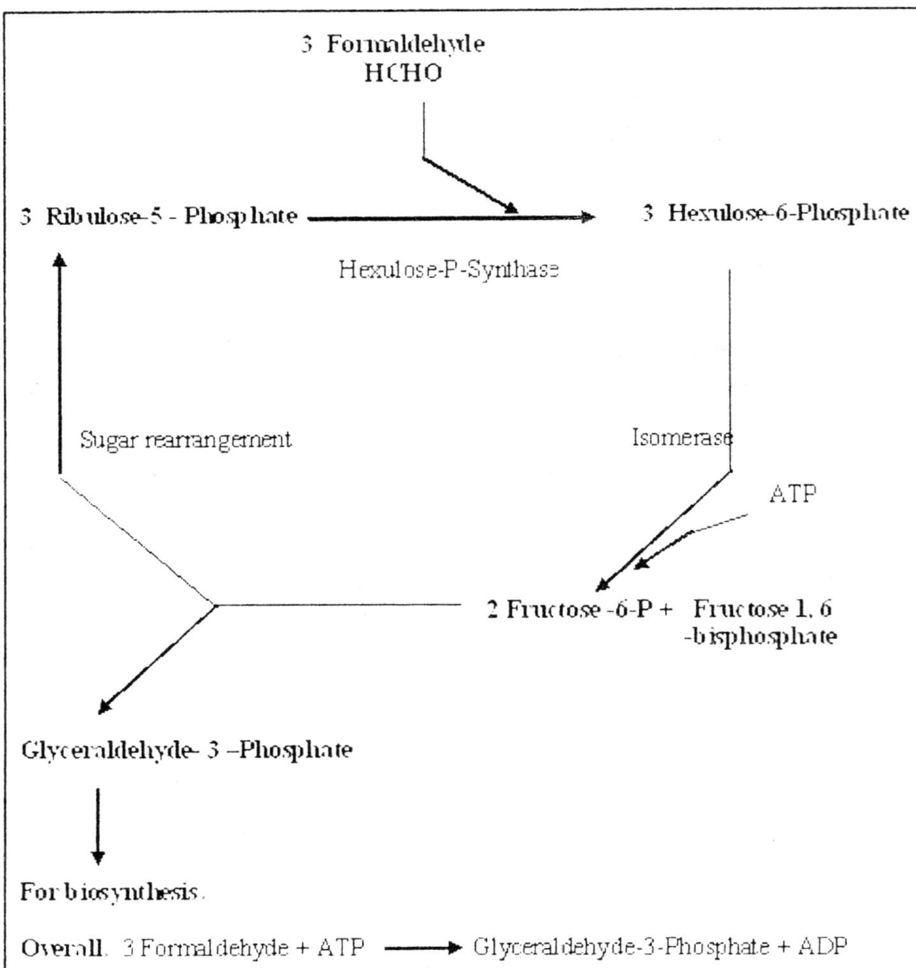

FIG. 2. An overview of the RUMP pathway in methylotrophs

3. SIGNIFICANCE OF METHYLOTROPHIC BACTERIA

3.1 Role of PPFMs (Pink Pigmented Facultative Methylotrophs) in Plant Growth Promotion

Among the various groups of methylotrophic bacteria that have been known till date the Pink Pigmented Facultative Methylotroph (PPFM's) are most unique and interesting. They are ubiquitous in nature and frequently reported on various plant species, where they form a substantial part of the aerobic, heterotrophic microflora of the surface of young leaves. They were first isolated as covert contaminants from the tissue cultures of liverwort, *Scapania nemorosa* (Basile *et al.*, 1969), but later identified as belonging to the genus *Methylobacterium mesophilicum* (Green and Bousfield, 1982). The presence of such a system in seemingly healthy tissue cultures has been a problem for sometime (Corpe and Basile, 1982). This is because-

(i) Their presence is not obvious though their numbers may be up to 10^8 cfu/g fresh weight.

(ii) They do not grow well on the plant tissue culture media and remain confined to the mass of cultured cells; though in old cultures the bacteria appear as a puddle around the senescing callus.

(iii) Regularly adapted sanitizing procedures for tissue cultures do not get rid of the bacteria.

PPFM's are classified as Alpha-proteobacteria (Patt *et al.*, 1976; Lidstorm, 1992). These bacteria are commonly found in soils, as well as on the surfaces of leaves, seeds and in the rhizosphere of a wide variety of plants, with highest numbers present on actively growing and meristematic tissues, sometimes averaging 10^4-10^6 colony forming units (cfu) per leaflet (Corpe, 1985; Dunleavy, 1988; Corpe and Rheem, 1989; Holland 1997).

Corpe and Basile (1982) showed that the PPFM cultures when inoculated to the culture of bryophytes without exogenous hormones could stimulate the growth of the cultures. The first evidence for the beneficial association of PPFM's was obtained from the studies of Basile *et al.* (1985) who found that PPFM's produced vitamin B_{12} and the addition of this vitamin to the cultures of the bryophyte *Jungermannia* and *Gynocola* could stimulate the plant growth and development.

PPFM's have been reported to influence seed germination and seedling growth by producing plant growth regulators like zeatin and related auxins and cytokinins (Holland and Polacco, 1992; Ivanova *et al.*, 2001; Omer *et al.*, 2004). PPFM's have also been observed to alter agronomic traits like branching, seedling vigour, rooting and heat /cold tolerance (Freyermuth *et al.*, 1996; Holland, 1997, Madhaiyan *et al.*, 2004 a). They have also been known to indirectly reduce or prevent the deleterious effects of pathogenic microorganisms, through induced systemic resistance (ISR) as found in rice and ground nut (Madhaiyan *et al.*, 2004b, 2006b). Their inoculation was found to increase the photosynthetic activity by enhancing the number of stomata, chlorophyll concentration and malic acid content of crops (Cervantes-Martinez *et al.*, 2004).

The studies on seed germination and development of plants treated with PPFM's and PPFM spent media showed that inoculated plants performed better than the uninoculated plants. Reduction in the population of the PPFM's in the seed coat also led to reduced germination levels speculating the role of PPFM's on seed germination (Holland and Pollaco, 1994). They have also been reported to produce enzymes like 1-aminocyclopropane-1-carboxylate deaminase that lowers the ethylene levels, which is a growth inhibitor at higher concentrations and thus promote root elongation in *Brassica compestris* seedlings (Madhaiyan *et al.*, 2006a).

3.1.1 PPFM's and Cytokinins

Cytokinins are a class of compounds that are defined by their ability to stimulate cell division in plants. They are found throughout the plant, but are most abundant in actively growing tissues. Their isolation from corn, following the discovery of their effect on plants cell in culture, led to assumption that cytokinins are endogenously produced plant growth regulators. This assumption gained the status of fact, although it was never proven. Conversely, cytokinins production by plant-associated microorganism is well documented (Freyermuth *et al.*, 1996; Holland, 1997a). Among the various groups of plant growth associated microorganisms, PPFM's have been widely implicated in the production of cytokinins in plant tissues.

3.1.2 Proposed model of cytokinin production by PPFM's (Holland, 1997b)

Holland (1997b) has proposed a speculative model for the production of cytokinins by plant associated PPFM's. This theory assumes that cytokinin production by PPFM's occurs in the growing tissues which are colonized by PPFM's. Cytokinins are said to act as signal molecules which initiate plant cell division leading to the demethylation of pectin and release of methanol, which is consumed by PPFM's as nutritional substrate. This speculative theory doubts the origin of cytokinins which were formerly thought to be of plant origin. This theory gains credit due to the fact that cytokinins are not systemic in nature and have only localized effect, and actively growing cells produce comparatively higher amounts of methanol due to faster rate of cell division, which in turn supports larger population of PPFM's. The microbial symbionts of plants are supposed to have coevolved along with the plants, and they have the means, the motive, and the opportunity to generate the cytokinins.

3.1.3 PPFM's and plant nitrogen metabolism

It has become apparent that PPFM's have access to plant metabolites and they are able to degrade ureides (a major transport mode for nitrogen in nodulated soy bean) and urea. PPFM's probably use ureides while living on soybean because they either do not take up arginine (the other logical source of metabolic urea) or do not have an arginase. When embryos lacking both plant and bacterial urease activity develop on a plant of the same genotype urea accumulates in seeds. If bacterial urease alone is active in the embryos, urea does not accumulate. Similarly during soybean germination, urea accumulation in the embryonic axis is mitigated by the activity of bacterial urease (Holland and Polacco, 1994).

3.2 Nitrogen Fixation by Methylotrophs

Methylotrophs are known nitrogen fixers in rice cultivars as non-legume forming symbiotic bacteria (Lee *et al.*, 2006). However one of the methylotrophs *viz.*, *Methylobacterium nodulans* has been reported to fix nitrogen in symbiotic relationship with some rare leguminous plants of the genus *Crotalaria*, it is highly specific to *Crotalaria* and forms nodules similarly as other rhizobia (Sy *et al.*, 2001). The expression and role of bacterial methylotrophy were investigated during symbiosis between *M. nodulans*; strain ORS 2060T, and its host legume, *Crotalaria podocarpa*. Using *lacZ* fusion to the *mxaF* gene, it was found that the methylotrophic genes are expressed in the root nodules, suggesting methylotrophic activity during symbiosis. In addition, loss of the bacterial methylotrophic function significantly affected plant development. Inoculation of *M. nodulans*, nonmethylotrophic mutants in *C. podocarpa* decreased the total root nodule number per plant up to 60%, decreased the whole-plant nitrogen fixation capacity up to 42%, and reduced the total dry plant biomass up to 46% compared with the wild-type strain. In contrast, inoculation of the legume *C. podocarpa* with nonmethylotrophic mutants complemented with functional *mxa* genes restored the symbiotic wild phenotype. These results demonstrate the key role of methylotrophy during symbiosis between *M. nodulans* and *C. podocarpa*. (Jourand *et al.*, 2005).

4. COMMERCIAL APPLICATIONS OF METHYLOTROPHS

Most of the earlier attempts in the commercial exploitation of methylotrophs were in the area of single cell protein (SCP) production, and production of heterologous proteins and fine chemicals. Methylotrophic organisms are most likely to be of commercial utility when they can either synthesize a novel product with desirable properties or a unique catalytic potential *e.g.* methane monooxygenase. Table 2 lists some of the commercially important products obtained from methylotrophs.

TABLE 2. Commercially important chemicals from methylotrophs (Lindstorm and Stirling, 1990)

Product	Patent holder	Year
Acetaldehyde	Agcy. Ind. Sci. Tech.	1986
Acetic acid	Shimixu Construction	1987
Acetone	Agcy. Ind. Sci. Tech.	1985
Adenosine	Toray. Ind. Inc.	1986
Aldehyde	Amer. Biogenetics. Co	1987
Aminoacids	Alinomoto k k	1987
^{13}C labeled biochemicals	Calgene Corp	1987
Ethanol	Agcy. Ind. Sci. Tech.	1985
Glycerol	Mitsubishi Gas Chem	1986
Isopropenol	Agcy. Ind. Sci. Tech.	1985
Vitamin B12	Nagai S	1988
Liposurfactant	Agcy. Ind. Sci. Tech.	1984
Propionic acid	Hoechst A G	1986

4.1 Single Cell Proteins (SCP's)

Methanol utilizing bacteria have been used as choice organisms for SCPs production, rather than the methylotrophic organism that have feed stock quality problems and explosion hazards. Some commercially successful ventures were the ICI "Pruteen" process and Philips "Prouesta" process. Due to the vagaries of the total world protein market, the interest in the production of cellular protein by methylotrophs has diminished and the attention has been turned towards higher value-added materials.

4.2 Enzymes and Specialty Proteins

The primary methylotrophic enzymes of interest have been the enzyme alcohol oxidase which holds promise in the surfactant area. Ever since a number of patents have been issued for the commercial production of enzymes from methylotrophs. List of enzymes produced by methylotrophs at commercial level is given in Table 3.

TABLE 3. Commercially produced enzymes from methylotrophs (Lindstrom and Stirling, 1990)

Product	Patent holder	Year
Alcohol dehydrogenase	Hoechst AG	1986
Alcohol oxidase	Philips Petroleum	1985
L-aminoacylase	Sauratu Ocean	1980
Glucose phosphate dehydrogenase	USSR Biochem	1983
Glutamate dehydrogenase	Akad Wissenschaft	1983
Heterologous proteins and host vector systems	Hoechst AG	1988

4.3 Recombinant Proteins

The *Methylobacterium* sp. has received special attention as a potential source of industrially pertinent natural and recombinant proteins, the industrial potential of *M. extorquens* as a producer of recombinant proteins is due to (i) the simplicity of its growth requirements and its utilization of a comparatively inexpensive substrate (methanol), (ii) the development and optimization of protocols for high-cell-density cultivation, (iii) the fact that the *M. extorquens* AM1 genome has been sequenced and annotated, and (iv) the development of genetic tools specifically for *M. extorquens* comprising novel cloning and expression vectors, efficient transposon mutagenesis, and simple and efficient electroporation protocols (Bourque *et al.*, 1995; Marx and Lidstorm, 2004). The potential of *M. extorquens* and other pink-pigmented facultative methylotrophs as "cell factories" is further enhanced by their inherent abilities to produce natural products of great importance, including pyrroloquinoline quinone, vitamin B_{12}, poly-ß-hydroxybutyric acid (PHB), carotenoids and phytohormones. By using the molecular tools and high-cell-density fermentation technologies mentioned above, it has been possible to overexpress a variety of recombinant proteins in *M. extorquens* strains ATCC 55366 and AM1. Representative proteins overexpressed in these two *M. extorquens* strains include the esterase from *Lactobacillus casei*, catechol 2, 3-dioxygenase from *Pseudomonas putida*, enterocin P from *Enterococcus faecium*, and haloalkane dehalogenase from *Xanthobacter autotrophicus*. Production of several of these recombinant proteins was achieved in high-cell-density fed-batch 20-liter and 150-liter fermentation processes (Choi *et al.*, 2006).

The production of green fluorescent protein (GFP) in *Methylobacterium extorquens* was studied by creating four different constructs using pJB3KmD, pRK310 and pVK101 vectors, as well as p*Lac* and soluble methane monooxygenase (sMMO) promoters. Plasmids were introduced into the cells by electroporation. Expression of GFP by selected clones was evaluated by growing cells in complex or defined media. The use of pRK310 as an expression vector containing the *lacZ* promoter resulted in a 100-fold increase of GFP production when compared to cells containing the p*Lac*-GFP-pJB3KmD construct. Higher production of GFP was also observed in cells containing p*Lac*-GFP-pRK310 and p*mmoX*-GFP-pVK101 constructs. While the transcriptional regulation of the *smmo* gene in *Methylosinus trichosporium* OB3b is known to be copper-dependent, expression of GFP by *M. extorquens* clones harboring p*mmoX*-promoters was not strongly controlled by the presence of copper in the medium. The production of GFP was generally constant throughout the growth of *M. extorquens* carrying the p*Lac*-GFP-pRK310 construct. GFP yields varied between 850 and 1000 μg of GFP/g biomass (Figueira *et al.*, 2000).

4.4 Bioplastics Production

Poly-ß-hydroxybutyric acid (PHB) and similar bacterial polyesters are promising candidates for the development of environment-friendly, totally biodegradable plastics. *Methylobacterium extorquens,* was found to be the most promising culture among 118 methylotrophic isolates studied by Bourque (1992). It was selected based on rapidity of growth and PHB accumulation. Being a methylotroph it utilizes methanol, one of the cheapest noble substrates available, which reduces the cost of producing such bioplastics. This isolate was obtained from a soil contaminated regularly with used oil products for some 40 years. Concentrations of methanol greater than 8 g/l affected growth significantly and the optimal methanol concentration was 1.7 g/l. PHB concentrations averaged 25–30% (w/v) of dry weight under non-optimized conditions. Controlling methanol concentration, using an open-loop configuration, led to biomass levels of 9–10 g/l containing 30–33% PHB while preventing methanol accumulation. This isolate was also able to produce the co-polymer PHB/poly-β-hydroxyvalerate (PHV) using the mixture methanol + valerate. The PHV-to-PHB ratio was about 0.2 at the end of the fermentation. An average molecular mass varying between 2 and 3 × 10^5 Da was obtained for three PHB samples using two different measurement methods (Borque *et al.*, 1992).

4.5 Bioremediation

The application of methylotrophic organisms for the bioremediation of polluted habits is one of the major areas of commercial exploitation. The ability of the enzyme methane monoxygenase to dehalogenate haloalkanes and haloalkenes has been known for long. The methane monooxygenase has been found to show its non-specific activity to affect oxidations, dechlorinations, condensations, and rearrangements of *meta*-chlorotoluene, phenol, chlorofluorobenzenes, and mono and dichlorobiphenyls via its non-specific activity (Murrell, 1992.) The methylotrophs have the ability to grow and mineralize mono and dihalo methanes. C1 utilizers are physiologically versatile in their ability to exist in a variety of habitats and live in hostile environments having a wide range of pH, temperature, heavy metal concentrations, oxygen concentrations, barometric pressures, salinity and radiation. The methylotrophs are more amenable to genetic manipulation than the methanotrophs and commercial biodegradation process could be developed soon.

5. CONCLUSION

Methylotrophic bacteria are a group of mysterious and interesting microorganisms which possess a distinctive and unique energy metabolism, they are promising candidates for further exploration to understand them better and exploit their different capabilities in various fields, and they might also prove to be an important part of puzzle of evolution as they can be a link between Archea and Eubacteria. They possess immense commercial significance due to their ability to produce hormones, enzymes and fine chemicals. Novel applications of methylotrophs include their ability to be used as heterologous hosts for the production of proteins and bioremediation agents in polluted environments. Future studies should focus more on the commercial utilization of this interesting group of microorganisms for the welfare of mankind.

REFERENCES

Anthony, C. (1986).Bacterial oxidation of methane and methanol. *Adv. Microbial. Physiol.* 26:113-203.

Basile, D.V., Basile, M.R., Li, Q.Y. and Corpe, W.A. (1985). Vitamin B12 stimulated growth and development of *Jungermannia leiantha* and *Gymnocolea. Bryologist.* 88(2):77-81.

Basile, D.V., Slade, L.L. and Corpe, W.A. (1969).An association between a bacterium and liverwort *Scapania nemorosa. Bull. Torrey Bot. Club.* 96(6):711-714.

Baxter, N. J., Hirt, R. P., Bodrossy, L., Kovacs, K. L., Embley, T. M., Prosser, J. I. and Murrell, J. C. (2002).The ribulose-1, 5-bisphosphate carboxylase/oxygenase gene cluster of *Methylococcus capsulatus* (Bath). *Arch Microbiol.* 177:279–289.

Bourque, D., Pomerleau, Y., and Groleau, D. (1995). High cell density production of poly-beta-hydroxybutyrate (PHB) from methanol by *Methylobacterium extorquens*: production of high-molecular-mass PHB. *Appl. Microbiol. Biotechnol.* 44:367-376.

Bourque, D., Ouellette, B., André, G. and Groleau. D. (1992). Production of poly-â-hydroxybutyrate from methanol: characterization of a new isolate of *Methylobacterium extorquens*. *Appl. Micro. Biotech.* 37(1):7-12.

Brocks, J.J., Buick, R., Summons, R.E., and Logan, G.A. (2003). A reconstruction of Archean biological diversity based on moleculer fossil from the 2.78 to 2.45 billion year old Mount Bruce Supergroup, Hamersley Basin, Western Australia. *Geochim. Cosmochim. Acta.* 67: 4321-4335.

Cervantes -Martinez, J., Lopez-Diaz, S., and Rodreguez-Garay, B. (2004). Detection of the effect of *Methylobacterium* in *Agave tequilina* Weber var. azul by laser induced fluorescence. *Plant Sci.* 166:889-892.

Chistoserdova, L., Jenkins, C., Kalyuzhnaya, M.G., Marx, C.J., Lapidus, A., Vorholt, J.T. Staley, J.A., and Lindstorm, M.E. (2006). The enigmatic planctomycetes may hold a key to the origin of methanogenesis and methylotrophy. *Mol. Biol. Evol.* 21 (7):1234-1241.

Chistoserdova, L., Vorholt, J.A., Thaver, R.K. and Lidstorm, M.E. (1998). C1 transfer enzymes and coenzymes linking methylotrophic bacteria and methanogenic Archea. *Science.* 281:99-102.

Choi, Y. J., Bourque, D., Morel, L., Groleau, D. and Míguez, C. B. (2006). Multicopy Integration and Expression of Heterologous Genes in *Methylobacterium extorquens* ATCC 55366. *Appl. Environ. Microbiol.* 72(1):753-759.

Corpe, W.A. (1985). A method for detecting methylotrophic bacteria on solid surfaces. *J. Microbial Methods.* 3:215-221.

Corpe, W.A. and Basile, D.V. (1982). Methanol-utilizing bacteria associated with green plants. *Dev. Indus. Microbiol.* 23: 483-93.

Corpe, W.A. and Rheem, S. (1989). Ecology of methylotrophic bacteria on living leaf surfaces. *FEMS Microbiol. Ecol.* 62:243-250.

Damaste, S., Fuerst, J.A., Strous, M., Rijpstra, W.I., Hopmans, E.C., Geenevasen, A.C., van Duin, J.A., van Niftrik, L.A. and Jetten, M.S. (2002). Linearly concatenated cyclobutane lipids form a dense bacterial membrane. *Nature,* 419:708-712.

De Marco, P. (2004). Methylotrophy versus heterotrophy: a misconception. *Microbiology,* 150:1606-1607.

Dunleavy, J.M. (1988). *Curtobacterium plantarum* sp. nov. is ubiquitous in plant leaves and is seed transmitted in soybeen and corn. *Int. J. Syst. Bacteriol.* 39:240-249.

Figueira, M.M., Laramée, L., Murrell, J.C., Groleau, D. and Miguez, C.B. (2000). Production of green fluorescent protein by the methylotrophic bacterium *Methylobacterium extorquens. FEMS Microbiology Lett.,*193(2):195-200.

Freyermuth, S.K., Long, R.L.G., Mathur, S. and Polacco, J.C. (1996).Microbial growth on C1 compounds. Kluwer Academic Publishers, Dordrecht, The Netherlands, Pp 277-284.

Fuerst, J.A. (1995). The planctomycetes: emerging model for microbial ecology, evolution and cell biology. *Microbiology,* 141:1439-1506.

Glockner, F.O., Kube, M., and Bauer M. (2003). Complete genome sequence of marine planctomycete *Pirulella* sp. strain 1. *PNAS USA,* 100:8298-8303.

Green, P.N. and Bousefield, I.J. (1982). A taxonomic study of some gram-negative facultatively methylotrophic bacteria. *J. Gen. Microbiol.* 128:623-38.

Harder, W. and Attwood, M.M. (1975). Biology, physiology and biochemistry of *Hyphomicrobia. Adv. Microb. Physiol.* 17:303-359.

Holland, M.A. (1997a). *Methylobacterium* and plants. *Recent Res. Devel. In Plant Physiol.* 1:207-21.

Holland, M.A. (1997b). Ocam's Razor applied to hormonology. *Plant Physiol.* 115:865-868.

Holland, M.A. and Polacco, J.C. (1992). Urease null and hydrogenase null phenotypes of a phylloplane bacterium reveal altered nickel metabolism in two soybean mutants. *Plant Physiol.* 98:942-948.

Holland, M.A. and Polacco, J.C. (1994). PPFM's and other covert contaminants: Is there more to plant physiology than just plant. *Annu. Rev. Plant Physiol. PlantMol.Biol.* 45:197 -209.

Ivanova, E.G., Doronina, N.V., Trotsenko, Y.A. (2001). Aerobic methylobacteria are capable of synthesizing auxins. *Microbiology* 70:392-397.

Jourand P., Renier A., Prin Y., Galiana A., Giraud E. and Dreyfus B. (2005). Role of methylotrophy during symbiosis between *Methylobacterium nodulans* and *Crotalaria podocarpa. Mol Plant Microbe Interact.* 18(10):1061-8.

Large, P.J., Peel, D. and Quayle, J.R. (1961). Microbial growth on C1 compounds: Synthesis of cell constituents by methanol and formate grown *Pseudomonas* AM1 and methanol grown *Hyphomicrobium vulgare. Biochem. J.* 81:470-480.

Lawrence, A.J., and Quayle, J.R. (1970). Alternative carbon assimilation pathways in methane utilizing bacteria. *J. Gen. Microbiol.* 63:371-374.

Lee, H.S., Madhaiyan, M., Kim, C.W., Choi, S.J., Chung, K.Y., Sa, T. (2006). Physiological enhancement of early growth of rice seedlings (*Oryza sativa* L.) by production of phytohormones of N_2-fixing methylotrophic isolates. *Boil. Fertil. Soils,* 42:402-408.

Lindsay, M.R., Webb, R.I., Strous, M., Jetten, M.S. Butler, M.K., Forde, R.J. and Fuerst, J.A. (2001). Cell compartmentalisation in planctomycetes: novel types of structural organization for the bacterial cell. *Arch. Microbiol.* 175:413-429.

Lidstorm, M.E. and Stirling, D.I. (1990). Methylotrophs: Genetics and commercial applications. *Annu. Rev. Microbiol.* 44:27-58.

Madhaiyan, M., Poonguzhali, S., Ryu, J., and Sa, T. (2006a). Regulation of ethylene levels in canola (*Brassica compastris*) by 1-aminocyclopropane-1-carboxilate deaminase containing *Methylobacterium fujisawaense. Planta.* 224:268-278.

Madhaiyan, M., Poonguzhali, S., Senthilkumar, M., Sundaram, S., Chung, H., Yang, A., and Sa, T. (2004b). Growth promotion and induction of systemic resistance in rice cultivers Co-47 (*Oryza sativa* L.) by *Methylobacterium* spp. *Bot. Bull. Acad. Sin.* 45:315-324.

Madhaiyan, M., Reddy, S .B. V., Anadham, R., Senthilkumar, M., Poonguzhali, S., Sundaram, S.P., Sa, T. (2006b). Plant growth promoting *Methylobacterium* induces defense responses in groundnut (Arachis *hypogia* L.) compared with rot pathogen. *Curr. Microbiol.* 53:270-6.

Madhaiyan, M., Sundaram, S.P., Kannaiyan, S. (2004a). Influence of pink pigmented facultative methylotrophic bacteria on seedling vigour of maize (*Zea mays* L.). *J. Microbe World,* 4:117-121.

Marx, C. J. and Lidstrom, M. E. (2004). Development of an insertional expression vector system for *Methylobacterium extorquens* AM1 and generation of null mutants lacking *mtdA* and/or *fch. Microbiology,* 150:9-19.

Murrell, J.C. (1992). The genetics and molecular biology of obligate methane-oxidizing bacteria, J.C. Murrell and H. Dalton (ed.), Methane and methanol utilizers. *Plenum Press, New York.* p.115-148.

Newaz, S.S., and Hersh, L.B. (1975). Reduced nicotinamide adenine dinucliotide-activated phosphoenolpyruvate carboxylase in *Pseudomonas* MA: potential regulation between carbon assimilation and energy production. *J. Bacteriol.* 124:825-833.

O'Connor, M.L. (1981). Extension of the model concerning linkage of genes coding for C-1 related functions *Methylobacterium organophilum. Appl. Environ. Microbiol.* 41:437-441.

Omer, Z.S., Tombolini, R., Broberg, A. and Gerhardson, B. (2004). Indole-3-acetic acid production by pink pigmented facultative methylotrophic bacteria. *Plant Growth Regul* .43:93-96.

Patel, R.N., Hau, C.T., Laskin, A.I. and Felix, A. (1978). Microbial oxidation of methane and methanol: isolation of methane utilizing bacteria and characterization of a facultative methane-utilizing isolate. *J. Bacteriol.* 136:352-358.

Patt, T.E., Cole, G.C. and Hanson, R.S. (1976). *Methylobacterium,* a new genus of facultative methylotrophic bacteria. *Int. J. Syst. Bacteriol.* 26:226-229.

Quayle, J. R. and Ferenci, T. (1978). Evolutionary aspects of autotrophy. *Microbiol Rev.* 42:251–273.

Quayle, J.R. (1961). Metabolism of C1 compounds in autotrophic and heterotrophic microorganisms. *Annu. Rev. Microbiol.*, 15:119-152.

Quayle, J.R. (1980). Aspect of the regulation of methylotrophic metabolism. *FEBS Lett.*117:K16-K27.

Sy, A., Giraud, E., Jourand, P., Garcia, N., Willems, A., Lajude, D.P., Prin, Y., Boivin-Masson, C., and Dreyfus, B. (2001). Methylotrophic *Methylobacterium* bacteria nodulate and fix nitrogen in symbiosis with legumes. *J. Bact.* 183(1):214-220.

Taylor, S., Dalton, H. and Dow, C. (1980). Purification and initial characterisation of ribulose 1, 5-bisphosphate carboxylase from *Methylococcus capsulatus* (Bath). *FEMS Microbiol Lett* 8:157–160.

Whittenbury, R., and Dalton, H. (1981). The methylotrophic bacteria. In M. P. Starr, H. Stolp, H.G. Truper, A. Balows, and H.G. Schlegel (ed), The Prokaryotes. Springer-Verlag KG, Berlin, Federal Republic of Germany.

Whittenbury, R., Phillips, K.C. and Wilkinson, J.F. (1970). Enrichment, isolation and some properties of methane utilizing bacteria. *J.Gen.Microbiol.* 61:205-218.

CHAPTER **5**

Recent Advances in Plant Growth Promotion Mechanisms of Agriculturally Important Bacteria

R. Muthukumarasamy*, V. S. Saravanan[2]
V. Ramesh Kumar[2], S. Natarajan[3] and G. Revathi[1]

[1]Main Bio-control Research Laboratory (Unit of Tamilnadu Co-operative Sugar Federation) Good Will Avenue, Venpakkam, Chengalpattu 603 111, Tamil Nadu, India;
[2]School of Bio-Engineering and Bio-Sciences, Vellore Institute of Technology (VIT) Vellore 632 014, Tamil Nadu, India
[3]Department of Plant Biology and Biotechnology, Gurunanak College, Velacherry, Chennai 600028, Tamil Nadu, India

ABSTRACT

Plant growth promoting (PGP) bacteria play an immense role in promoting crop growth mainly by direct mechanisms such as N_2 fixation, phosphate solubilization and phytohormone production and indirectly may help to suppress pathogens or contribute to enhance absorption of nutrients leading to increase yields. The N_2-fixers are grouped based on their niche adaptation as endophytic nodular symbionts (*Rhizobium* spp.), free living (*Azotobacter* spp.), associative symbionts (*Azospirillum* spp.) and endophytic non-nodular symbionts (*Azoarcus* and *Gluconacetobacter diazotrophicus*). In the recent past, phosphate solubilizing and phytohormone synthesizing bacterial communities also gained momentum.

Key words: plant growth promotion, agriculturally important bacteria, symbionts, *Rhizobium*

1. INTRODUCTION

Soil bacteria are the central components in earth's biodiversity. According to Lovell (2003), the estimated described species were 5,500 (<1%) whereas the undescribed species account to 1,00,000 (>99%) which need to be identified

*Email: mutindia@yahoo.com

in future. In recent years, identification of bacterial community has become rapid by 16S rDNA gene analysis. India is rich in biodiversity comprising of 12 mega diversity regions. Western Ghats is one such region rich in biodiversity and 17% of species diversity being of microbial origin. The bacteria in soil are diverse and active making the soil ecosystem as dynamic both in natural and managed agricultural soils mediating various beneficial as well as harmful processes to plants. The beneficial roles include formation of soil structure, decomposition of organic matter, removal of toxin, release of mineral nutrients and bio-geochemical nutrient cycling. These microbes also promote plant growth directly by N_2-fixation, phytohormone production, P-solubilization, organic mineralization, soil aggregation by polysaccharide production, siderophore production, vitamin synthesis, proton-extrusion mediated mineral uptake, degradation of potential xenobiotics, production of organic volatile compounds and soil bound and plant growth modulating enzymes. Further, they indirectly influence the plant growth by suppressing soil borne pathogens through mechanisms like antibiosis, hyperparasitism, induction of systemic induced resistance (ISR) resulting in production of pathogenesis-related (PR) proteins and lastly, enhance the stress (biotic and abiotic) resistance in plants.

Thus, soil bacteria play an invaluable role in the present scenario of sustainable agriculture and precision farming. This chapter discusses the agriculturally important bacterial species and their role in plant growth promotion.

2. PLANT GROWTH PROMOTING N_2-FIXERS

N_2-fixing bacterial numbers are ever increasing and it is outside the scope of this chapter to discuss about them. We discuss only N_2-fixers that have the ability to influence plant growth positively. N_2-fixation is the sole property of certain eubacteria and archebacteria and this trait is not discovered in any other domain of life.

2.1 Endophytic Nodular Symbiont

i. Rhizobium

The term "*rhizobium*" was viewed as a common name and it refers to any bacterium irrespective of the genus that can nodulate and fixes nitrogen in legumes. This was one of the oldest known symbiotic bacterial associations with legume plants known to the mankind. A plethora of references are available on the role of rhizobia in legume nodulation (Hirsch *et al.*, 2001; Gage, 2004; Prell and Poole, 2006). The first species of rhizobium namely *R. leguminosarum* was isolated by Frank (1889) later emended by Young *et al.* (2001). However, the taxonomy of *rhizobium* is under state of rapid change in the recent decades. The older classifications of rhizobium were mainly on the basis of host plants (Hiltner and Stormer., 1903). Further, description was based on nodulation range of *rhizobium* on the host plants (Fred *et al.*, 1932). In turn Norris (1965) grouped *rhizobium* based on biochemical properties. Vincent (1974) and Burton (1980) also developed certain classification based on cross-inoculation groups. After the advent of molecular techniques, a much clear understanding on the *rhizobium* phylogeny was obtained (Sawada *et al.*, 2003).

As per the recent classification, 57 validly published species distributed across 12 genera. Most belong to rhizobiales, a monophyletic group of *Proteobacteria* (Fig. 1). Formerly, *rhizobium* was known to the β-Proteobacterial group but the recent study confirms that it also occurs in the *β-Proteobacteria* group. Though they are within the *Proteobacteria*, they are scattered in different families. The genera *Rhizobium* or *Bradyrhizobium* are also found to colonize various non-legumes, acting as non-nodular symbiotic endophytes under favorable environments. A few studies documented their free living state in soil or association with plants like rice, sweet potato and maize as non-nodular symbionts (Yanni *et al.*, 1997; Chaintreuil *et al.*, 2000; Rosenblueth and Martinez-Romero, 2004). The study by Chaintreuil *et al.* (2000) focuses on the occurrence of endophytic non-nodular *Bradyrhizobium* in African wild rice (*Oryza breviligulata*) grown in Senegal and Guinea. It colonized the root surface and intercellular regions. It was isolated as an endophytic bacterium was also obtained from water in the rice fields. Thus it appears that the previous views of *Rhizobium* as a strict nodular symbiont may not be true. Under favorable conditions, the soil dwelling rhizobium could efficiently colonize the plants and could live inside plants as non-nodular endophytic bacteria.

2.2 Free Living N_2-Fixers

i. *Azotobacter*

Azotobacter is a common free living, aerobic, N_2-fixing, high carbon requiring bacterium that commonly occurs in soil. The genus *Azotobacter* was coined by Beijerinck (1901). The family *Azotobacteriaceae* includes two genera namely, *Azomonas* (cyst formers) comprising of three species (*A. agilis*, *A. insignis* and *A. macrocytogenes*) and *Azotobacter* (non-cyst formers) comprising of seven validly published species that commonly occurred in soil and rhizosphere, which include: *A. chroococcum* (Beijerinck, 1901), *A. vinelandii* (Lipman, 1903), *A. beijernckii* (Lipman, 1904), *A. nigricans* (Krasi¾nikov, 1949), *A. paspali* (Döbereiner *et al.*, 1966 and Thompson and Skerman, 1980), *A. armenicus* (Thomson and Skerman, 1981), *A. salinestris* (Page and Shivprasad, 1991). The agricultural importance of this bacterium was much studied by

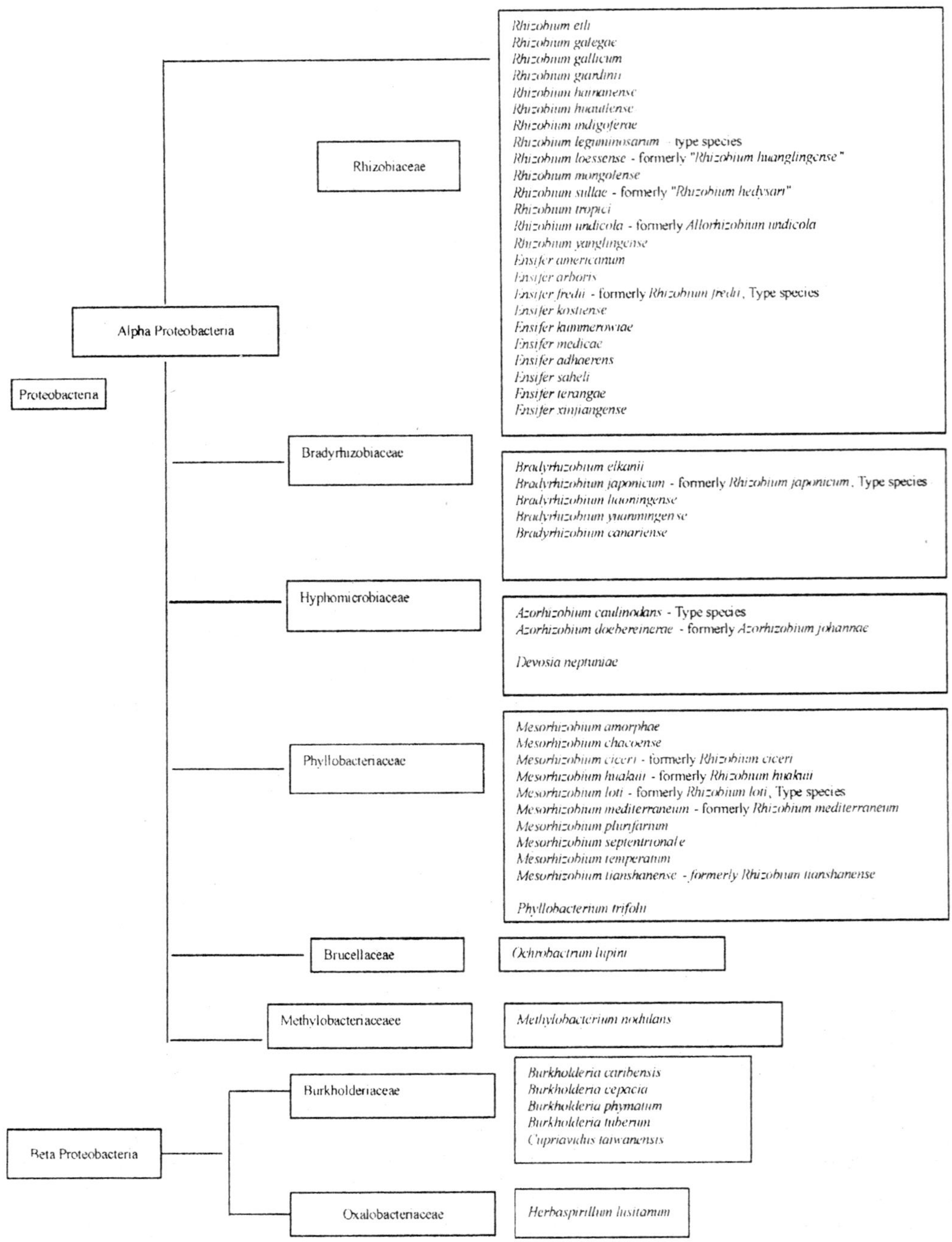

FIG. 1. Rhizobium taxonomy (*source:* http://www.rhizobia.co.nz/taxonomy/rhizobia.html)

Mrkovacki and Miliae (2001) and Behl *et al.* (2006). Several properties of *Azotobacter* are considered to be responsible for plant growth promotion effects. They are- a) ability to fix nitrogen to ammonia (Newton *et al.*, 1953; Bishop *et al.*, 1982); b) produce vitamins and other growth promoting substances that enhance seed germination (Brown, 1975; Narula *et al.*, 1981); c) production of auxins like IAA and other growth promoting hormones like gibberellins and cytokinin. This influences the root growth, which in turn helps in nutrient absorption; d) production of siderophore, that could help in efficient iron scavenging from the environments (Suneja *et al.*, 1996); e) inhibition of phytopathogenic fungi using antimicrobial compounds (Verma *et al.*, 2001); f) solubilization of tri-calcium phosphate and Mussoorie rock phosphate (Kumar and Narula, 1999) which helps in P- nutrition for plants.

2.3 Associative symbiotic N_2-fixer

Among the associative N_2-fixing bacteria, *Azospirillum* belongs to *β-Proteobacteria* and it has been recognized as a plant growth promoting rhizobacterium for a long period. It was isolated from various grasses and cereals around the world. It used as an efficient bacterium in numerous field and green-house studies conducted in India (Bashan *et al.*, 2004). This is a gram negative motile bacterium with much potential to establish in rhizosphere (strain Sp7) or inside the plants as an endophyte (Sp245). It has the ability to fix nitrogen which makes it as an elite bio-inoculant for agriculture. Currently, it has ten validly published species that occur in various plants (Table 1).

Azospirillum promotes plant growth by different mechanisms like N_2-fixation, production of phytohormone (auxin, gibberellins and cytokinin) and certain amino acids, P- solubilization, increased nutrient uptake, enhanced stress resistance, induction of chlorophyll synthesis in plant, vitamin synthesis in culture broth and siderophore production. It also possesses certain indirect plant growth promoting mechanisms like biocontrol and antibiosis. It is the only plant growth promoting rhizobacterium that has been so extensively studied for various plant growth promotion mechanisms in association with different crops (Steenhoudt and Vanderleyden, 2001; Seshadri *et al.*, 2000; Dobbelaere *et al.*, 2003; Thuler *et al.*, 2003; Rodriguez *et al.*, 2004; Gonzalez-Lopez *et al.*, 2005; Bashan *et al.*, 2006; Chowdhury *et al.*, 2007). However, it was reported that ACC deaminase, the bacterial enzyme that helps in plant growth promotion was absent in *A. brasilense* (Holguin and Glick, 2003).

2.4 Endophytic Non-Nodular Symbionts

For many years, it was thought that rhizobium-legume interactions are the foremost symbiotic relations with the plant systems. However, with the advent of molecular techniques, presence of bacteria inside the plant without causing any disease symptom is recognized. Endophytes have certain ecological advantages over the rhizosphere bacteria and they also differ from phytopathogenic bacteria in the following aspects:

TABLE 1. List of *Azospirillum* species identified in various crop plants

S. No.	*Azospirillum* sp.	Isolation source	Reference
1.	*A. brasilense*	Sp7 - Rhizosphere soil of *Digitaria decumbens*, Brazil Sp245 - Surface sterilized wheat roots, Brazil	Tarrand *et al.*, 1978 Baldani *et al.*, 1986
2.	*A. lipoferum*	Roots of wheat and maize, Brazil	Tarrand *et al.*, 1978
3.	*A. amazonense*	Roots of rhizosphere soil of graminaceous plants, Amazon region, Brazil	Magalhaes *et al.*, 1983
4.	*A. halopraeferens*	Roots of Kallar grass grown in saline soils, Pakistan	Reinhold-Hurek *et al.*, 1987
5.	*A. irakense*	Roots of rhizosphere of rice, Iraq	Khammas *et al.*, 1989
6.	*A. doebereinerae*	Roots of Gramineous plant Miscanthus, Germany	Eckert *et al.*, 2001
7.	*A. largimobile*	Fresh water, Australia	Skerman *et al.*, 1983; Ben Dekhil *et al.*, 1997; Sly and Stackebrandt (1999)
8.	*A. oryzae*	Roots of rhizosphere of rice, Japan	Oyaizu-Masuchi and Kamagata, (1988); Xie and Yokota (2005)
9.	*A. melinis*	As an endophyte in molasses grass (*Melinis minutiflora*), China	Peng *et al.*, 2006
10.	*A. canadense*	Rhizosphere of corn plants, Canada	Mehnaz *et al.*, 2007

i. The endophytes are in better niches (intercellular/intracellular/xylem) that may help in effective C utilization. Internal localization would help the diazotrophic endophytes to acquire a better environment for N_2-fixation, as they prefer partially oxygenated (microaerophilic) condition, which is absolutely necessary for the expression and functioning of nitrogenase system.

ii. Being endophytic, they are better safeguarded from certain potential bacterial predators like protozoa rather than rhizospheric population.

iii. Although the nitrogen-fixation and phytohormone production are common among the rhizosphere bacteria, products of these processes including ammonia are not totally available to the plants, owing to a possible degradation. However, endophytic mediated production inside plants may be fully available to the plant system.

iv. Endophytic bacteria are more tolerant to man made organic compounds (xenobiotics) and inorganic metals than the rhizospheric bacteria (Idris *et al.*, 2005). The high resistance for these compounds by endophytic bacteria helps them to survive easily inside the plants and potentially degrade them.

The features that differentiate endophytes from plant pathogens are:

i. The colonizing numbers of endophytic bacteria are always less than the phytopathogens. For example, *Clavibacter michiganensis* sub sp *sepedonicus* causing bacterial wilt disease accounted for $1x10^7$ cfu g^{-1} fresh weight. Conversely, the size of the artificially introduced endophytes ranged between $1x10^3$ - $1x10^5$ cfu g^{-1} of tissue for most investigated plant species (Lodewyckx *et al.*, 2002). The large numbers of phytopathogens could result in blocking of conducting vessels leading to wilting symptom. Whereas, the endophyts are probably regulated by plant system and maintained in a level significant to the plant system. The recent findings show that host plant may be actively involved in the establishment with beneficial endophytes (Vinagre *et al.*, 2006)

ii. The other major difference exists in the production level of certain cell wall degrading enzymes that could help in the development of endophytic habitat. Endophytes produce certain pectic and cellulolytic enzymes that are mild and more localized to attack only the cell wall polymers. But pathogenic *Ralstonia solanacearum* (the wilt pathogen) produce high quantity of these enzymes which is deleterious to plant development.

iii. Generally endophytes are unable to utilize common carbohydrates, but plant pathogens can efficiently use.

Though large number of bacterial endophytes was isolated from different sources (Rosenblueth and Martínez-Romero, 2006), only *Azoarcus* sp. BH72 and *G. diazotrophicus* that occur in Kallar grass and sugarcane plants respectively were studied (Reinhold-Hurek *et al.*, 1993; Cavalcante and Döbereiner, 1988).

i. Azoarcus

Recently, the genome of *Azoarcus* sp. strain BH72 has been completely sequenced (Krause *et al.*, 2006). This report insight into the endophytic life style of this bacterium. It is a potential N_2-fixer in rice plants but present inside the plant in an unculturable state (Hurek *et al.*, 2002). Commonly this bacterium does not excrete substantial amount of fixed nitrogen under laboratory conditions. However, it contributes significantly for the N nutrition of plants. Siderophore production by *Azoarcus* sp. is not documented so far and there is no evidence for the presence of common pathways for siderophore production in this bacterium, but the genomic analysis revealed that it has certain putative receptors to uptake a few groups of siderophores. At present, the reason for expression of such siderophore uptake systems is not understood. But high numbers of putative siderophore receptors may be useful to develop this bacterium as a rhizosphere competent bacterium and also a successful bio-control agent (Krauze *et al.*, 2006).

ii. Gluconacetobacter diazotrophicus

It is the first acetic acid endophytic bacterium discovered with the trait of N_2-fixation from sugarcane plants growing in Brazil (Cavalcante and Döbereiner, 1988). In India also, it was identified in sugarcane plants cultivated with low N-fertilized fields (Muthukumarasamy *et al.*, 1999; 2002a). Its association was documented in several other crop plants also including certain vegetables like carrot, raddish and beetroot (Muthukumarasamy *et al.*, 2002b; Madhaiyan *et al.*, 2004). This bacterium was recently reported in association with certain Indian and Korean rice varieties (Muthukumarasamy *et al.*, 2005; 2007) and was found inside the root and stem tissues apart from rhizosphere soils. Muthukumarasamy *et al.* (2005 and 2007) focused soil as a possible mode of transmission for this bacterium. *G. azotocaptans* another species of *Gluconacetobacter* was detected only in the rhizosphere of corn plants and but not inside the plant tissues (Mehnaz *et al.*, 2006).

Gluconacetobacter diazotrophicus performs a pivotal role in the growth promotion of sugarcane. It influences on growth mainly by N_2-fixation under favorable plant niches. A study based on N dilution/N balance revealed that up to 60-80% of sugarcane plant N (equivalent to over 200 kg N ha^{-1} y^{-1}) could be N_{2-}fixation and *G. diazotrophicus* was predicted as the responsible bacterium for this huge BNF activity in sugarcane (Boddey *et al.*, 1991). Among the diazotrophs available, *Azoarcus* sp. strain BH 72 and *G. diazotrophicus* are alone having a conclusive proof for N_2-fixation under *in planta* conditions. Several plant growth promoting traits are present in *G. diazotrophicus* including, phytohormone production, P and Zn solubilization, biocontrol of pathogens (Bastián *et al.*, 1998; Mahesh Kumar *et al.*, 1999; Muthukumarasamy *et al.*, 2000; Madhaiyan *et al.*, 2004; Blanco *et al.*, 2005; Saravanan *et al.*, 2007).

iii. Phosphobacteria

Phosphorus is an essential nutrient and its deficiency severely restricts the growth of the plants mainly in the early periods of crop growth. Tropical

TABLE 2. List of phosphate solubilizing bacteria and the type of organic acids involved in the solubilization (modified from Khan *et al.*, 2006)

Bacterial source	Type of organic acids produced	Reference
Enterobacter intermedium	2- keto gluconic acid	Hwangbo *et al.*, 2003
Bacillus amyloliquefaciens, *B. licheniformis*, *B. atrophaeus*, *Paenibacillus macerans*, *Vibrio proteolyticus*, *Xanthobacter agilis*, *Enterobacter aerogenes*, *E. taylorae*, *E. asburiae*, *Kluyera cryocrescens*, *Pseudomonas aerogenes*, *Chryseomonas luteola*	Lactic, itaconic, isovaleric, isobutyric, acetic acid	Vazquez *et al.*, 2000
Pseudomonas cepacia	Gluconic, 2-keto gluconic acid	Bar-Yosef *et al.*, 1999
Bacillus polymyxa, *B. licheniformis*, *Bacillus* sp	Oxalic acid, citric	Gupta *et al.*, 1994
Pseudomonas striata	Malic, glyoxalic, succinic, fumaric, tartaric, ? -ketobutyric acid	Gaur, 1990
Arthrobacter sp.	Oxalic, malonic	Banik and Dey, 1982
Bacillus firmus	2-ketogluconic, succinic	Banik and Dey, 1982
Micrococcus spp.	oxalic	Banik and Dey, 1982
Bacillus subtilis, *Bacillus* sp.	Oxalic, succinic, citric, 2-ketogluconic acid	Banik and Dey, 1982
Chryseobacterium, *Gordonia*, *Arthrobacter ureafaciens*, *Phyllobacterium myrsinacearum*, *Rhodococcus erythropolis* and *Delftia* sp	citric acid, gluconic acid, lactic acid, succinic acid, propionic acid and three unknown organic acids	Chen *et al.*, 2006

Contd...

Table 2. Contd...

Pantoea agglomerans		Son *et al.*, 2006
Burkholderia cepacia CC-Al74	Gluconic acid	Lin *et al.*, 2006
Paenibacillus mucilaginosus		Hue *et al.*, 2006
Gluconacetobacter diazotrophicus	5-ketogluconic acid	Saravanan *et al.*, 2007

and subtropical soils are acidic in nature and phosphorus gets fixed in these soils leading to deficiency. It is generally available as organic phosphorus found in the living tissues which reaches soil as dead organic matter after death of an organism. The P available through this source can be made bio-available by action of certain bacterial phosphatases and phytases leading to P mineralization (conversion of organic phosphate to mineral phosphate). The other source of P is fixed in the soil through chemical reactions. A research data showed that 75-90% of the added phosphatic fertilizer was precipitated in soil (Stevenson, 1986). This kind of accumulated phosphate of time has the potential to sustain crop yield in agricultural crops for longer period if properly managed (Goldstein *et al.*, 1993). In this context, the use of efficient phosphorus mobilizing bacteria can augment the P nutrition of plants on a sustainable basis.

The insoluble form of P can be solubilized by bacteria and fungi by different mechanisms. However, a microbial solubilization is frequently linked to the excretion of organic acids produced through aerobic metabolism (tricarboxylic acid cycle and extracellular aldose oxidation pathway) or by anaerobic pathway (lactic acid production). The solubilization may be accomplished by reduction in pH, organic acids secreted by the microbes causing direct dissolution of mineral phosphate.

3. PHYTOHORMONE PRODUCTION

Phytohormone refers to plant growth regulating compounds that have a regulatory role in the plant growth and development. The hormone levels are critical for the characteristic change in the plant growth and development. The role of microbial hormonal effects on plant growth was first recognized in *Azospirillum brasilense,* which produced different classes of enzyme like IAA, gibberellins and cytokinins. Since then numerous reports appeared on phytohormone production by plant growth promoting (PGP) bacteria that are summarized in Table 3. Moreover, the enhancement of plant growth observed by *Azospirillum* spp. in most of the inoculation studies was mainly due to its phytohormone producing ability (Okon and Kapulnik, 1986). This was also demonstrated in several other PGP bacteria using *nif*$^-$ mutants, which were unable to fix nitrogen but were able to produce phytohormones and other metabolites. By using the *nif*$^-$ mutants, the phytohormone production and plant growth promotion was demonstrated in plants inoculated with *G. diazotrophicus* and *Azoarcus* sp. (Sevilla *et al.*, 2001; Hurek *et al.*, 2002). Recent study showed that *Serratia proteamaculans* has the ability to transform certain cytokinins and adenine derivatives and has the ability to utilize a secondary metabolite as a carbon source (Taylor *et al.*, 2006) and which could help for its rhizosphere competency. It is a rare property among the soil culturable bacteria. Recently, *Methylobacterium* was demonstrated as a prolific cytokinin producer (Koenig *et al.*, 2002). The endophytic occurrence, prolific cytokinins producing ability and ubiquitous presence would make this bacterium, as an elite inoculant for crop production in future.

TABLE 3. List of plant growth promoting bacteria producing phytohormones

S. No	Hormone type and the producing bacteria		
	Auxins	**Gibberellins**	**Cytokinin**
1	*Pseudomonas fluorescens* NJ-101	*Gluconacetobacter diazotrophicus*	*Azospirillum brasilense*
2	*Pseudomonas fluorescens* EM 85	*Herbaspirillum seropedicae*	*Azotobacter beijerinckii*
3	*Pseudomonas* spp.	*Azospirillum lipoferum*	*A. chroococcum*
4	*Bacillus* spp.	*A. brasilense*	*A. vinelandii*
5	*Acinetobacter* spp.	*Bacillus licheniformis*	*Paenibacillus polymyxa*
6	*Xanthomonas maltophilia*	*B. pumilus*	*Rhizobium leguminosarum*
7	*Bacillus thuringiensis*	*Rhizobium phaseoli*	*Bacillus subtilis*
8	*Aeromonas veroniii*	*Azotobacter chroococcum*	*Methylobacterium* spp.
9	*Enterobacter cloacae*	*A. vinelandii*	*Pseudomonas fluorescens* G20-18
10	*Enterobacter* sp.	*A. paspali*	*Bacillus megaterium*
11	*Agrobacterium* sp.	*Bacillus macroides*	*Proteus mirabilis*
12	*Alcaligenes piechaudii*		
13	*Azospirillum brasilense*		
14	*Bradyrhizobium* sp.		
15	*Rhizobium leguminosarum*		
16	*Comamonas acidovorans*		

Contd...

Table 3. Contd...

17	*Paenibacillus polymyxa*		
18	*Pseudomonas putida* GR-12-2		
19	*Rhizobium phaseoli*		
20	*Azotobacter chroococcum*		
21	*A. paspali*		
22	*A. vinelandii*		
23	*Flavobacterium* spp		
24	*Stenotrophomonas* sp.		
25	*Pantoea* sp.		
26	*Paracoccus* sp.		
27	*Burkholderia* sp.		
28	*Herbaspirillum seropedicae*		
29	*Klebsiella pneumoniae*		
30	*Azospirillum lipoferum*		
31	*Azotobacter beijerinckii*		
32	*Azoarcus* sp.		
33	*Gluconacetobacter diazotrophicus*		

4. CONCLUSION

It is clear from the previous studies that utmost importance has been given only to a few PGP traits like N_2-fixation, phytohormone production and P-solubilization. However, in future, bacterial cell to cell communication and its interaction with plant be extensively studied. Another area that needs to be addressed would be the basic metabolic and genetic potential of the PGP bacteria. This seems to play a profound role in the expression of each and every PGP characteristic of a bacterium. At present, lack of knowledge in this area may be a reason for the inconsistent performance of bacterial inoculants. In future, it is important to study other plant growth promoting factors such as ACC deaminase role in plant growth promotion and also discovering certain novel plant growth promotion molecules of microbial origin. Another important area to be critically studied is the plant genotype versus bacterial strain interactions and the mechanism of inoculated bacteria which maintain more numbers than native numbers of the same group in plant micro-environments.

Acknowledgements

R.M and G.R thank Tamilnadu Co-operative Union for partly funding our research programs through its Co-operative Research and Development grant. VSS Thanks Management, Vellore Institute of Technology for the financial support extended to the research activities.

REFERENCES

Baldani, V.L.D., Alvarez, M.A. deB., Baldani, J.I. and Döbereiner, J. (1986). Establishment of inoculated *Azospirillum* spp. in the rhizosphere and in roots of field grown wheat and sorghum. *Plant Soil* 90:35-46.

Banik, S. and Dey, B.K. (1982). Available phosphate content of an alluvial soil as influenced by inoculation of some isolated phosphate solubilizing microorganisms, *Plant Soil* 69:353-364.

Bar-Yosef, B., Rogers, R.D., Wolfram, J.H. and Richman, E. (1999). *Pseudomonas cepacia*-mediated rock phosphate solubilization in Kaolinite and Montmorillonite suspensions. *Soil Sci. Soc. Am. J.* 63:1703-1707.

Bashan, Y., Bustillos, J.J., Leyva, L.A., Hernandez, J.P. and Bacilio, M. (2006). Increase in auxiliary photoprotective photosynthetic pigments in wheat seedlings induced by *Azospirillum brasilense. Biol. Fertil. Soils* 42:279-285.

Bashan, Y., Holguin, G. and Bashan, L.E.D. (2004). *Azospirillum*-plant relationships: physiological, molecular, agricultural, and environmental advances (1997–2003). *Can. J. Microbiol.* 50: 521-577.

Bastián, F., Cohen, A., Piccoli, P., Luna, V., Baraldi, R. and Bottini, R. (1998). Production of indole-3-acetic acid and gibberellins A1 and A3 by *Acetobacter diazotrophicus* and *Herbaspirillum seropedicae* in chemically defined culture media. *Plant Growth Regul.* 24: 7-11.

Behl, R.K., Narula, N., Vasudeva, M., Sato, A., Shinano, T. and Osaki, M. (2006). Harnessing wheat genotype x *Azotobacter* strain interactions for sustainable wheat production in semi arid tropics. *Tropics* 15: 121-133.

Beijerinck, M.W. (1901). Ueber oligonitrophile Mikroben. *Zentr. Bakt. Parasitenk* II, 7: 561-582.

Ben Dekhil, S., Cahill, M., Stackebrandt, E. and Sly, L.I. (1997). Transfer of *Conglomeromonas largomobilis* sub sp. *largomobilis* to the genus *Azospirillum* as *Azospirillum largomobile* comb. nov., and elevation of *Conglomeromonas largomobilis* subsp. *parooensis* to the new type species of *Conglomeromonas. Syst. Appl. Microbiol.* 20: 72-77.

Bishop, P. E., Jarlenski, D. M. and Heterington, D.R. (1982). Expression of an alternative nitrogen fixation system in *Azotobacter vinelandii. J. Bacteriol.* 150:1244-1251.

Blanco, Y., Blanch, M., Piñón, D., Legaz, M.E. and Vicente, C. (2005). Antagonism of *Gluconacetobacter diazotrophicus* (a sugarcane endosymbiont) against *Xanthomonas albilineans* (pathogen) studied in alginate-immobilized sugarcane stalk tissues. *J. Biosci. Bioeng.* 99: 366-371.

Boddey, R.M., Urquiaga, S., Reis, V.M. and Döbereiner, J. (1991). Biological nitrogen fixation associated with sugarcane. *Plant Soil* 37: 111-117.

Brown, M.E. (1975). Rhizosphere microorganisms opportunist bandits or benefactors. In: *Soil Microbiology*. N. Walker (Ed). Buterworths pp. 21-38.

Burton G. C. (1980). New developments in inoculating legumes. In: Recent Advances in *Biological Nitrogen Fixation*. Ed. N. S. Subba Rao. pp 380-405. Edward Arnold, Great Britain.

Cavalcante, V.A. and Döbereiner, J. (1988). A new acid-tolerant nitrogen-fixing bacterium associated with sugarcane. *Plant Soil* 108: 23-31.

Chaintreuil, C., E. Giraud, Y. Prin, J. B. Lorquin, M. Gillis, P. de Laudie and B. Dreyfus. (2000). Photosynthetic bradyrhizobia are natural endophytes of the African wild rice *Oryza breviligulata*. *Appl. Environ. Microbiol.* 66: 5437–5477.

Chen, Y.P., Rekha, P.D., Arun, A.B., Shen, F.T., Lai, W.A. and Young, C.C. (2006). Phosphate solubilizing bacteria from subtropical soil and their tricalcium phosphate solubilizing abilities. *Appl. Soil Ecol.* 34 : 33-41.

Chowdhury, S.P., Nagarajan T., Tripathi R., Mishra M.N., Rudulier, D.L. and Tripathi, A.K. (2007). Strain-specific salt tolerance and osmoregulatory mechanisms in *Azospirillum brasilense*. *FEMS Microbiol Lett.*, (in press) DOI:10.1111/j.1574-6968.2006.00540.x.

Dobbelaere, S., Vanderleyden, J. and Okon, Y. (2003). Plant growth-promoting effects of diazotrophs in the rhizosphere. *Crit. Rev. Plant Sci.* 22:107-149.

Döbereiner, J. (1966). *Azotobacter paspali* sp. nov., uma bacteria fixadora de nitrogenio na rhizosfera de Paspalum. *Pesquisa Agropecuaria Brasileira*, 1: 357-365.

Eckert, B. Weber, O.B., Kirchhof, G., Halbritte, A., Stoffels, M. and Hartmann, A. (2001). *Azospirillum doebereinerae* sp. nov., a nitrogen-fixing bacterium associated with the C4-grass *Miscanthus*. *Int. J. Syst. Evol. Microbiol.* 51: 17-26.

Frank, B. (1889). Ueber die Pilzsymbiose der Leguminosen. *Ber Dtsch Bot Ges* 7: 332-346.

Fred, E. B., Baldwin, I. L. and McCoy, E. (1932). Root nodule bacteria and leguminous plants, University of Wisconsin Press, Madison, Wis., p. 343.

Gage, D.J. (2004). Infection and invasion of roots by symbiotic, nitrogen-fixing rhizobia during nodulation of temperate legumes. *Microbiol. Mol. Biol. Rev.* 68: 280-300.

Gaur A.C. (1990). Phosphate solubilizing microorganisms as biofertilizers, Omega Scientific Publisher, New Delhi, India. p. 176.

Goldstein, A.H., Rogers, R.D. and Mead, G. (1993) Mining by microbe, *Bio.Technol.* 11: 1250–1254.

González-López, J. Rodelas, B., Pozo, C., Salmerón-López, V., Martínez-Toledo, M.V. and Salieron, V. (2005). Liberation of amino acids by heterotrophic nitrogen fixing bacteria. *Amino Acids* 28: 363–367.

Gupta R.R., Singal R., Shanker A., Kuhad, R.C. and Saxena, R.K. (1994). A modified plate assay for screening phosphate solubilizing microorganisms, *J. Gen. Appl. Microbiol.* 40: 255-260.

Hiltner, L., and K. Störmer. (1903). Neue Untersuchungen über die Wurzelknöllchen der Leguminosen und deren Erreger. Arbeiten aus der Biologischen Abteilung für Land- und Forstwirthschaft, Kaiserlichen Gesundheitsamte, Berlin 3:151-307

Hirsch, A.M., Lum, M.R. and Downie, J.A. (2001) What makes the rhizobia-legume symbiosis so special? *Plant Physiol.* 127: 1484-1492.

Holguin, G. and Glick, B.R. (2003). Transformation of *Azospirillum brasilense* Cd with an ACC deaminase gene from *Enterobacter cloacae* UW4 fused to the *Tetr* gene promoter improves its fitness and plant growth promoting ability. *Microb. Ecol.* 46:122-133.

Hue, X., Chen, J. and Guo, J. (2006). Two phosphate- and potassium-solubilizing bacteria isolated from Tianmu Mountain, Zhejiang, China. *World J. Microbiol. Biotechnol.* 22:983-990.

Hurek, T., Handley, L.L., Reinhold-Hurek B. and Pich Y. (2002). *Azoarcus* grass endophytes contribute fixed nitrogen to the plant in an unculturable state. *Mol. Plant-Microbe Interact.* 15 : 233-42.

Hwangbo, H., Park, R.D., Kim, Y.W., Rim, Y.S., Park, K.H., Kim, T.H., Suh, J.S. and Kim, K.Y. (2003).2-ketogluconic acid production and phosphate solubilization by *Enterobacter intermedium. Curr. Microbiol.* 47: 87-92.

Idris, R. Trifonova, R., Puschenreiter, M. Walter W. Wenzel, W.W. and Sessitsch, A. (2005). Bacterial communities associated with flowering plants of the Ni hyper accumulator *Thlaspi goesingense. Appl. Environ. Microbiol.* 70: 2667-2677.

Khammas, K.M., Ageron, E., Grimont, P.A.D. and Kaiser, P. (1989). *Azospirillum irakense* sp. nov., a nitrogen-fixing bacterium associated with rice roots and rhizosphere soil. *Res. Microbiol.* 140: 679-693.

Khan M.S., Zaidi, A. and Wani, P.A. (2006).Role of phosphate-solubilizing microorganisms in sustainable agriculture – A review. *Agron. Sustain. Dev.* 26: 1-15.

Koenig, R.L., Morris R.O. and Polacco J.C. (2002). tRNA is the source of low-level *trans*-zeatin production in *Methylobacterium* spp. *J. Bacteriol.* 184: 1832-1842.

Krasi¾nikov, N.A. (1949).Opredelitel' bakterii i aktinomitsetov (Index manual of bacteria and actinomycetes).- Izd. AN SSSR (Published by the Academy of Sciences of the USSR), Moscow-Leningrad.

Krause, A. Ramakumar, A., Bartels, D., Battistoni, F., Bekel, T., Boch, J., Böhm, M., Frauke F., Hurek, T., Krause, L., Linke, B., McHardy, A.C., Sarkar, A., Schneiker, S., Syed, A.I., Thauer, R., Vorhölter, F.J., Weidner, S., Pühler, A., Reinhold-Hurek, B., Kaiser, O. and Alexander, G. (2006) Complete genome of the mutualistic, N_2-fixing grass endophyte *Azoarcus* sp. strain BH72. *Nat. Biotechnol.* 24, 1385-1391.

Kumar, V. and Narula, B. (1999). Solubilization of inorganic phosphates and growth emergence of wheat as affected by *A. chroococcum* mutants. *Biol. Fertil. Soils* 28: 301-305.

Lin T.F., Huang H.I., Shen F.T. and Young, C.C. (2006). The protons of gluconic acid are the major factor responsible for the dissolution of triclacium phosphate of *Burkholderia cepacia* CC-A174. *Bioresour. Technol.* 97: 957-960.

Lipman J.G. (1903). Report on the New Jersey Agricultural Experiment Station, 24: 217-285.

Lipman J.G. (1904). Report on the New Jersey Agricultural Experiment Station, 25: 237- 289.

Lodewyckx, C., Vangronsveld, J., Porteous, F., Moore, E.R.B., Taghavi, S. and van der Lelie, D. (2002). Endophytic bacteria and their potential applications. *Crit. Rev. Plant Sci.* 21: 1-24

Lovell C.R. (2003). Microbial diversity In: Desk Encyclopedia of Microbiology ed. Moselio Schaechter, Academic press, pp. 326-339.

Madhaiyan, M., Saravanan, V.S., Bhakiya Shilba Sandal Jovi, D., Lee, H.S., Thenmozhi, R., Hari, K. and Sa, T.M. (2004). Natural endophytic occurrence of *Gluconacetobacter diazotrophicus* in tropical and subtropical plants of Western Ghats, India. *Microbiol. Res.* 159:233-243.

Magalhães, F.M., Baldani, J.I., Souto, S.M., Kuykendall, J.R. and Döbereiner, J. (1983). A new acid-tolerant *Azospirillum* species. *An. Acad. Bras. Ciên.* 55: 417-429.

Mahesh Kumar, K.S., Krishnaraj, P.U. and Alagawadi, A.R. (1999) Mineral phosphate solubilizing activity of *Acetobacter diazotrophicus*: a bacterium associated with sugarcane. *Curr Sci.* 76: 874-875.

Mehnaz, S., Weselowski, B. and Lazarovits, G. (2006). Isolation and identification of *Gluconacetobacter azotocaptans* from corn rhizosphere. *Syst. Appl. Microbiol.* 29: 496-501.

Mehnaz, S., Weselowski, B. and Lazarovits, G. (2007). *Azospirillum canadense* sp. nov., a nitrogen-fixing bacterium isolated from corn rhizosphere. *Int. J. Sys. Evol. Microbiol.* (in press).

Mrkovaèki, N. and Miliæ, V. (2001). Use of *Azotobacter chroococcum* as potentially useful in agricultural application *Ann. Microbiol.* 51:145-158.

Muthukumarasamy, R., Revathi, G and Lakshminarasimhan, C. (1999) Influence of N-fertilization on the ioslation of *Acetobacter diazotrophicus* and *Herbaspirillum* spp. from Indian sugarcane varieties. Biol. Fertil. Soils. 29: 157 – 164.

Muthukumarasamy, R., Revathi, G. and Vadivelu, M. (2000). Antagonistic potential of N_2 fixing *Acetobacter diazotrophicus* against *Colletotrichum falcatum* Went, a causal organism of red-rot of sugarcane. *Curr Sci.* 78:1063-1065.

Muthukumarasamy, R., Revathi, G. and Loganathan, P. (2002a). Effect of inorganic N on the population, *in vitro* colonization and morphology of *Acetobacter diazotrophicus* (Syn. *Gluconacetobacter diazotrophicus*). *Plant Soil*. 243: 91-102.

Muthukumarasamy, R., Revathi, G., Seshadri, S. and Lakshminarasimhan C. (2002b). *Gluconacetobacter diazotrophicus* (syn. *Acetobacter diazotrophicus*), a promising diazotrophic endophyte in tropics. *Curr. Sci.* 83 : 137-145.

Muthukumarasamy, R., Cleenwerck, I., Revathi, G., Vadivelu, M., Janssens, D., Hoste, B., Gum, K.U., Park, K.D., Son, C.Y., Sa, T.M. and Caballero-Mellado, J. (2005). Natural association of *Gluconacetobacter diazotrophicus* and diazotrophic *Acetobacter peroxydans* with wetland rice. *Syst. Appl. Microbiol.* 2:505-508.

Muthukumarasamy, R., Kang, U.G., Park, K.D., Jeon, W.T., Park, C.Y., Cho, Y.S., Kwon, S.W., Song, J., Roh, D.H. and Revathi, G. (2007). Enumeration, isolation and identification of diazotrophs from Korean wetland rice varieties grown with long-term application of N and compost and their short-term inoculation effect on rice plants. *J. Appl. Microbiol.* (In press) doi:10.1111/j.1365-2672.2006.03157.x

Norris, D.O. (1965). Acid production by *Rhizobium* a unifying concept. *Plant Soil* 22: 143-166.

Narula,N., Lakshminarayana, K.L and Tauro, P. (1981). Ammonia excretion by *Azotobacter chroococcum*. *Biotechnol. Bioeng.* 23: 467-470.

Newton, J.W., Wilson, P.W. and Burris, R.H. (1953). Direct demonstration of ammonia as an intermediate in nitrogen fixation by *Azotobacter*. *J. Biol. Chem.* 204: 445-451.

Okon, Y. and Kapulnik, Y. (1986). Development and function of *Azospirillum* inoculated roots. *Plant Soil* 90: 3-16.

Oyaizu-Masuchi, Y. and Komagata, K. (1988). Isolation of free-living nitrogen-fixing bacteria from the rhizosphere of rice. *J. Gen. Appl. Microbiol.* 34:127-164.

Page, W.J. and Shivprasad, S. (1991). *Azotobacter salinestris* sp. nov., a sodium-dependent, microaerophili, and aeroadaptive nitrogen-fixing bacterium. *Int. J. Syst. Bacteriol.* 41: 369-376.

Peng, G., Wang, H., Zhang, G., Hou, W., Liu, Y., Wang. E.T. and Tan, Z. (2006). *Azospirillum melinis* sp. nov., a group of diazotrophs isolated from tropical molasses grass. *Int. J. Sys. Evol. Microbiol.* 56: 1263-1271.

Prell, J. and Poole, P. (2006).Metabolic changes of rhizobia in legume nodules. *Trends Microbiol.*14:161-168.

Reinhold, B., Hurek, T., Fendrik, I., Pot, B., Gillis, M., Kersters, K., Thielemans, S. and De Ley, J. (1987). *Azospirillum halopraeferens* sp. nov., a nitrogen-fixing organism associated with roots of Kallar Grass (*Leptochloa fusca* (L.) *Kunth*). *Int. J. Syst. Bacteriol.* 37: 43-51.

Reinhold-Hurek, B., Hurek, T., Gillis, M., Hoste, B., Vancanneyt, M., Kersters, K. and De-Ley, J. (1993). *Azoarcus* gen. nov., nitrogen-fixing proteobacteria associated with roots of Kallar grass (*Leptochloa fusca* (L.) Kunth), and description of two species, *Azoarcus indigens* sp. nov. and *Azoarcus communis* sp. nov. *Int. J. Syst. Bacteriol.* 43:574-584.

Rodríguez, H. Gonzalez, T. Goire, I. and Bashan, Y. (2004). Gluconic acid production and phosphate solubilization by the plant growth-promoting bacterium *Azospirillum* spp. *Naturwissenschaften* 91:552–555.

Rosenblueth, M. and Martínez-Romero, E. (2004). *Rhizobium etli* maize populations and their competitiveness for root colonization. *Arch. Microbiol.* 181:337-344.

Rosenblueth, M. and Martínez-Romero, E. (2006). Bacterial endophytes and their interactions with hosts. *Mol. Plant-Microbe Interact.* 19: 827-837.

Saravanan, V.S., Madhaiyan, M. and Thangaraju, M. (2007). Solubilization of zinc compounds by the diazotrophic, plant growth promoting bacterium *Gluconacetobacter diazotrophicus. Chemosphere* 66:1794-1798.

Sawada, H. Kuykendall, L.D. and Young J.M. (2003). Changing concepts in the systematics of bacterial nitrogen-fixing legume symbionts. *J. Gen. Appl. Microbiol.* 49: 155-179.

Seshadri, S., Muthukumarasamy, R., Lakshminarasimhan, C. and Ignacimuthu, S. (2000). Solubilization of inorganic phosphates by *Azospirillum halopraeferans. Curr. Sci.* 79: 565-566.

Sevilla, M., Burris, R.H., Gunapala, N. and Kennedy, C. (2001). Comparison of benefit to sugarcane plant growth and $^{15}N_2$ incorporation following inoculation of sterile plants with *Acetobacter diazotrophicus* wild. *Mol. Plant-Microbe Interact.* 14: 358-366.

Skerman, V. B. D., Sly, L.I. and Williamson, M. (1983). *Conglomeromonas largomobilis* gen. nov., sp. nov., a sodium-sensitive, mixed-flagellated organism from freshwater. *Int. J. Syst. Bacteriol.* 33: 300-308.

Sly, L.I. and Stackebrandt, E. (1999). Description of *Skermanella parooensis* gen. nov., sp. nov. to accommodate *Conglomeromonas largomobilis* subsp. *parooensis* following the transfer of *Conglomeromonas largomobilis* subsp. *largomobilis* to the genus *Azospirillum. Int. J. Syst. Bacteriol.* 49: 541-544.

Son, H.J., Park, G.T., Cha, M.S. and Heo, M.S. (2006). Solubilization of insoluble inorganic phosphates by a novel salt- and pH-tolerant *Pantoea agglomerans* R-42 isolated from soybean rhizosphere. *Bioresour. Technol.* 97: 204–210.

Steenhoudt, O. and Vanderleyden, J. (2001). *Azospirillum*, a free-living nitrogen-fixing bacterium closely associated with grasses: genetic, biochemical and ecological aspects. *FEMS Microbiol. Rev.* 24:487-506.

Stevenson, F.J. (1986) Cycles of soil carbon, nitrogen, phosphorus, sulphur micronutrients, Wiley, New York.

Suneja, S., Narula, N., Anand, R.C. and Lakshminarayana, K. (1996). Relationship of *Azotobacter chroococcum* siderophore with nitrogen fixation. *Folia Microbiol.* 40: 154-158.

Tarrand, J.J., Krieg, N.R. and Döbereiner, J. (1978). A taxonomic study of the *Spirillum lipoferum* group, with descriptions of a new genus, *Azospirillum* gen. nov. and two species, *Azospirillum lipoferum* (Beijerinck) comb. nov. and *Azospirillum brasilense* sp. nov.. *Can. J. Microbiol.* 24: 967-980.

Taylor, J.L. L. Zaharia, I., Chen, H., Anderson, E. and Abrams, S.R. (2006). Biotransformation of adenine and cytokinins by the rhizobacterium *Serratia proteamaculans*. *Phytochemistry* 67: 887-1894.

Thompson, J.P. and Skerman, V.B.D. (1980). *Azotobacteraceae*: the taxonomy and ecology of the aerobic nitrogen-fixing bacteria. Academic Press, London.

Thompson, J.P. and Skerman, V.B.D. (1981).Validation list No6. *Int. J. Syst. Bacteriol.* 31: 215-218.

Thuler, D.S., Floh, E.I.S., Handro, W. and Barbosa, H.R. (2003).Plant growth regulators and amino acids released by *Azospirillum* sp. in chemically defined media. *Lett. Appl. Microbiol.* 37: 174-178.

Vazquez, P., Holguin, G., Puente, M.E. Lopez-Cortes, A. and Bashan, Y.(2000).Phosphate solubilizing microorganisms associated with the rhizosphere of mangroves in a semi arid coastal lagoon. *Biol. Fert. Soils,* 30: 460-468.

Verma, S., Kumar, V., Narula, N. and Merbach, W. (2001). Studies on *in vitro* production of antimicrobial substances by *Azotobacter chroococcum* isolates/ mutants. *J Pl. Dis. Protect.* 108:152-165.

Vinagre, F., Vargas, C., Schwarcz, K., Cavalcante, J., Nogueira, E.M., Baldani, J.I., Ferreira, P.C.G. and Hemerly, A.S. (2006). SHR5: a novel plant receptor kinase involved in plant-N_2-fixing endophytic bacteria association. *J. Exp. Bot.* 57: 559-569.

Vincent J. M. (1974). Root-nodule symbiosis with *Rhizobium*. In: The Biology of Nitrogen Fixation. Ed. A Quispel. Pp 265–341. North Holland Publishing Company, Amsterdam.

Xie, C.H. and Yokota, A. (2005). *Azospirillum oryzae* sp. nov., a nitrogen-fixing bacterium isolated from the roots of the rice plant *Oryza sativa*. *Int. J. Sys. Evol. Microbiol.* 55: 1435-1438.

Yanni, Y.G., Rizk, R.Y., Corich, V., Squartini, A., Ninke, K., Philip-Hollingsworth, S., Orgambide, G., de Bruijn, F., Stoltzfus, R., Buckley, D., Schmidt, T., Mateos, P.F., Ladha, J.K. and Dazzo, F.B. (1997). Natural endophytic association between *Rhizobium leguminosarum* bv. *trifolii* and rice roots and assessment of its potential to promote rice growth. *Plant Soil*, 194: 99-114.

Young, J.M., Kuykendall, L.D., Martínez-Romero, E., Kerr, A. and Sawada, H. (2001). A revision of *Rhizobium* Frank 1889, with an emended description of the genus, and the inclusion of all species of *Agrobacterium* Conn 1942 and *Allorhizobium undicola* de Lajudie *et al.* 1998 as new combinations: *Rhizobium radiobacter, R. rhizogenes, R. rubi, R. undicola* and *R. vitis. Int. J. Syst. Evol. Microbiol.* 51: 89-103.

□□□

CHAPTER **6**

Microbial Biocontrol Agents for Insects and Phytopathogens

S. Indira Devi[1*] and Ram Lal[2]

[1]Institute of Bioresources and Sustainable Development, Takyelpat, Imphal, India
[2]S.B.S. (P.G) Institute of Biomedical Sciences and Research, Balawala, Dehradun, India

ABSTRACT

Soil borne fungal pathogen causing root diseases of vegetables are difficult to control by chemical methods. The existing biocontrol bacteria are not very effective in control of soil borne fungal pathogens. As such very less number of bacterial biopesticide is available commercially to control the fungal pathogens. Increasing concerns in recent years about human health and environmental problems due to chemical pesticides, couple with increasing pest resistant to chemical, have led to a reappraisal of our approach to control pest and disease in crops. The trends now are towards a management concept in which the best combination of tools is used to manage plant diseases below economic thresholds with minimal risk to human health and environment. The modern management concept includes extensive use of biocontrol agents which are harmless to non-target organisms. Development of effective technology for control of pest and disease by using microbial resources of a region to enhance the quality of ecosystem is a potential alternative approach to chemical control of plant diseases. In such strategy, search for microorganisms and biological system is an important prerequisite prior to development and practical exploitation of biocontrol. Rhizosphere microorganisms are shown to be perspective candidate for biocontrol agent as they provide the front line of defense for plant roots against the attack by various plant pathogens causing the most diseases of vegetables and crops. Chemical control of soil-borne fungal pathogen had been the subject of growing concern because of their adverse effect on the environment and emergence of chemical pesticide resistant to plant pathogen, moreover, soil borne pathogens are very difficult to control. Crop rotation, breeding for resistant plant varieties and application of pesticides are insufficient to control root diseases of important plants because they survive as sclerotia and has got wide host range.

Key words: rhizosphere microorganisms, antibiosis, antifungal compound, biopesticide

*Email: sidevi1@yahoo.co.in

1. INTRODUCTION

Agricultural crops are always liable to attack by a variety of unwanted organisms that may destroy the crop and thus reduce the yield and value of crops. These include insect pests and plant pathogenic microorganisms (bacteria, fungi, viruses). Therefore, in present day agricultural practices synthetic chemicals are widely used to control the population if these pests so that the agricultural produce remain economically viable. Those chemicals include herbicides, insecticides, fungicides and nematocides which are used to kill weeds, insects, fungi and nematodes respectively. Use of these chemicals is the only reliable and economic method for control of pests and plant diseases. The estimated crop losses in India due to weeds and pests are about 30% of the total agricultural production (Thakur, 2006) and worldwide there is about 15-20% reduction in crop due to insect damages (Aggrawal, 2000). Farmers prefer these synthetic pesticides because they involve simple application method, these have broad spectrum of activity, success rate and cost benefit ratio is high.

Halogenated hydrocarbons were the first group of insecticides introduced for controlling of insect, pests and disease vectors (Chapalamadugu and Chaudhury, 1992). However, their intensive use in 1930 resulted in environmental problems like contamination of soil, ground water and even their metabolites were found in humans (Adeshima and Todd, 1990; Barquet *et al.*, 1981; Chaudhury and Chapalamadugu, 1991). This led to the development of carbonates and phosphorous compounds as alternate pesticides, which gradually replaced most of the chlorinated compounds. At present, most of the pesticides and many of the herbicides contain carbonates and organophosphates as active ingredients. In spite of high potency at low doses and low persistence of the carbonates and organophosphates, the wide spread and indiscriminate use has also caused environmental and health problems (Chapalamadugu and Chaudhury, 1992). Majority of the organic pesticides undergo mineralization through biochemical processes alone or in combination with purely chemical reactions within the time of one growing season or less (Alexander, 1981; Thakur, 2006). However, some of them have novel chemical structure and therefore resist physical, chemical and biological degradation and persist in air, water and soil and thus constitute an environmental burden of considerable magnitude. At higher concentrations or prolonged exposure several adverse effects have been observed in humans and other organisms. These include toxicity, carcinogenesis and mutagenesis. Moreover, some substances that are present at very low concentration in environment and are not considered pollutants may undergo bioaccumulation in food chain to high concentrations (Chapalamadugu and Chaudhury, 1992;

Thakur, 2006). The residues of pesticides and their metabolites that accumulate in the food chain have been reported to cause short as well as long term human health problems (Cohan *et al.*, 1984; Belluk and Benjamin, 1995; Kaufman, 1984).

The recalcitrance of chemical pesticides in some cases may lead to severe ecological damage. Concentration of chemicals can lead to sterility of species far removed in size and location from the target pests. The pesticide that is used to control a particular pest act on a far broader spectrum of organisms. A typical insecticide also kills not only the specific insect pest on a plant but many harmless other species. Thus use of chemical insecticides may seriously affect the long term survival of different ecosystems as they disturb predator-prey relationships and cause loss of biodiversity (Aggrawal, 2000; Thakur, 2006). The build up of toxic levels of residues in soil also disrupts the activities of soil microflora. All those adverse effects of pesticides have cast doubts about the long term sustainability of chemical control of agricultural pests. Another disadvantage of use of chemical pesticides is the acquisition of resistant to these agents by increasing number of insect pests (O'Brien, 1967; Vanden Heuvel). During the period from 1920 to 1990 about six-fold increase in number of crop damaging pest has been observed in India (Thakur, 2006). As a result higher concentrations of pesticides are required to achieve the desired level for insect control thus resulting in greater risks of environmental pollution, ecological imbalance, human health, soil microflora and fauna and other non-target organisms all warrant the development of alternate safe and ecofreindly methods of insect pest control. Biological control agents offer an attractive and potential alternative to the chemical pesticides.

2. BIOCONTROL AGENTS

Biological control, in simple terms, has been defined as use of one organism to eliminate or control another (Whipps, 1997). However, in more elaborative terms it may be defined as eradication or suppression of insects and plant pathogens either by encouragement of their natural enemies such as parasites, predators and disease causing organisms or by artificial introduction of these agents. In microbiological terms it is the control of macroorganisms (pests) or microorganisms (fungi) with infecting microorganisms.

The idea of using one living species to control another is quite old. Agostino Baggi was the person to advance the idea of using one organism for control of another. In 1834, Baggi showed that the fungus *Benveria bassiana* responsible for causing a disease 'muscardine' in silkworms. He suggested that the microorganism might be used for controlling purposes. Towards the end of the nineteenth century Japanese silk industry was struck by another disease of silkworm named 'flacherie' which ruined the industry. An aerobic spore forming bacterium was isolated by Japanese bacteriologist Ishiwata and was shown to be responsible for the disease. He named the bacterium, *Bacillus sotto*. Later, Japanese bacteriologists made an important observation that old sporulated culture of *B. sotto* was highly infective to healthy silkworm but young cultures fed to the caterpillar before sporulation were ineffective. Thus it was concluded that it was the spores of the organism that were pathogenic to silkworm not the vegetable cells. This Japanese work attracted little attention and the work was forgotten. However, in 1915 Berliner, a German microbiologist isolated a similar bacterium from *Anagasta kuhniella* a pest of stored grain and floor and in 1927 another similar organism was isolated by another German microbiologist from the same insect. He named it as *B. thuringensis*. Both these researchers made an important observation that as the vegetative cell grew and produced spore at one end, another body is also produced at the other end of the bacterial cell which they referred to as the 'Restekorper' and this lies inside the sporangium along with the spore. This body was released as the sporangium wall broke down. Later in 1930s extensive field trials were carried out with spore dust of Matte's bacterium against European corn borer. These were mostly successful and a commercial product called 'Sporiene' was produced in France around Second World War.

Since 1930 research on biocontrol agents have been a subject of great interest as an alternate method to chemical control of pests. This interest has mainly stemmed from increasing public concern over the environmental pollution, ecological imbalance increase pest resistance and increasing cost of factory-produced pesticides. Each year almost one million people have been reported to suffer from toxic residues of chemical pesticides and numerous deaths have been attributed to these chemicals (Deshpande, 1999). The use of biopesticide is an ecological approach that is based on the natural interactions among populations of organisms and use of one or more organisms to control the other. The biopesticides are ecofreindly, target specific and less vulnerable to the development of resistance by insects. Generally, biological control uses specific organisms that attack or interface with activity of specific insects and pathogens that have no effect on non-target organisms. Therefore, these agents have no or least effect on biodiversity.

During past 70 to 80 years an intense research in this area has generated many valuable information and technological know how on different aspects of biological control agents and several formulations have been commercialized. The current biological control agents is estimated to be about 10% of the world total market of agrochemicals and it is expected to grow at the rate of 10-15% per annum in coming years (Thakur, 2006). Biological control agents fall into two categories: those used for control of agricultural pest and those used for plant pathogens.

2.1 Biocontrol of Agricultural Pests (Bio-insecticides)

There are about 1500 naturally occurring microorganisms or their products have been identified that hold promise to be developed as biological pesticides (Aggarwal, 2000). The microorganisms that affect insects are called entomopathogens. These agents can be used alone to infect the pests or in combination with chemical insecticides. These insect damaging agent belong to all major groups of microorganisms i.e. bacteria, fungi, viruses and protozoa (Cook and Baker, 1983; Aggrawal, 2000). However, bacteria, fungi and viruses have been developed as commercial insecticides (Powell and Faull, 1989).

2.1.1 Bacteria

Many bacteria have been found to be obligate or opportunistic entomopathogens and they belong to families- *Pseudomonadaceae, Entemobacteriaceae, Lactobacillaceae, Micrococcaceae* and *Baullaceae*. Obligate entomopathogens are generally fastidious and grow only within living host insect. About 100 bacteria have been reported to infect insects but only four viz *Bacillus thuriengensis, B. papillilliae, B. lentimorbus* and *B. sphaericus* has been investigated for use as biocontrol agents. Most bioinsecticides are base on *B. thuriengensis* and *B. sphaerica* both of which produce a toxin (Cry protein) which is insecticidal (Chernin and Chet, 2003). *B. popillae* and *B. lentimorbus* kill the host by virtue of their ability to multiply extensively in the body of the target insect. Both these bacteria are pathogens of various beetles (Table 2). The killing effect of the bacterium is attributed to the removal of nutrients and essential growth factors by the bacterium. However, toxins are also thought to be involved as cell free filtrates of culture are lethal when small quantity is infected into larvae (Bulla *et al.*, 1978b). Larvae of the Japanese beetle (*Popillia japorica*) are infected by *B. popilliae*. The larvae ingest the spores which germinate in the digestive cavity of the larvae and multiply extensively in the haemolymph. Infected larvae develops a milk appearance before death hence the name 'milky disease' (Bulla *et al.*, 1978a). Before death of the insect large mass of spore accumulates in the body that is ultimately released. The spores are eaten by newly hatched beetle larvae thus initializing the infection process again. One application of these bacteria continue the infection for many reasons, however, persistence is affected by several factors like properties of the soil, climatic conditions, agricultural practices and larval populations.

Strains of *B. sphaericus* are highly toxic to larvae of mosquitoes (Priest, 1992). It shows infectivity of varying degree against different genera of mosquito. Experimental commercial preparations of this bacterium have been produced at prices competitive with those of chemical insecticides.

Bacillus thuringensis, so far has been the most successful BCA. Products based on this bacterium contribute the largest share of the total bio-pesticides that are marketed today (Chernin and Chet, 2003). It has been found applications in controlling insect pest of agricultural, veterinary and medical importance (Van Frankenhuyzen, 1993). *B. thuringiensis* is a gram-positive endospore forming bacterium that is widely distributed in nature (Prabhagaran *et al.*, 2003; Mahalakshmi *et al.*, 2005). It is pathogenic to about 150 species belonging to orders Lepidoptera, Diptera and Coleoptera (Beegle and Yamamoto, 1992) that include some of the most important agricultural pests. Some of these insect are cabbage hoper, tobacco hornworm, tobacco budworm, Gipsy moth, European cora borer, Grape leaf folder, sugarcane borer, cotton bollworm, and Indian meal moth (stored grain). The larvae of the insects become paralyzed within an hour of ingestion when bacterial spore culture is fed; however, vegetative cells from a young culture have no effect. Microscopic examination of a sporulating cell shows a large crystal is called parasporal body. Indeed, the production of this parasporal body is the main diagnostic feature of this organism which otherwise closely resemble *B. cereus* which is ubiquitously found in nature. The crystal is composed of an alkali soluble protein called δ-endotoxin (Cry protein). This protein is responsible for killing of the insect larvae. The crystalline protein in fact is a protoxin which is broken down into smaller active toxin by proteases in the gastric juices. The crystalline protein is not toxic to man and other mammals. Useful insects such as bees are resistant to low doses of toxin. It shows high specific toxicity to Lepidoptera larvae. Some preparations of toxic protein give LD_{50} values as low as 0.9μg/g of larvae, a value that is comparable with some widely used chemical insecticides. Crude preparation of parasporal body also contains a small molecular weight metabolite called θ-toxin. It is a basic compound with a molecular weight of about 900 and is an adenine nucleotide with an unusual sugar derivative called allomucic acid. This toxic is active against Diptera (flies and mosquitoes), Coleopteran (beetles), Orthoptera (termites) as well as Lepidoptera (butterflies and moths). δ-endotoxin or Cry protein of different strains of bacteria have molecular weight of about 130 - 140 k Da. More than 100 cry genes have been sequenced and some of them have been cloned and expressed in *E.coli* (Rajamohan and Dean, 1998). The action of the cry protein have been reported to involve the following steps (i) the solubilization of the crystal albumin pH of the midgut and the digestion of the proteolytic enzymes into an active protein fragment (ii) binding of the activated toxin to the receptors present on the epithelia cells of the midgut

(iii) insertion of the activated toxin into the cell membranes of the epithelial cells thus creating ion- channels or pores (Hill and Pinnock, 1988). Proteolytic cleavage of δ-endo toxin involves the removal of a stretch of C-terminal of the protoxin thus leading to the activation of N-terminal fragment (Tojo and Aizawa, 1983). In some species the initial toxicity is followed by extensive germination of the spores and multiplication of vegetative cells in the body of the caterpillar resulting in another round of sporulation and parasporal body synthesis in the cadaver.

2.1.2 Commercial Production

The production of a biocontrol agent on commercial scale is determined by several factors. Some of the factors are (1) the biocontrol agent should be equally or more effective than the existing pesticides; (2) it should be easy to produce involving cost effective technologies in its production; (3) it should be produced in the form of products that can be easily applied in the fields; (4) the organisms contain in the biocontrol agent should not lose its potency or effectiveness during storage. *B. thuringensis* fulfills almost all these criteria. At present it is the most accepted environment-friendly alternate method to agrochemical pesticides. It is also one among the nodal organisms that can be used as a gene pool for introduction of insect resistance in microorganisms and transgenic plants (Prabagaran *et al.*, 2003). *B. thuringiensis* has been found suitable organism that can be produced large scale by fermentation as it has simple nutritional requirements and can be produced using standard fermentation equipments (Pearson and Ward, 1988; Powell and Faull, 1989). Production on industrial scale is comparatively easier as the organism is amylolytic and grows on a mineral salt medium containing starch. Semi solid media based on bran or submerged cultures are used for growing the organism. In the latter case the product is dried by vacuum drying. The fermentation procedures are aimed at maximizing the concentration of insecticidal parasporal crystalline protein. Since, 1930s onwards commercial products containing *B. thuringiensis* have been produced world wide and used widely for pest control agricultural crops, fruit and vegetable crops and forestry (Falcon, 1971). Among the biocontrol pesticides of Bacillus group namely *B. thuringiensis, B. sphaerius, B. Popilliae, B. lentimorbu,* the first two are produced by fermentation process whereas the last two are produced exclusively in living insects. Third Japanese beetle (*Papillia japonica*) larvae are infected with spores of *B. popilliae* and *B. lentimorbus* and in abated in soil mixed with rye so that larvae may feed on it. Infected larvae are harvested after 16-21 days, pulverized, and stabilizing agents are added to form slurry. The slurry is dried and packed as a dry powder. Attempts to produce spores by cultivation of these bacteria *in vitro* have been largely unsuccessful. Field trials show that at least 10^4 - 10^7 spores per gram of soil are required to infect

50% of natural population of larvae (Falcon, 1971). However, use of such a product for spreading on soil is expensive. *B. thuringiensis, B. popillae* and *B. lentimorbus* have been found to have no harmful effects on humans and animal. At least *B. thuringiensis* at present is well accepted as eco-friendly biocontrol agent to use in agriculture and forestry (Prabagaran *et al.*, 2003; Mahakshmi *et al.*, 2005). Insecticides based on *B. thuringiensis* are used to obtain short term control of one reason. The organism does not persist from year to year and also does not spread extensively in the insect population. To be effective the insect must ingest a toxic dose of the organism, hence spraying should be sufficient enough to cover the surface of the leaves. Also the sprayed toxic should be protected from sunlight as it shows a degree of sensitivity to ultraviolet rays. Hence use of emulsifying agents, stickers and ultraviolet protectants are important ingredients of formulations.

Since the inception of the biological control different bioinsecticide formulations has been used in field trials and small commercial scale and there has been no discernible resistance development in target insects against these agents. Therefore, it is a general view that biocontrol agent does not suffer from pest resistance.

The genes responsible for production of endotoxin (cry genes) are found on chromosome, plasmids or on both in different strains of *B. thurindiensis* (Powell and Faull, 1989). This has made possible the genetic manipulation and construction of transgenic microbes and plants. Transgenic microorganisms of different genera like *Pseudomonas, Bacillus, Rhizobium* and a few others have been developed that can be applied either as foliar spays or in power or granule formulation in soil (Prabagaran, 2003). Important development, however, was the insertion and expression of *B. thuringiensis* toxin gene in plants using T-DNA vector from *Agrobacterium tumifaciens* (Vaeck *et al.*, 1987). This has made possible the development of several transgenic crops that are resistant to insect pests. First such transgenic crop Bt-cotton was first commercialized in 1996 by Monsanto. This commercial crop along with Bt-corn is presently grown in about total area of more than 80 million ha worldwide (Bates *et al.*, 2005). Indian government has also permitted growing of Bt-cotton and this crop is currently being grown in considerable area of Maharastra and Gujrat and some other areas (Thakur, 2006). However, controversies have also erupted due to studies in India that indicate that Bt-cotton planted in India are not as efficient in killing bollworms (*Helipcoverpa armigera*) as it is in USA or China (Kranthi, 2005; Jayaraman, 2005).

2.1.3 *Fungi*

A large number of fungi have been found to infect insects and there are about 150 fungal species distributed in approximately 100 genera (Chernin and Chet, 2004). The idea of using fungi as biological control agents was conceived at the time of Louis Pasteur. During 1880, *Metarhizium anisopliae* was mass produced to use against wheat cockchafer (*Anisoplia anstricata*) and sugar beet curculionid (*Cleonus punctiventris*) by Metchukoff and Krassilstschik respectively (Lumsdeu and Lewis, 1989). Around the end of 19th century and beginning of 20th century *Beauveria beongniartii* was used against European Cockchafer (*Melocontha melolontha*) in France and Ascherronia against white fly in USA. However, most researchers during these early days of development of mycoinsecticides considered their application as impractical due to adverse environmental factors that are prevalent in the field. But as the work in this progress factors responsible for establishment and maintenance of epizooties were revealed and presently several entomopathogenic fungi have been developed as biocontrol agents which includes fungi belonging to genera *Metarhizium beauveria, Verticillium, ascharsonia, Nomuraea entomophthora* and *Zeozygites* (Lumsden and Lewis, 1989, Deshpande, 1999).Many fungi are being developed for the biocontrol of diseases (Table 1&2).

TABLE 1. Fungi developed or being developed for the biological control of diseases (Burges, 1998; Butt *et al.*, 1999, 2001).

Product	Fungus	Target	Producer
Entomogenous fungi			
Mycotal	*Verticillium lecanii*	Whitefly and thrips	Koppert, The Netherland
Vertalec	*Verticillium lecanii*	Aphids	Koppert, The Netherland
BIO1020	*Metarhizium anisopliae*	Vine weevil	Licenced to Taensa, USA
Biogreen	*Metarhizium anisopliae*	Scarab larvae on pasture	Bio-care Tech., Australia
Metaquino	*Metarhizium anisopliae*	Spittle bugs	Brazil
Bio-path	*Metarhizium anisopliae*	Cockroaches	Ecoscience,USA
Bio-blast	*Metarhizium anisopliae*	Termites	Ecoscience,USA
Cobican	*Metarhizium anisopliae*	Sugercane spittle bug	Probioagro,Venezuala
Candida	*Beauveria bassiana*	Coffee berry borer	Live system Tech. Colombia
Ostrinil	*Beauveria bassiana*	Corn borer	NPP, France
Cornguard	*Beauveria bassiana*	European corn borer	Mycotech, USA
Mycotrol GH	*Beauveria bassiana*	Grass hoppers, locust	Mycotech, USA
Mycotrol WP & Botaniguard	*Beauveria bassiana*	Whitefly,aphids, thrips	Mycotech, USA
Naturalis L	*Beauveria bassiana*	Cotton pest , ballworm	Troy Bioscience,USA
Proecol	*Beauveria bassiana*	Army worm	Probioagro, Vanezuala
Boverin	*Beauveria bassiana*	Colorado beetle	Former USSR
Boverol	*Beauveria bassiana*	Colorado beetle	Czechoslovakia
Boverosil	*Beauveria bassiana*	Colorado beetle	Czechoslovakia
Engerlingspilz	*Beauveria brongniartii*	Cockchafer	Andermatt, Switzerland
Schweizer Beauveria	*Beauveria brongniartii*	Cockchafer	Eric Schweizer, Switzerland
Melocont	*Beauveria brongniartii*	Cockchafer	Kwizda, Austria
Green muscle	*Metarhizium flavoviride*	Locust, grasshoppers	CABI-Bioscience, USA
PFR-97	*Paecilomyces fumosoroseus*	Whitefly	ECO-tek, USA
Pae-Sin	*Paecilomyces fumosoroseus*	Whitefly	Agrobionsa, Maxico
LagiNix	*Lagenidium gigantium*	Mosquito larvae	Agra Quest, USA

TABLE 2. Microbial pesticides registered in the United States against insects (as of 1993)

	Species of Microbes	Pest control
Bacteria	*Bacillus popilliae* Ductky 1 *Bacillus lentimorbus* Dutky *Bacillus thuringiensis* Berliner subsp. kurstaki *Bacillus thuringiensis* subsp. Israelensis *Bacillus thuringiensis* var. san diego *Bacillus thuringiensis* subsp. Tenebrionis *Bacillus thuringiensis* subsp. Kurstaki strain EG2348 *Bacillus thuringiensis* subsp. Kurstaki strain EG2424 *Bacillus thuringiensis* subsp. Kurstaki strain EG2371 *Bacillus sphaericus* Neide. *Bacillus thuringiensis* subsp. aizawai *Bacillus thuringiensis* subsp. aizawai	Japanese beetle larvae Lepidopteran larvae Dipteran larvae Coleopteran larvae Coleopteran larvae Lepidopteran larvae Lepidopteran/Coleopteran larvae Lepidopteran larvae Dipteran larvae Lepidopteran larvae Lepidopteran larvae
Fungi	*Phytophthora citrophthora* (Smith and Smith) Leonian *Colletotrichum gloeosporioides* (Penzig) and Saccardo in Penzig f. sp. aeschynomene ATCC 20358 *Lagenidium giganteum* Couch *Metarhizium anisopliae* (Metchnikoff) Sorokin strain ESF1 Protozoa	Citrus strangler vine Northern joint vetch Mosquito larvae Cockroach and fly control

During recent years entomopathogenic fungi are receiving increasing attention as environment friendly insect control agent as their mode of infection is different than the bacterial biocontrol agents. Fungi act on the target insects by contact and do not depend on the ingestion by the insect. Cuticle degrading enzymes play an important in the insecticidal activity of the entomopathogenic fungi (Deshpande, 1999; Charnley and St.Lerger, 1991). In addition, the mass production of many of the biocontrol fungi is easy and fungi are relative more specific with respect their target insects (Deshpande, 1999; Aggrawal, 2000). With a few exceptions, fungi with high virulence for insects do not cause disease in animals and they do not harm the beneficent insects. Only *Conidiolobus coronatus* has shown possible infectivity to mammals. However, fungi also have from some disadvantages. Pathogenicity of most insecticidal fungi is depended on the environmental factors such as temperature and humidity (Lumusden and Lewis, 1989). Spores of some fungi may cause allergy in humans and metabolites of some like those of *Aspergillus flavus* that infect insects may exert toxic effect.

Endopathogenic fungi have two categories based on their nutritional requirements. Some have comparatively simple nutritional requirement and these can be easily cultured on artificial media. Others are obligate biotrophs and can grow only in living hosts. Most studies have been carried on the exploitation of the culturable fungi. In most cases the spreading of the infection agent among target insects occur by spores. Infection of insects involves adhesion and germination of spore as prerequisite steps. There is formation of germ tube and aspersoria. These structures and cuticle degrading enzymes make the infection possible. Penetration in some cases is largely due to enzymatic degradation and lesser extent to mechanical pressure (Samuels and Paterson, 1995; Bidockka *et al.*, 1997). Conidial germination and growth on the cuticle is dependent on moisture and temperature. In Denteromycetes like *Beauveria bassiona, Metarhizium anissopliae, Paecelomyces* spp and *V. lecanu* conidia germinate in relative humidity (RH) range of 92-100%. Conidia of *Aschersonia akysodis* can infect white fly larvae at R.H of 50%. Optimum temperature for mycelial growth of *M. anisopliae, Beauveria bassiona* and *M. anisopliae* is 20-30°C and considerable reduction in growth occurs below 15°C and above 35°C (Lumusdan and Lewis, 1989).

After the fungus has crossed the cuticle barrier, it starts growing in the body of the insect and the fungus undergo a change in morphology to hyphal bodies or yeast like cells that circulate in the haemolymph and multiply by budding. The insect is killed in 3-14 days. After death of the insect the fungus reverts back to the mycelial form and reproductive bodies are formed on the external surface of the cadaver. The killing mechanisms used by different entomopathogenic fungi are mechanical damage due to tissue degradation, depletion of nutrients, production of cuticle degrading enzymes, antibiotics and toxins etc (El-Sayed *et al.*, 1989; Charnley and St.Leger, 1991). The enzymes that are responsible for cuticle degradation include proteases, lipases and chitinases. Among these chitinases are thought to contribute significantly in the degradation of cuticle. Chinolytic enzymes have been demonstrated in several entomopathogenic fungi including *Aspergillus flavus, B. bassiana, M. anysopliae, Nomurea rileyi, P. farinosus, V. lecanii, Zoophthora* spp and *Entomophthora* spp (St.Leger *et al.*, 1993; Chernia and Chet, 2004). Proteolytic enzymes have been found to be excreted early during the infections which degrade the proteins thus exposing the chitin which then is attached by chitinases (El-Sayed *et al.*, 1993). In some entomopathogenic fungi the enzymes are secreted in sequential manner. Esterase and proteolytic enzymes appear first followed by chitobiase, chitinase and lipase (Charnley and St.Leger, 1991).

3. CONTROL OF SOIL-BORNE PATHOGENS

Soil borne plant pathogens viz. bacteria, fungi and nematodes; among these, fungal diseases are the major cause of crop losses. These soil-borne pathogens are difficult to control and methods like crop rotation, breeding for resistant varieties and the use of chemical fungicides are used to control them. However, these methods are not effective enough and farmers still suffer from heavy losses due to fungal diseases. Moreover, considering the limitations of chemical fungicides, the biological control of soil-borne pathogens is emerging as important attractive or supplement to the chemical fungicides. Numerous microorganisms, both bacteria and fungi have been investigated for biocontrol activity and a few of them have been used effectively in green house crops and field trials (Chernin and Chet, 2002). However, biological control of plant pathogens at present is not as well established as biocontrol of insect pest. The biological pest control has been comparatively more successful on commercial scale and has now emerged as rapidly developing area of research. During past a few years several agrochemical and biotechnological companies are coming up to invest in research and development of biocontrol of plant diseases and about its commercial preparation of microbial antagonists have been developed. These are all based on seven species of bacteria and about ten species of fungi. The bacterial preparations are based on *Agrobacterium radiobacter, Bacillus subtilis, Burkholdaria cepacia, Pseudomonas fluorescence, P. syringae, Stroptomyces grieseoviridis* and *S. lydicus* and fungal species include *Ampelomyces quisqualis, Candida oleophila, Conidiothyrium urinitaus, Fusarium oxysporum, Gliocladium virens, Phlebia gigantean, Pythium oligandrium, Trichoderma hazianum* and other *Trichoderma* spp (Chernin and Chet, 2002).

The idea of using antagonistic microorganisms dates back to 1930s. In 1932 Weindling noted that *Trichoderma* and *Glicladium* secrete metabolites that were toxin to other fungi (Weindling, 1932). More interest in this field was generated in subsequent years and the subject was discussed at 1963 international symposium entitled "Ecology of Soil-borne Plant Pathogens" prelude to biocontrol held in USA (Haas and Defago, 2005). Since that time a lot of research has been carried out on this subject. However, only a few successful biocontrol agents for soil-borne pathogens has been developed commercially (Whipps, 1997). There are several reasons for it but one important reason is the inconsistency in the performance of the biocontrol agent. Large number of antagonist had been reported but only a few have emerged successful in field trials. The reason for this failure is erroneous screening method. Ecological considerations were not taken into account *in vitro* screening procedures involving agar plate tests. Therefore, the organisms viewed as potential and successful *in vitro* screens generally do not perform well in larger contain glasshouse or field trials. *In vitro* tests do not take into consideration the environmental and microbiological factors that a particular biocontrol agent has to interact in the field trials. Another reason is that earlier researchers were unaware of the fact that the rhizosphere suitable habitat for isolation of potential biocontrol agents. Recent successes in biological control in contrast to the earlier attempts, is partly due to greater understanding of the rhizosphere and isolation selection of the strains more adapted to the rhizosphere environment. During past 20 to 25 years greater progress has been made in different aspects of biological control like properties of the antagonist that determine the effectiveness of the agent, mechanisms of the biocontrol and the interactions of biocontrol agent with plant and the pathogen (Weller, 1988; Harman *et al.,* 2004; Haas and Detago, 2005).

4. PROPERTIES THAT DETERMINE THE EFFECTIVENESS OF BIOCONTROL AGENTS

To be an effective biocontrol agent the selected microbe should have several ecological properties that the persistence of the organism in the habitats and compete successfully with the other microorganism. Some of the properties that play crucial role are:

4.1 Ecological Competence

One important requirement for a successful biocontrol agent is that it should successfully grow and persist in the ecological niche where it is intended to bring about the control of the plant pathogens. The ecological competence is ability of the biocontrol agent to survive and successfully compete with the microorganisms present in the habitat (Scroth *et al.,* 1984). Many traits of an organism had been reported to contribute in ecological competence most of which are unknown. Any of these traits may be lost by the organism. For example, many bacteria are surrounded by a capsular exopolysacharide (EPS) when first isolated, however, on subsequent sub-culturing EPS deficient mutants may appear spontaneously and over time they may predominate as they multiply faster. Such mutants may survive poorly when used as biocontrol agents (Weller, 1988). The organism may also be affected by the interference or completion from the non-target pathogens.

4.2 Root Colonization

Another useful trait of the biocontrol agent is that it should be an aggressive root colonizer. To have beneficial effect on plant growth and efficient disease control the biocontrol agent must be present in sufficient high number on the roots. Welder (1988) has defined a root colonizer as "a root colonizer is a bacterium that when introduced becomes distributed along the root in the natural soil, propagates, and survives for several weeks in the presence of competition from the indigenous rhizosphere microflora". Scher *et al.* (1984) defined a root colonizer in terms of the number per gram of root as bacteria that attain cfu of > $5x10^3$ per gram of root of corn. In case of *Pseudomonas* species it has been found that for effective protection of plants *Gaeumannomyces tritici* or *Pythium* spp. a colonization level of 10^5-10^6 cfu g^{-1} of root must be attained by the bacterium (Bull *et al.*, 1991). Colonization of plant roots by fungi is also an important mechanism of biological control by fungal antagonists. It has been found that some *Trichoderma* strains can colonize only local sites on roots (Metcalf and Wilson, 2001), but rhizosphere competent strains colonize entire root surfaces for several weeks or months (Harman, 2000; Thrane *et al.*, 1997). The root colonization by biocontrol bacteria is affected by many abiotic and biotic factors such as rhizosphere potential, water percolation, temperature, plant genotype and indigenous flora (Weller, 1988).

4.3 Rhizosphere Competence

In addition to favorable rhizosphere environment, long term root colonization is also determined by rhizosphere competence traits that are important in attachment, distribution growth and survival of the biocontrol agent in the rhizosphere. The structure of rhizosphere community is influenced by plant species and by the type and quality of root exudates (Goddard *et al.*, 2001). The exudates are rich in carbon compounds like organic acids (citrate, malate, succinate, pyruvate, fumarate, maltose, sucrose, galactose and ribose) are the main. In addition, variable amounts of Ü-amino acids, nitrogen bases and vitamins (such as thiamin and biotic are also excreted (Rovira, 1969). The ability to utilize organic acids like malate and succinate is considered as rhizosphere competent trait (Goddard *et al.*, 2001; Lugtenberg and Bloemberg, 2004). Rhizosphere competence can also be envisaged as the relative root colonizing ability of a strain and can be quantified by measuring the population microorganism attain on the root surface, or by the length or the number of root colonized. Bacterial traits that determine the rhizosphere competence are largely unknown that some of these traits include surface polysaccharides, fimbril, flagella, and the outer membrane protein of *F. chemotaxis*, osmotolerance and ability to utilize complex carbohydrates such as cellulose, hemicellulose and pectin (Luytenberg and Bheemberg, 2004).

5. MECHANISMS OF BIOCONTROL

The suppression of pathogenic microorganisms in soil by biocontrol agents is brought about by rhizosphere colonization, antibiosis, iron-chelation by siderophore production, production of volatile compounds, induction of systemic resistance (ISR) and competition for nutrients (Weller; 1988; Haas and Defayo, 2005).

5.1 Substrate Competition and Niche Exclusion

The first mechanism that is thought to be applied by the biocontrol agent to suppress the plant pathogens is the competition for limited resources such as nutrient and space in the rhizosphere and root surfaces. The antagonist capable of more efficient utilization of essential nutrients such as carbon and nitrogen compounds, volatile organic materials, plant residues, iron, trace elements etc. will deprive the pathogen of these nutrients. Also the antagonist will compete with pathogens for an ecological niche and colonization of root surfaces and rhizosphere thus making the condition adverse for growth of the pathogen. Competition for nutrient is thought to occur in most of interaction between bacteria and pathogen and has been found responsible at least partially for observed biocontrol by introduced bacteria (Elad and Baker, 1985; Elad and Chat, 1987). Large populations of bacteria established on planting material and roots has been found to utilize nutrient in the rhizosphere resulting in reduction in carbon and nitrogen required to stimulate the growth of fungal spores and subsequent colonization of the roots (Elad and Baker, 1985). Organisms like *fluorescent pseudomanads* because of their versatile nutritional abilities and rapid growth are especially suited for nutrient depletion in rhizosphere (Weller, 1985).

5.2 Antibiosis

The fact that the production of antibiotics may play a crucial role in suppression of competing microorganisms by the producer, was envisaged since a long back, however there were little evidence to prove that the antibiotics are actually produced in the soil. The first evidence was provided by Wright in 1956, when he reported that *Trichoderma viride* produced detectable amount of the antifungal antibiotic in wheat straw buried in soil. However, the antibiotic substances contributed to the biocontrol activity remained doubtful until recently. The direct proof to prove this fact was lacking because the technique to detect small quantities of antibiotics were not available. During recent years, by using improved analytical techniques and complementation studies with mutants of biocontrol microorganisms defective in antibiotic production, it has been proved that antibiosis contribute to the biocontrol activity. Mutants lacking antibiotic production *in vitro* had decreased biocontrol activity *in vivo* bioassays. However, complementing the gene restored antibiotic production as well biocontrol activity (Thomashow and Weller, 1988; Keel *et al.*, 1990; Kraus and Loper, 1995). Antibiotic production by some bacteria of genera like Agrobacterium, Pseudomonas and Bacillus has been found to play a major role in disease suppression (Weller, 1988; Yu *et al.*, 2002). *A. radiobacter* produces a type of antibiotic called Agrocin 84 that brings about suppression of crown gall caused by *A. tumifaciens* (Kerr, 1980). *Fluorescent pseudomonads* produce phenazine compound that play role in suppression of take-all disease of wheat caused by *Gaeumannocyces graminis* var *tritici*. Phenazine-1-carboxylate is produced by *P. fluorescen*2-29, which suppresses the take-all in fields when applied as seed treatment (Weller and Cook, 1983). The fact that the phenazine plays role in biocontrol activity has been proved by mutation and complementation studies (Thomashow and Weller, 1987).

5.3 Siderophores

Some bacteria, mostly *fluorescent pseudomonads* produce some low molecular weight substances that chelate Fe^{3+} ions and then transport the iron into the cell thus enabling the producer to grow under iron-deficient conditions. These siderophores are produced to grow under iron-deficient conditions and they act as carriers of iron (Leong, 1986; Haas and Defago, 2005). *Fluorescent pseudomonads* produce an extracellular diffusible, yellow-green, fluorescent pigment called pyoverdin. This pigment has high affinity for Fe^{3+} which it binds and forms a complex (Ferripyoverdin). This complex is recognized by a receptor that is present on the outer membrane of the producer bacteria (Buyer and Leong, 1986; Hohnadel and Meyer, 1986). Subsequently the Fe^{3+} is transported into the cell. The role of siderophores in biological control was first of all demonstrated by Kloepper *et al.* (1990). Later, many studies have reported siderophore mediated suppression of *Fusarium oxysporum* (Elad and Baker, 1985a, b), *G. graminis* var tritici (Kloepper *et al.*, 1980; Weller *et al.*, 1988) and *Pythium* spp. (Loper, 1988).

Siderophores bind the small quantity of Fe^{3+} present in the rhizosphere thus limiting the availability to the pathogen and leading to the suppression of its growth (Scroth and Hancock, 1982; Kloepper *et al.*, 1980). The pathogens suffer suppression because either it does not produce siderophores or it produces siderophores with less affinity to Fe^{3+} than those of the antagonist. Evidence has been produced that support the role of sederophores in biocontrol. *P. putida* strain B10 suppresses *Fusarium* wilt and take-all disease, but this effect is lost if soil is supplemented with the iron as the latter represses siderophore production (Kloepper *et al.*, 1980). Moreover, pyoverdin deficient mutants of *fluorescent pseudomonads* have been found to be less effective in biocontrol than the parent strain (Keel *et al.*, 1989; Loper and Buyer, 1991).

5.4 Mycoparasitism

This is a sort of predation shown by some fungi in which one fungus feeds directly on or inside the other. Many saprophytic non-pathogenic fungi have been found to show this type of antagonism towards plant pathogenic fungi. When one fungus parasitizes the other it is called mycoparasitism. This process involves the growth of the antagonist in the vicinity of pathogen, recognition and contact with the host, secretion of cell-wall degrading enzymes and entry into the host (Elad *et al.*, 1982; Haran *et al.*, 1998). This mechanism is followed by *Trichoderma* spp. parasitize many other fungi and these have been widely used as biocontrol agents. The mycoparasitic *T. harzianum* has been reported to use living host mycelial mass for its own growth, however, another mycoparasitic fungus *Pythium nunn* primarily used organic substrates for its growth but not the living mycelium of its host *P. ultimum* (Deshpande, 1999). This indicates that *P. nunn* uses different mechanisms for antagonizing its host than those of *Trichoderma* spp. In general the mechanisms involved are competition, antibiosis and mycoparasitism. Exoenzymes like chitinases, glucanases, cellulases and proteases are supposed to play crucial role in mycoparasitism (Chernin and Chet, 2002).

5.5 Induced Systemic Resistance (ISR)

Some localized or systemic defense like mechanisms are activated in peanuts under some conditions like attack by pathogenic microorganisms, physical damage caused by insects and other factors, treatment of plants with some chemical and by the presence of some non-pathogenic rhizobacteria. In this process some chemical substances like salicylic acid or jasmonic acid are produced at the site of infection (for example roots) that induces synthesis of antimicrobial metabolites at the other part of the plant (for example leaves). This is called induced systemic resistance (ISR) (Kuc, 2001; Oostendorp *et al.*, 2001). These plants defense responses result in induction and accumulation of low molecular weight proteins designated as pathogenesis-related (PR) proteins, deposition of some polymeric substances like callose and lignin and plant enzymes like β-1,3 glucanases, chitinases and peroxidases (Kombrink and Somssich, 1992; Dlisay and Kuc, 1995).

Induced systemic resistance is also elicited by non-pathogenic root colonizing rhizobacteria that are used as biocontrol agents against soil-borne plant pathogens. This resistance is sometimes also called as rhizobacteria-induced resistance (RISR) (Harman *et al.*, 2004). However, induced resistance in these cases differ in the fact that PR-proteins do not produced in detectable amounts and accumulation of salicylic acid does not occur, but plant respond by developing increased resistance if it is attacked by a pathogen. The inoculation of such plants with a leaf pathogen such as *P. sysingae,* shows that plant becomes 'primed',that is it shows faster and stronger response to the pathogen. Many biocontrol rhizobacteria have been reported to elicit ISR irrespective of the fact that they produce antibiotic or not (Zehnder *et al.*, 2001). Similarly, different *Trichoderma* spp.that are used as biocontrol agents have been reported to ISR in plants (De Meyer *et al.*, 1998; Harman *et al.*, 2004).

6. MICROORGANISM WITH BIOCONTROL POTENTIALS

6.1. Bacteria

Bacteria belonging to several genera have been found to possess biocontrol activity (Table 3). Some of these genera are *Actinoplanes, Agrobacterium, Alaligenes, Arthrobacter, Azotobacter Bacillus, Burkholdaria, Enterobacter, Cellulomonas, Erwinia, Flavobacterium, Micromonaspora, Pseudomonas, Rhizobium, Bradyrhizobium, Serratia, Streptomyces* and *Xanthomonas* (Weller, 1988). Among these the species belonging to genera *Agrobacterium, Bacillus, Pseudomonas, Enterobacter* and *Streptomyces* have been tested in large scale in green houses and fields. Some of them has been developed on commercial scale or are in the process of registration (for example *A. radiobacter, B. subtilis, P. fluorescens, P. syringae, B. cepacia, S. griseoviridis, S. lydicus* (Whipps, 1997; Chernin and Chet, 2002).

Agrobacterium radiobacter is one among the initial successful biocontrol agents is routinely used. It is used for the control of crown gall which is caused by *A. tumifaciens* (Kerr, 1972; Whipps, 1997). Many species of genus Bacillus have been investigated as biocontrol agent against plant pathogenic fungi (Broadbent *et al.*, 1977; Yuen *et al.* 1985; Yu *et al.*, 2002). Commercial preparations based on *Bacillus* spp. like *B. subtilis, B. pumilus, B. licheniformis* and *B. amyloliquefaciens* are currently in use in USA and China (Haas and Defago, 2005). *Bacillus* species being spore forming bacteria are more tolerant to adverse environmental factors like heat and desiccation, there the commercial products are easy to formulate and they have better shelf-life. Antibiosis and ISR are the mechanisms that play role in biocontrol activity of *Bacillus* spp. (Asaka and Shoda, 1995; Yu *et al.*, 2002; Schister *et al.*., 2004).

Another bacterial genus with great biocontrol potential is *Pseudomonas.* During pass few years extensive studies have been carried out on various *Pseudomonas* species with biocontrol potentials against plant pathogenic fungi (Weller, 1998; Haas and Defago, 2005). Most of these agents are fluorescent Pseudomonads belonging mostly to species *P. aeruginosa, P. fluorescens, P. putida* and *P. syringae. Pseudomonas* species are versatile aerobic gram negative bacteria with inherent capability to grow in nutritionally habitats as they can obtain energy by oxidizing diverse substrates. Most biocontrol *Pseudomonas* strains use different mechanisms like production of siderophore antibiotics and eliciting ISR. These *Pseudomonas* strains produce many antibiotic substance, some of these that have been found to play role in biocontrol activity includes: phenazines, fluoroglucicols, pyoluteorin, cyclic lipopeptides and hydrogen cyanide (HCN) (Haas and Defago, 2005).

TABLE 3. Examples of some bacteria applied to seeds or roots providing biocontrol of fungal plant pathogens

Bacterial antagonist	Fungal pathogen	Plant
Actinoplanes spp.	*Pythium ultimum*	Table beet
Bacillus spp.	*Rhizoctonia solani*	Wheat
Bacillus subtilis GB03	*Fusarium oxysporum*	Chick pea
Bulkholderia cepacia A3R	*F.graminearum*	Wheat
	Pythium spp.	Maize
Pseudomonas aureofaciens	*P.ultimum.*	Tomato
Psedomonas fluorescens	*F.oxysporum*	Radish
	P.ultimum	Lotus
	R.solani	Rice
Pseudomonas putida	*F.oxysporum*	Radish
	R.solani	Rice
	Pyricularia grisea	Rice
	P.aphanidermatum	Cucumber
Serratia plymathica	*P.ultimum*	Cucumber
Streptomyces spp.	*F.oxysporum*	Cucumber

6.2 Characteristics of an Ideal Biocontrol Agent

6.2.1 Bacteria

Despite the numerous reports of successful experiments where biological agents have been used to control plant diseases, there have been limited commercial successes. In general, the field results have been found to be inconsistent in several studies. However, few have focused on the mechanisms with respect to the spatial-temporal relationship of the biocontrol bacteria of interest.

Though selection criteria for a potential biocontrol agent have developed from different bases, but this task becomes very difficult to find organisms that can function effectively under variable conditions prevailing during the period of plant growth. For selecting ideal biocontrol agents of various plant diseases, some criteria are taken into considerations. They include (a) genetic stability (b) high consistent efficacy, (c) ability to survive under adverse conditions, (d) effectiveness against a wide range of pathogens on a variety of fruits and vegetables (e) amenability for growth on an inexpensive medium in fermenters, (f) stability of the end products during storage, (g) non production of secondary metabolites that might be toxic to humans, (h) strain resistant to standard fungicides, (i) compatibility with other chemical and physical treatments of the commodity such as heating and waxing.

6.2.2 Fungi

Many fungi can show antagonism towards other fungi and such fungi have been exploited for controlling the soil-borne diseases of plants (Table 4 & 5). Such fungi belong to genera *Ampelomyces, Candida, Coniothyrium, Fusarium, Gliocladium, Penicillium, Phoma, Phlebia, Pythium, Laetisaria, Sporidesmium, Talaromyces,* and *Verticillium* (Whipps, 1997; Chernin and Chet, 2002). However, most promising among these are *Trichoderma* spp, *Gliocladium* spp and non-pathogenic *Fusarium* species (Chernin and Chet, 2002; Herman *et al.*, 2004). Many species of fungi contain both pathogenic and non-pathogenic races and strains as well as non-pathogenic strains. For example, genus contains pathogenic as well as non-pathogenic strains. Non-pathogenic *Fusarium* strains possess strong biocontrol activities against pathogenic strains of the same species as well as other pathogens like *P. ultimum* (Fucks *et al.*, 1997; Benhamou *et al.*, 2002). Likewise, genera *Pythium* and *Rhizoctonia* have been found to contain both pathogenic and non-pathogenic strains. Non-pathogenic strains possess biocontrol activities (Hwang and Benson, 2003; Weller, 1997). Biocontrol activity of all these antagonists involves mechanisms like direct antagonism through competition, antibiosis, mycoparasitism and ISR (Benhamou *et al.*, 2002; Howell, 2003).

Trichoderma spp are common saprophytic fungi that are found in soil and rhizosphere. Soils from temperate and tropical regions contain about 10^1-10^3 culturable propagates per gram. These fungi have also been found to colonize roots of woody and herbaceous plants. During past few years *Trichoderma* spp. has attracted considerable attention as potential biocontrol agents for soil-borne pathogens (Samuels, 1996; Harman *et al.*, 2004). Several species of *Trichoderma* and *Gliocladium* like *G. virens, G. roseum, T. viride. T. harzianum* and *T. hamatum* have been tried as biocontrol agents like *Botrytis cinerea, Fusarium* spp., *Phytophthora cactorum, Pythium* spp, *R. solani* and *Sclerotinia* spp. The *Trichoderma* spp has been found to colonize plant root and posses several properties that make them highly rhizosphere competent. These strains colonize entire root surfaces for several weeks and even for months (Thrane *et al.*, 1997; Herman, 2000). They successfully compete with other rhizosphere microorganisms for space and key nutrients present in root exudates thus making them unavailable for propagules plant pathogenic fungi.

These nutrients required to stimulate germination of the propagules. These *Trichoderma* species produces lytic enzymes like chitinases and β 1,3-glucanases and several metabolites that are fungicidal or fungistatic in action. Different strains produce more than 100 different metabolites with antibiotic activities (Sivasithamparam and Ghisalberti, 1998; Herman and Kubicek, 1998). These agents execute their biocontrol activities through competition, antibiosis, mycoparasitism and eliciting ISR in plants.

TABLE 4. Examples of fungal-fungal interactions examined in the spermosphere and rhizosphere associated with biocontrol disease control

Fungal antagonist	Fungal Pathogen	Plant
Cladorrhinum foecundissimum	*Pythium ultimum, R.solani*	Egg plant, pepper
Glomus intraradices	*F.oxysorum*	Carnation
Idriella bolleyi	*Bipolaris sorokiniana*	Barley
Penicillum oxalium	*F.oxysporum*	Tomato
Talaromyces flavus	*Verticillium dahliae*	Eggplant
Trichoderma harzianum	*R.solani, S.sclerotiorum*	Soyabean
	P. ultimum	Cucumber
T.longibrachiatum	*R.solani*	Bean
T.viride	*P.ultimum,*	Bean
T.virens	*P.ultimum , R.solani*	Eggplant
T.virens GL-1,21	*R.solani*	Cucumber, pea,
T.virens GL-3	*F.graminearum*	Eggplant
	P. arrhenomanes , P. ultimum	Maize

Trichoderma spp and other root colonizing fungi have been found to enhance plant growth and productivity. Several commercial and academic trials on the effect of *T. harzianum* T-22 on maize have shown that there is an average increase of about 5% yields. Increase plant growths are also observed in other plants in greenhouse and field trials (Harman, 2004). *Trichoderma* spp. increase the uptake of many nutrients like copper, phosphorous, iron, manganese and sodium. They also increase the efficiency of nitrogen use in plants and generalized increase in uptake of elements like arsenic, cobalt, cadmium, chromium, nickel, lead, magnesium, manganese and many others. *T. harzianum* T-22 has been found to solubilize rock phosphate, Fe^{3+}, Cu^{2+}, and Mn^{4}. Other *Trichoderma* spp are also believed to possess similar capability (Harman *et al.*, 2004).

7. CONCLUSIONS

Several rhizosphere microorganisms have the potential to control various root, foliage and post harvest diseases of agricultural crops caused by plant pathogenic fungi, bacteria and viruses. These microorganisms are not only

TABLE 5. Fungi developed or being developed for the biological control of diseases (Burges, 1998; Butt *et al..*, 1999, 2001)

Product	Fungus	Target	Producer
Mycoparasites			
Rotstop	*Phlebiopsis gigantea*	*Heterobasidium annosus*	Kemira Agro
Primastop	*Gliocladium catenulation*	Several plant diseases	Oy,Findland
SoilGard	*Gliocladium virens*	Several plant diseases	Kemira Agro
		Damping off & root	Oy,Findland
CotansWG	*Coniothyrium minitans*	pathogen	Thermo Trilogy, USA
AQ10Biofungicide	*Ampelomyces quisqualis*	*Sclerotina* species	
YIELDPLUS	*Cryptococcus albidus*	Powdery mildews	Prophyta, Germany,
Aspire	*Candida oleophila*	*Botrytis* spp., *Penicillium*	KONI
Fusaclean	*Fusarium oxysporium*	spp.	Ecogen Inc. USA
Biofox C	*Fusarium oxysporium*	*Botrytis* spp., *Penicillium*	Anchor Yeast, S. Africa
		spp.	Ecogen Inc. USA
Polygandron,	*Pythium oligandrum*	*Fusarium oxysporium*	Natural Plant
Polyversum		*Fusarium oxysporium*	Protection,France
Trichoderma 2000	*Trichoderma harzianum*	*Fusarium moliliforme*	SIAPA, Italy
		Pythium ultimum	
Trichpel	*Trichoderma harxianum*		Plant ProtectionInst.
		Rhizoctonia solani, Sclerotium	Slovak Republic
T-22 and T-22 HB	*Trichoderma harxianum*	*rolfsii, Pythium*	Mycontrol (Ef A1) Ltd.
Bio-		Wide range of fungal	Israe
Trek,RootShield	*T.harzianum & T.viride*	diseases	
Trichodowels,			Agrimm Technologies
Trichoject,	*T. harzianum, T.*	*Pythium, Rhizoctonia,*	Ltd., New Zealand
Trichosel & others	*polysporum*	*Fusarium, Sclerotina*	BioWorks, Geneva, USA
Binab T		*Chondrostereum purpureum*	
		& other soil & foliar	
	T. harzianum		Agrimms Biologicals,
Trichodex		pathogens	new Zealand & others
		Fungi causing Wilt, Wood	Bio-Innovation, Sweden
		decay and take all	
		Fungal diseases ; eg.	Makhteshim-Agan
		Botrytis cineria	Several European
			Companies
			e g. De Ceuster, Belgium.

the appealing candidates for biocontrol of plant diseases, but also many of these play an important role in promotion of plant growth and yield. Being environmentally safe and because of hazardous effects of agrochemicals the interest in biocontrol of plant diseases has increased in the recent years. Several manufacturing companies are involved in the development of various bioproducts and some commercial bioproducts are available in the market today. However, the number of bioproducts is much smaller than the traditional pesticides for the control of various plant diseases but, the purposeful introduction of these bioagents where the commercial pesticides are less effective. An understanding of ecological requirements for survival, colonization and activity of biocontrol agents is essential to achieve commercial success. Numerous other factors such as efficacy, standard performance, cost, storage and methods of application of these bioproducts are also important for their commercial production.

Further improvements can be achieved by developing precise screening techniques to find highly compatible bioproducts under different environmental conditions. Thus a combination of byproducts and cultural control may offer efficacy nearly equal to that of agrochemicals. This review has focus on resent research concerning interaction between biocontrol agents and pathogens in the rhizosphere. Greatest interest recently has concerned the phenomenon of induced resistance as the molecular tools for its clearly justified. Several other areas requiring further work have been highlighted in the texts and in addition, there are a few other topics that could develop further in the future. For example, the concept of breeding plants to improve the effectiveness of biocontrol agents or plant growth promoting microbes appears to be a novel approach (Smith *et al.*, 1997; Smith and Goodman, 1999). Integration of biocontrol and chemical control systems may also be an approach that receives more attention in the future. If pesticide tolerant isolates of biocontrol agents could be used to reduce the application of pesticides then an environmental benefit would ensure.

Another area that has been largely neglected is the role of soil fauna in the interaction between biocontrol agents and plant pathogens in the rhizosphere. Often soil fauna are studied in isolation and the significance of this is only just beginning from microorganisms to be appreciated. For instance, Sclerotia of *Sclerotina sclerotiorum* which had been grazed by the larvae of the fungus *Bradysia coprophila* were degraded rapidly in soil infested with *T. hamatum*. Larval damage altered the sclerotia both physically and chemically, enhancing the activity of *T. hamatum*. In other studies, both collembolans and mites were shown to transmit conidia of the mycoparasite, *Coniothyrium minitane* between sclerotia of *Sclerotinia sclerotiorum* in soil (Williams *et al.*, 1998 a, b)

Finally, perhaps the greatest interest in the future lies with the application of modern molecular techniques and their integration with conventional experimental procedures to understand and utilize soil-plant-microbe interactions. The significance of these techniques has already been described with the monitoring of biocontrol agents and their impact on microbial populations, in the construction of *A. radiobacter* K1026, and in understanding the modes of action of biocontrol agents, particularly with induced resistance in plants. Further, much effort has gone into developing transgenic bacteria and fungi expressing genes that provide enhanced biocontrol activity, and in transgenic plants expressing genes that provide disease resistance, while also allowing a greater understanding of the mechanisms operating in the rhizosphere. With environmental concerns and existing legislation, it remains to be seen whether transgenic micro-organisms and plants for disease control become universally accepted both as research tools and as commercial products.

REFERENCES

Adeshima, E. and Todd, E.L. (1990). Organochlorine compounds in human adipose tissue from north Texas. *J. Toxicol. Env. Health.* 23: 147-152.

Aggrawal, S.K. (2000). Environmental Biotechnology, A.P.H Pub. Corp. New Delhi, Pp.317-336.

Alexander, M. (1981). Biodegradation of Chemicals of environmental concern. *Science,* 211:132-138.

Asaka, O. and Shoda, M.(1996). Biocontrol of *Rhizoctonia solani* damping off of tomato with *Bacillus subtilis* RB 14. *Appl. Environ. Microbiol.* 62:4081-4085.

Bajan, C., Kalalova, S., Kimitova, K., Samsinakova, A. and Wojciechowska, M. C. (1979).The relationship between infections activities of entomophagus fungi and their production of enzymes.Bull. Acad. Pol. Sci. Cl2 Ser. *Sci. Biol.* 27:963-968.

Barquet, A., Morgade, C. and Paffenberger, C.D. (1981). Determination of organochloride and metabolites in drinking water, human blood serum and adipose tissue. *J. Toxicol. Environ. Health,* 7:469-472.

Bates, S.L., Zhao, Z-H., Roush, R.T. and Shelton, S.M. (2005). Insect resistance management in GM crops: past, present and future. *Nat. Biotechnol.* 23:57-62.

Beegle, C.C. and Yamamoto, T. (1992). History of *Bacillus thuringiensis* Berliner research and development. *Can. .Entamol.* 124:587 - 616.

Belluck, D.A. and Benjamin, M.S. (1990). Pesticides and environmental health. *J. Environ. Health,* 53:11-19.

Benhamou, N., Garanci, C. and Goulet, A. (2002). Ability of non-pathogenic *Fusarium oxysporum* strain FO47 to induce resistance against *Pythium ultimum* infection in cucumber. *Appl. Environ. Microbiol.* 68: 4044-4060.

Bidockka, M.J., St Lager, R.J. and Roberts, D.W. (1997). Mechanisms of deuteromycetes fungal infections in grasshoppers and locusts: An overview. Memoirs *Entomol. Soc. Can.* 171:213-224.

Broadbent, P., Baker, K. F., Franks, N. and Holland, J. (1977). Effect of *Bacillus* spp. On increased growth of seedlings in steamed and in not treated soil. *Phytopathol.* 67:1027-34.

Bull, CT., Weller, D.N. and Thomashow, L.S. (1991). Relationship between root colonization and suppression of *Gaeumannomyces graminis* var *tritici* by *Pseudomonas fluorescens* strain 2-79. *Phytopathol.* 81: 954-959.

Bulla, L.A., Costilow, R.N. and Sharpe, E.S. (1978a). Biology of *Bacillus popillae*. *Adv. Appl. Microbiol.* 23:1 - 18.

Bulla, L.A., Rhodes, R.A. and St.Julien, G. (1978b). Bacteria as insect pathogens. *Ann. Rev. Microbiol.* 29:163 - 190.

Burges, H. D. (1998). Formulation of microbial pesticides. Kluwer Academic Publishers. Dordrecht, The Netherlands.

Butt, T. M., Goettel, M. S. and Papierok, B. (1999). Directory of specialists involved in the development of fungi as biocontrol agents. Colin Butt Design & Print, Warley, West Midlands.

Butt, T. M., Jackson, C. W. and Magan, N. (2001). Fungal biological control agents: Progress, Problems and Potential. CABI International, Wallingford, Oxon, U.K. plant growth-promoting and plant deleterious *Pseudomonas* strains. *J. Biol. Chem.* 261:791-794.

Chapalamadugu, S. and Choudhury, G.S. (1992). Microbiological and Biotechnological aspects of metabolism of carbonates and organophosphates. *Crit. Review Biotech.* 12:357-389.

Charnley, A.K. and St. Leger, R. (1991). The role of cuticle degrading enzymes in fungal in fungal pathogenesis in insects. In the Fungal spore and Disease Initiation in Plants and Animals. Gole G.T et al (Eds). Pleunum Press New -York.

Chaudhury, G.R. and Chapalamadugu, S. (1991). Biodegradation of chlorinated compounds. *Microbial Rev.* 57:59-61

Chernin L. and Chet, I. (2002). Microbial enzymes in the biocontrol of plant pathogens and pest. In Enzymes in The Environment: Activities, Echology and applications, Burus R.G and Dick, R.P (Eds). pp 171-226, Marcel-Dekker Inc. New-York.

Cohen, S.Z, Creegers, S.M., Carsel, R.F. and Enfield, G.G. (1984). Potential pesticide contamination of ground water from agricultural uses. In treatment and disposal of pesticide wastes. Krueger R.F and Scimber, J.N (Eds), pp 29 -34, American Chemical Society Washington, D.C.

Cook R.J. and Baker K.F. (1983). The Nature and Practice of Biological Control of Plant Pathogens. *American Phytopathol.* Society, St.Paul Minnesota USA.

Dalisay, R.F. and Kuc, J.A. (1995). Persistence of induced resistance and enhanced peroxidase and chitinase activities in cucumber plants. *Physiol. Mol. Plant. Pathol.* 47:315-327.

De Meyer, Ge Bigirimana, J., Elad, Y. and Hofte, M. (1998). Induced systemic resistance in *Trichoderma harzianum* T39 biocontrol of *Botrytis cinerea*. *Eur. J. Plant Pathol.* 104:279-286.

Deshpande, L. (1999). Mycopesticide production by fermentation potential and challenges. *Critical remains in Microbiol.* 25:229-243.

Elad, Y. and Backer, R. (1985). Influence of trace amounts of cation and siderophore producing pseudomonads on Chlamydosphore germination of *Fusarium oxysporum*. *Phytopathol.*75:1047-1052.

Elad, Y. and Baker, R.C. (1985b). The role of competition for iron and carbon in suppression of Chlamydospore germination of *Fusarium* spp. By *Pseudomonas* spp. *Phytopathol.* 75:1053-1059.

Elad, Y. and Chet, I. (1987). Possible role competition for nutrients in biocontrol of *Pythium* damping off by bacteria. *Phytopathol.* 77:190-195.

Elad, Y., Chet, I. and Henis, Y. (1982). Degradation of plant pathogenic fungi by *Trichoderma harziaanum*. *Can. J .Microbiol.*. 28:719-725.

El-Sayed, G.N., Ignoffo, C.M., Leathers, T.D. and Gupta S.C. (1993). Effect of cuticle source and concentration on expression of hydrolytic enzymes by an entomopathogenic fungus. *Nomuraea rileyi*. *Mycopathol.* 122: 149-152.

El-Sayed, G.N., Coudron, T.A. and Ignaffo, C.M. (1989). Chitinolytic activity and virulence associated with native and mutant isolates of an entomopathogenic fungus *Nomuraea rileyi*. *J. Invertbr. Pathol.* 54:394-403.

Falcon, L.A. (1971). Use of bacteria for microbial control. In Microbiol control of Insects and Mites. B urges, H.D. and Hussey, N.W (Eds) pp 67-95. Academic Press, London and New York.

Fucks, J.G., Moenne-Loccoz, Y. and Defago, G. (1997). Non-pathogenic *Fusarium oxysporum* strain FO47 induces resistance to *Fusrium* wilt of tomato. *Plant. Dis.* 81:492-496.

Goddard, V.J., Baily, M.J., Darran, P., Lily, A.K. and Thompson I.P. (2001). Monitoring temporal and spatial variation in rhizosphere bacterial population diversity: a community approach for the improved selection of rhizosphere competent bacteria. *Plant Soil,* 232: 181-183.

Gubriel, B.P. (1968). Enzymatic activities of some entomopathogenic fungi. *J. Invertebr. Pathol.* 11:70-81.

Haas, D. and Defago, G. (2005). Biological control of soilborne pathogens by *fluorescent pseudomonads*. *Nat. Rev. Microbiol.,* 3:307-319.

Haran,S., Benhamn,N. and Chet,I. (1998). Mycoparasitic enzymes and lytic enzymes. In *Trichoderma harzianum*. *Can.. J. Microbiol.* 28:719-725.

Harman, G.E. (2000). Myths and dogmas of biocontrol changes in perceptions desired from research on *Trichoderma harzianum* T22. *Plant Dis.* 84:377-393.

Harman, G.E., Howell, C.R., Viterbo, A., Chet, I. and Lorito, M. (2004). *Trichoderma* species-opportunistic, a virulent plant symbionts. *Nat. Rev. Microbiol.* 2: 43-56.

Hill, C.A. and Pinnock, D.E. (1998). Histopathological effect of *Bacillus thuringiensis* on the alimentary canal of the sheep louse *Bovicola ovis*. *J. Inverbr. Pathol.*72:9: 20-22.

Hohnadel, D. and Meyer, J.M. (1986). Pyoverdine-facilitated iron uptake among *fluorescent pseudomonds*. In iron, siderophores and plant diseases, pp 119-129, Swinburne, T.R. (Eds), Pleunum, New York.

Howell, C.R. (2003). Mechanisms employed by *Trichoderma* species in the biological control of plant diseases: the history and evolution of current concepts. *Plant Dis.* 87:4-10.

Huran, S.,Benhamon, N.and Chet, I.(1998).Mycoparasitic enzymes and lytic enzymes. In *Trichoderma* and *Gliocladium, Harma* G.E and Kubicek, C.K.(Eds). Vol. 2, pp 153-172

Hwang, J. and Benson, D.M. (2003). Expression of induced resistance in poinsietta cuttings against *Rhizoctonia* stem rot by treatment of stock plants with binucleate *Rhizoctonia*. *Biol. Control*.27:73-80.

Jayaraman, K.S. (2005). Monsantos Bollgard potentially compromised in India. *Nat. Biotechnol.*, 23:1326.

Kaufman, D.D. (1984). Degradation of pesticides by soil microorganisms. In Pesticides in Soil and Water, Mcenzi, W.D (Ed) pp 133 – 156. Soil Science Society of America, Medison WI.

Keel, C., Voisard, C., Berling, C.H., Kahr, G. and Defago, G. (1989). Iron sufficiency, *fluorescens* strain CHAO under gnobiotic conditions. *Phytopathology*, 79:584-689.

Keel, C., Schnider, V., Maurhoper. C. and Voisard, C. (1992). Suppression of root diseases by *Pseudomonas fluorescens*, CHAO: importance of the bacterial secondary metabolite 2,4, diacetylphosphoglucinol. *Mol. Plant-Microbe Interact.* 5: 4-13.

Kerr, A. (1980). Biological control of crown gall through production of agrocin 84. *Plant Dis.* 64: 25-30.

Kloepper, J.W., Leong, J., Teintze, M. and Schroth, M.N. (1980a). Enhaced plant growth by siderophores produced by plant-growth promoting rhizobacteria. *Nature*. 286:885-886.

Kloepper, J.W., Leong, J., Teintze, M. and Schroth, M.N. (1980b). *Pseudomonas* siderophores; A mechanism explaining disease suppressive soils. *Curr. Microbiol.* 4:317-320.

Kombrink,.E. and Somssich, I.E. (1992). Defence responses of plants to pathogens. *Adv. Bot.* Res. 21:1-34.

Kranthi, K,T., Naidu, S., Dhawad C.S., Tatwadi, A., Kate, M., Patil, E., Bharase, A.A., Behere G.T., Wadasker, R.M.. and Kranti., S. (2005). Temporal and interplant variability of cry Ac expression in Bt cotton and its influence on the survival of cotton bollworm *Helicovera armigera* (Hubner) (Noctuidae: Lepidoptera). *Curr. Sci.* 87:291-298.

Kraus,J. and Loper, J.E.(1995). Characterization of genomic region required for production of the anti biotic pyoluteorin by the biological control *agent Pseudomonas fluorescens* Pt-5. *Appl. Environ. Microbiol.* 61:849-854.

Kuc, J. (2001). Concepts and directions of induced systemic resistance in plants and its application. *Eur. J. Plant Pathol.* 107:7-12.

Leong, J. (1986). Siderophores; their biochemistry and possible role in the biocontrol of plant pathogens. *Ann. Rev. Phytopathol.* 24:187-190.

Loper, J.E. (1988). Role of fluorescent siderophore production in biological control of *Pythium ultimum* by a *Pseudomonas fluorescens* strain. *Phytopathol* 78: 166-172.

Loper, J.E., and Buyer, J.S. (1991). Siderophores in microbial interactions on plant surfaces. *Mol. Plant Microbe Intract.* 4:5-13.

Lugtenberg, B.J.J. and Bloemberg, G.V. (2004). In Pseudomonas, Vol 1, Ramos, J.L (Eds). pp 403-430, Kluwer Academic/ Pleunum Publishers New York, USA.

Lumusden, R..D. and Lewis, J.A (1989). Fungal control of agricultural and horticultural pests. In Biotechnology of Fungi for Improving Plant growth. Whipps, J.M. and Lumsden, R.D. (Eds) pp 171-193. Cambridge University, Cambridge.

Mahalakshmi, A., Shenbagrathai, R. and Sujatha, K. (2005). Identification of novel indigenous *Bacillus thuringiensis* isolates.*Indian J.Exp. Biol.* 43: 867-872.

Metcalf, D.A. and Wilson,C.R. (2001). The process of antagonism of *Sclerotium capivorum* in white net affected onion roots by *Trichoderma* . *Plant Pathol.* 50:249-257.

O' Brien, R.D. (1962). Insecticides Action and Metabolism, pp 290 - 319. Academic Press, New york.

Oostendorp, M., Kunz, W., Dietrich, B. and Staub, T. (2001). Induced disease resistance in plants by Chemicals. *Eur. J. Plant. Pathol.* 107:19-28.

Pearson, D. and Ward, O.P. (1988). Bioinsecticidal activity bacterial cell lysis and proteolytic activity in cultures of *Bacillus thuringiensis.* var isralnsis. *J.Appl. Bacteriol.* 65:195-202.

Powell, K.A. and Faull, J.L. (1989). Commercial Approaches to the use of biological control agents. Biotechnology of Fungi for Improving Plant Growth, J.M Whipps and R.D Lumsden (eds), Cambridge University Press, Cambridge, Pp 259 - 276.

Prabagaran, S.R., Rupesh, K.R., Nimal, S.J., Sudha Rani, S. and Jayachandran, S. (2003). Advances in pest control: The role of *Bacillus thuringiensis. Indian J. Biotech.* 2:302 - 321.

Priest, F.G. (1992). Biological Control of mosquitoes and other biting flies by *Bacillus sphaeericus* and *Bacillus thuringiensis. J. Appl. Bacterial.* 72:357 - 369.

Rajamohan, F. and Dean, D.H. (1996). Molecular biology of bacteria for the biological control of insects. In Molecular Biology of the Biological Control of Pests and Disease of Plants. Gunasekaran, M and Weber D.J (Eds), CRC Press. USA, Pp 105 - 122.

Raymond, J.C. (1993). Prospects and progress for *Bacillus thuringiensis* based pesticides. *Pestic. Sci.* 37:331-335.

Roberts, D.W and Frontiers. In Industrial Mycology, pp 144-159. Leatham G.F.C. (Eds). Chapman and Hall, New York.

Rovira, A.D. (1969). Plant root exudates . *Bot. Rev.* 35: 35-57.

Samuels, B.I. and Paterson, I.C.(1995). Cuticle degrading proteases from insect moulting fluid and culture filtrates of entomopathogenic fungi. *Comp. Biochem. Physiol.* 110B: 661-669.

Samuels, G.J. (1996). *Trichoderma*: a review of biology and systematics of the genus. *Mycol. Res.* 100:923-935.

Scher, F.M., Ziegle, J.S. and Kloepper, J.W. (1984). A method for assessing the root colonizing capacity of bacteria on maize. *Can. J. Microbiol.* 30:151-157.

Schister, D.A., Slininger, P.J., Behle, R.W. and Jackson, M.A. (2004). Formulation of *Bacillus* spp. For biological control of Plant diseases. *Phytopathology,* 94:1267-1271.

Schroth, M.N. and Haucock, J.G. (1982). Disease-suppressive soil and root colonizing bacteria. *Science,* 216:1376-1381.

Scroth, M.N., Lopper J.E. and Hildebrand, D.C. (1987). Bacteriaas biocontrol agents. In current perseptives in Microbiol Ecology, Klug M.J and reddy, C.A. (Eds). pp 362-369. ASM, Washington,D.C.

Sivasithamparam, K. and Ghisalberti, E.L. (1998). In *Trichoderma* and *Gliocladium. Kubicek* C.P. and Harman, G.E. (Eds.) Vol. 1, pp139-191. Tylor and Francis London.

Smith, K.P. and Goodman, R.M. (1999). Host variation for interections with beneficial plant-associated microbes. *Phytopathology,* 37: 473-491.

Smith, K.P., Handelsman, J. and Goodman, R.M. (1997). Modeling dose-response relationships in biological control; participating host responses to the pathogen and biocontrol agent. *Phytopathology,* 87:720-729.

St. Lager, R.J., Staples, R.J. and Roberts, D.W. (1993). Entomophytogenic isolates of *Metarhizium anisopliae, Beauvera bassiana, Aspergillus flavus* produce multiple extracellular chitinase isoenzymes. *J.Invert. Pathol.* 61: 81-84.

Thakur, I.S. (2006). Industrial Biotechnology problems and remedies. I.K International Pvt. Ltd., New Delhi, pp 184 – 212.

Thomashow, L.S. and Weller, D.M. (1988). Role of a phenazine antibiotic from *Pseudomonas fluorescens* in biological control of *Gaeunmanomyces gramines* var. *tritici. J.Bacteriol.* 170: 3499-3508.

Thrane, C. Tironsmo, A. and Jensen, D.F. (1997). Endo 1,2 ß-glucanase and cellulose from *Trichoderma harzianum:* Purification and partial characterization, induction and biological activity against plant pathogenic *Pythium* spp. *Eur. J. Plant Pathol.* 103: 331-334.

Tojo, A. and Aizawa, K. (1983). Dissolution and degradation of *Bacillus thuringiensis* ?-endotoxin by gut juice protease of the silk worm *Bombyx mori. Appl. Environ. Microbial.* 45:576 - 580.

Van den Heuvel, M.J. and Cochram, D.G. (1965). Cross resistance to Organophosphorus Compounds in malathion and diazinon-resistant strains of *Blatella germanica. J. Econ. Entamol.* 58:872 - 894.

Van Frankenhuygen, K. (1993). The challenges of *Bacillus thuringiensis.* In: *Bacillus thuringiensis* an environmental pesticide: Theory and Practice, pp 1, Wiley and Sons, New York.

Van Loon, L. C., Bakkes, P.A.H.M. and Pietarse, C.M.J. (1998). Systemic resistance induced by rhizosphere bacteria. *Ann. Rev. Phytopathol.* 36:453-483.

Weinling, R.(1932). *Trichoderma lignorum* as a parasite of other fungi. *Phytopathology,* 22:837-841.

Weller, D.M. (1988). Biological control of soilborne plant pathogens in the rhizosphere with bacteria. *Ann. Rev. Phytopathol.* 26:379-407.

Weller, D.M and Cook. R.J. (1983). Suppression of take-all of wheat by seed treatments with *fluorescent pseudomonds. Phytopathology,* 73:463-469.

Weller, D.M. (1985). Application of *fluorescent pseudomonads* to control root diseases. In Ecology and Management of soilborne Plant Pathogens, Parker, C.A; Rovira, A.D., Moore K.J., Wong P.T.W. and Kollmorgen, J.F.(Eds) St. Paul: *Am. Phytopathol. Soc.* pp 137-140.

Weller, D.M., Howie, W.J. and Cook, R..J. (1988). Relationship between *in vitro* inhibition of *Gaeumannomyces graminis* var tritici and suppression of take all of wheat by *fluorescent pseudomonads. Phytopathology,* 78:717-723.

Whipps, J.M. (1997). Ecological consideration involved in commercial development of biological control agents for soil-borne diseases. In Modern Soil Microbiology, van Ilsas et al. (Eds), pp 215 - 243. Marcel Dekker, Wellington EMS.

Williams, R.H., Whipps, J.M. and Cooke, R..C. (1998a). The role of mesofaur.a in dispersal of *Coniothyrium minitans;* transmission of sclerotia of *Sclerotina sclerotiorium. Soil Biol. and Biochem.* 30:1929-1935.

Williams, R.H., Whipps, J.M. and Cooke, R.C. (1998b). The role of mesofauna in dispersal of *Coniothyrium minitans;* mechanisms of transmission. *Soil Biol. Biochem.* 30:1937-1945.

Wraight, S.P. and Carruthers, R.I. (1999). Production, delivery and use of mycopesticides for control of insect pests on field crops. In Biopesticides: Use and Delivery, Pp 233-269. Humara Press.

Wright, J.M. (1956). The production of antibiosis in soil III. Production of gliotoxin in wheat straw buried in soil. *Ann. Appl. Biol.* 44:461-466.

Yu, G.Y., Sinclair, J.B., Hartman. G.L. and Bertagnolli, B.L. (2002). P of iturin A by *Bacillus amyloliquefaciens* suppressing. *Rhizoctonia solani. Soil. Biochem.* 34:955-963.

Yuen, G.Y., Scroth, M.N. and McCain, A.H. (1985). Reduction of *Fusarium* wilt of carnation with suppressive soils and antagonistic bacteria. *Plant Dis.* 69:1071-1075.

Zehnder, G.W., Murphy, J.F., Sikora, E. J. and Kloepper, J.W. (2001). Application of rhizobacteria for induced resistance. *Eur. J. Plant. Pathol.* 107:39-60.

❑❑❑

CHAPTER **7**

Natural Products from Endophytic Fungi

Margareth B. C. Gallo, Denise O. Guimarães
Luciano da S. Momesso, Mônica T. Pupo*
Faculdade de Ciências Farmacêuticas de Ribeirão Preto Universidade de São Paulo, Avenida do Café, s/n, 14040-903, Ribeirão Preto, SP, Brazil

ABSTRACT

Endophytes are microorganisms that live inside the host plant tissues without causing symptoms of disease. There has been a crescent interest in the prospecting of these microorganisms as sources of novel and bioactive natural products. In fact, a significant number of interesting natural products have been reported in the last years. In this chapter some aspects regarding the endophytes are detailed: the relationship with their host plants, isolation, preservation, taxonomy, cultivation, frequently isolated strains, tropical endophytes, novel structures isolated and also their biological activities.

Key words: natural product, endophytic fungi, plant fungus interaction, secondary metabolites.

1. INTRODUCTION

Plants, insects, microorganisms and marine organisms exhibit complex interactions with the environment and produce small molecules natural products useful for their survival. As a consequence of their biological role, these metabolites might exhibit a broad range of biological activities. In fact, natural products have played a fundamental role in drug discovery and development processes for a long time (Pupo *et al.*, 2006). In addition, the structural complexity of the natural products have inspired synthetic organic chemists and also provided useful research tools for the understanding of several biochemical pathways. However, one crucial aspect to be considered for a successful discovery of useful novel natural products is the selection of

*Email: mtpupo@fcfrp.usp.br

the source of the compounds to be studied. It is important to take into account that untapped sources of biological diversity are often related to new chemical diversity (Clardy and Walsh, 2004). Since the discovery of the penicillin, soil microorganisms have been extensively prospected in the search of bioactive small molecules. However, endophytes, marine microorganisms, microorganisms from extreme environments and microorganisms found in association with other microorganisms or insects have been less studied, and should be considered as under-explored sources of new useful natural products. Endophytes might be defined as microorganisms that can be detected at a given moment within the tissues of an apparently healthy plant host (Schulz and Boyle, 2005), and they have been found to produce a significant number of interesting natural products (Tan and Zou, 2001; Gunatilaka, 2006; Zhang *et al.*, 2006). In this chapter we will focus on endophytes as sources of natural products, including aspects of their interaction with the host plants, methods for isolation, preservation, identification and culturing, as well as tropical endophytes and recently isolated novel structures.

2. PLANT-FUNGUS INTERACTIONS

Fungi present a universal characteristic which is the impossibility of doing photosynthesis, hence they are completely heterotrophic. Also, fungi must acquire their nutrients from the environment, from living, dying, or dead organisms. They can attack many different hosts so virulently that they kill the hosts and then absorb the released nutrients (Mauseth, 2003). In addition, some fungi are able to develop a symbiotic relationship with their hosts varying from mutualism to parasitism (Schardl *et al.*, 2004).

Symbiosis between a fungus and a plant is a widespread and ancient phenomenon in nature that plays a major role in structuring plant communities by affecting colonization, competition, coexistence and soil nutrients dynamics (Pyrozynski and Hawksworth, 1998; Clay and Holah, 1999; Lemons *et al.*, 2005). Most, if not all, plants studied in natural ecosystems are infested by fungi without none external manifestation of disease. By definition, those fungi are called endophytes and occur in an astonishing diversity (Kogel *et al.*, 2006). The term 'endophyte' has been controversial since its appearance (Petrini, 1991) and there are various reports that endophytes can become parasites under certain conditions (Schulz and Boyle, 2005). Since there are lots of independent evolutionary events, the distinctions between mutualistic and parasitic fungi are unavoidably blurred. However, a good number of published studies are still based on the conventional wisdom that endophytic fungi are plant-defending mutualists (Salminen *et al.*, 2005; Grayer and Kokubun, 2001). It is known as mutualism, the interaction that is beneficial for both partners, being of foremost ecological and evolutionary importance. Into this context, it is necessary to have a range of fungus-plant adjustments to establish a balanced mechanism among, for example, the endophyte and host reproduction conflicts, energetic costs of harboring the endophyte, the nutrient demand of the endophyte, the suppression of the host immune systems that not only allow the endophyte to subsist but also might increase the susceptibility to other fungal pathogens, and host control of fungal growth within the plant (Saikkonen *et al.*, 2004). Therefore, this mutualistic relationship has been evolved during the great physical proximity between the interacting cells as a finely tuned balance of antagonisms with a great expenditure of energy to maintain a stable state that disadvantages neither partner. Under this view, disease is the exception in plant-microbe interaction, being regarded as an unbalanced status of a symbiosis (Kogel *et al.*, 2006). Currently, key elements for the evolution of the endophytic lifestyle involve multiple levels of causation, the whole network of species that interact with host plant directly or indirectly (multispecies interaction), are swayed by biotic and abiotic environmental conditions, influenced by genetic factors, and have traits related to mutual exploitation (Saikkonen *et al.*, 2004). The endophyte profits from host plant by receiving organic nutrients, shelter and guaranteed transmission to the next host generation whereas infected plants are more vigorous, drought resistant and toxic to herbivores, nematodes and pathogens than uninfected plants (Müller and Krauss, 2005; Schardl *et al.*, 2004), as there has been shown for the symbiotic interaction between *Neotyphodium* endophytes and certain grasses resulting in the synthesis of a number of alkaloids which defend the plants against herbivory (Salminen *et al.*, 2005). Sometimes this gain is extended to a third entity of the symbiosis, for instance a hemiparasitic plant. In this case, the endophyte also enhances the resistance of the hemiparasitic plant, which has an ameliorated performance that means a reducing in the growth of the host grass; straight away the mutualistic endophytic fungus becomes parasitic in the presence of the hemiparasitic plant (Lehtonen *et al.*, 2005). Further, the mutualistic association can suffer modifications in the chemical balance. It happens in the case of *Brachiaria plantaginea*, a weed grass species. It yields aconitic acid that stimulates the development of the endophytic fungus *Fusarium solani* into the seeds of other weed species competitor. However, the stimulation of that fungus has an adverse effect on germination of those competitors (Voll *et al.*, 2003; Voll *et al.*, 2005).

The molecular and biochemical basis for the switch from endophytic to parasitic lifestyle are characterized by an imbalance in nutrient exchange that can explain why colonization of different hosts can cause a fungus to adopt contrasting lifestyles. In this regard, *Colletotrichum magna*, a fungal pathogen which provokes anthracnose in cucurbit plants, exercises an endophytic lifestyle when growing asymptomatically on an assortment of non-cucurbit species (Hahn and Mendgen, 2001; Redman *et al.*, 1999).

As a result of these long-held interactions, it is feasible to arouse that some endophytic microbes may have devised genetic systems allowing for transfer of information between themselves and the host plant. As a consequence, the associated microorganisms can learn as much as teach biochemical pathways in order to produce substances common to their hosts or *vice versa* that may have application outside the host plant in which they normally reside (Strobel, 2002). In fact, the study of natural products from plants and their endophytes has shown the probability of that hypothesis. From the toxic locoweeds *Astragalus mollissimus*, *Oxytropis lambertii* and *O. sericea* were isolated endophytic fungi capable of producing the alkaloid swainsonine (1), also produced by their hosts. This compound causes locoism, a disease of ruminant animals (Braun *et al.*, 2003). 3-Nitropropionic acid (2), a potent antimycobacterial agent, was isolated from endophytes and several plants species (Chomcheon *et al.*, 2005). Interestingly, some useful plant-derived anticancer drugs have also been identified from endophytic fungal cultures. Paclitaxel (TaxolÒ) (3), one of the most important drugs available for the treatment of breast and ovarian cancers, was isolated from *Taxomyces andreanae*, an endophytic fungus associated with the Pacific yew *Taxus brevifolia* (Stierle *et al.*, 1993). An endophytic fungus from leaves of *Catharanthus roseus* was reported to biosynthesize the potent antileukemia agent vincristine (4) (Yang *et al.*, 2004). *Entrophospora infrequens*, an endophyte isolated from *Nothapodytes foetida* (Icacinaceae), was able to produce camptothecin (5) (Puri *et al.*, 2005), chemotherapeutic agent efficient against lung, ovarian and uterine cancer, which was first isolated from *Camptotheca acuminate* (Nyssaceae) (Amna *et al.*, 2006). Podophyllotoxin (6), a natural product precursor of useful anticancer agents, was found to be synthesized by *Trametes hirsute*, a novel endophyte from *Podophyllum hexandrum* (Puri *et al.*, 2006), and also by the endophytic fungus *Phialocephala fortinii*, associated with *Podophyllum peltatum* (Eyberger *et al.*, 2006). Figure 1 shows the structures of the compounds 1-6. Based on these examples, knowledge of plant-microorganisms interactions can direct the research of novel bioactive natural products for pharmaceutical and agrochemical industries (Rubini *et al.*, 2005; Tan and Zou, 2001; Strobel, 2006).

In addition to the previous examples, a significant number of interesting molecules has been found to be produced by endophytes, including alkaloids, steroids, terpenoids, isocoumarins, quinones, flavonoids, phenylpropanoids, lignans, peptides, phenolics, aliphatics, and volatile organic compounds, many of them biologically active (Tan and Zou, 2001; Gunatilaka, 2006; Zhang *et al.*, 2006). Therefore, endophytes are an outstanding source of small molecules.

3. ENDOPHYTIC FUNGI ISOLATION

The rationale for the host plant selection is crucial to increase the chances of isolating novel microorganisms and new bioactive compounds. Plants should be selected mainly on the basis of their unique environmental setting, ethnobotanical history, endemism, unusual longevity, and large areas of biodiversity (Strobel *et al.*, 2004).

Generally, the isolation of endophytes is first based on the principle that the selected plant is apparently healthy. Then, the following procedures are relatively simple and initially anchored in the sterilization of epiphytic microbes. One of the critical needs for isolating endophytes is the acquisition of fresh plant material. It is recommended the use of sturdy paper bags, wax paper bags, zip-locked or perforated bags designed for proper transport and temporary storage of most types of plant tissues (Bills, 1996). However, if the plants are originated from foreign countries, they can be stored in sterile sealed plastic bags and transported by air within a few days (Strobel *et al.*, 1999).

The surface treatment usually initializes with plant material being washed in running tap water, by means of detergent or not, to remove extraneous matter (Seena and Sridhar, 2004). Then it is used to subject the plant material under airflow (Comcheon *et al.*, 2005), sterile filter paper (Gao *et al.*, 2005) or sterile blotters under airflow to ensure complete drying (Tejesvi *et al.*, 2006). Subsequently a surfactant such as ethanol and/or Tween 80 is employed to rinse the plant material, followed by a sterilizing agent, such as sodium hypochlorite (Schulz and Boyle, 2005) or hydrogen peroxide (Gao *et al.*, 2005). The surfactant and sterilizing agent concentrations required for the sterilization varies with the kind of the plant tissue (Table 1). For especially delicate tissues, it is suggested serial washings in sterile distilled water and changes in the sterilization approach (Raviraja *et al.*, 2006; Bussaban *et al.*, 2001). On the other hand, if the leaves contain latex, it is not appropriated the use of ethanol as the rinse solution (Bayman *et al.*, 1998). The time into the sterilizing agent must be major for seeds (Mirlohi *et al.*, 2006). Later on, the plant tissues are excised with a sterilized scalpel into pieces of about 5 mm length and plated onto the culture medium, normally potato dextrose agar (PDA) (Suryanarayanan *et al.*, 2003) or malt extract agar (MEA) (Arnold *et al.*, 2000) supplemented or not with antibiotic agents such as chloramphenicol (Suryanarayanan *et al.*, 2003), streptomycin, tetracycline or penicillin (Otero *et al.*, 2002) to restrain bacterial growth until emergence of fungal colony from the plant segments. Afterward, the plates are incubated at temperatures ranging from 18 to 30 °C for several days until fungal growth (Table 1). Additionally, aliquots of the water from final rinse solutions can be plated on the same media employed for the endophyte culture to check the effectiveness of sterilization procedure (Cao *et al.*, 2004). Fungal outgrowth from the plant tissues is subcultured on fresh antibiotic-free medium for identification based on morphological examination by scrutinizing the fungal culture, the mechanism of conidia production and the conidial characters. Nevertheless, some isolates must be cultured on different media as banana leaf pieces impregnated on PDA (Tejesvi *et al.*, 2006), oatmeal agar (Bayman *et al.*, 1998), malt extract agar, plant-origin tissue fragments (Strobel *et al.*, 1999) or several other industrializing media (Shen *et al.*, 2006) to induce sporulation.

4. FUNGAL STRAINS PRESERVATION

Fungi have been used extensively in industrial processes. Moreover, many researchers have made use of fungi either in teaching, medical diagnosis, chemotaxonomic or investigation purposes. In any case, the preservation of fungal strains, as type material or reference stocks, becomes a strategic approach to acquire reproductive outcome. Consequently, all laboratories undertaking commercial biochemical production or research involving living fungi have to possess a culture collection with a compatible way of keeping fungi alive. Preferably, stored strains should remain viable over long periods but being dormant in order to prevent mutations which can cause morphological, physiological and metabolic alterations. Besides viability, it may be important to retain sporulation since in general the fungal conidia provide the propagule for spraying onto equipment or materials under test (Smith, 1993). However, the choice of preservation method depends on the service asked for, the laboratory availabilities and other factors. For instance, a teaching collection does not demand the guarantee of the stability and longevity of a national type culture depository. Yet, a taxonomic collection requires the morphological conservation of conidia form of the fungus whereas an industrial collection needs to retain a stable metabolic state.

Several methods for the storage of fungal cultures have been developed mainly during the past century until nowadays. They can be divided into two main groups based on the continuous or suspended metabolism of the fungus (Onions, 1983). The first group includes storage in: sterile water (Castellani, 1963), serial transfer in agar, cool storage in a standard refrigerator at 5-8°C, deep freeze at about -20 °C, saline, under mineral or paraffin oil, grain or soil (Douds Jr. and Schenck, 1991) at room temperature. The second

TABLE 1. Recently employed methods for endophytic fungi isolation

Washing	Rinse in ethanol solution	Surface disinfection	Rinsed in ethanol solution	Rinsed in sterile distilled water	Incubation (days, temperature)	Reference
Running tap water (RTW)	70%	3%, 3 min[a]	70%	twice	3-15, 28°C	Rubini *et al.*, 2005
Water and detergent[c]	70%, 1 min	15%, 1 min[b]	70%, 1 min	ni[c]	ni	Gao *et al.*, 2005
RTW [d]	70%, 1 min	5%, 5 min[a]	ni	twice, 1 min	30 °C[e]	Chomcheon *et al.*, 2005
RTW	75%, 1 min	6%, 3 or 5 min[a]	75%, 0.5 min	three times	30, 25 °C	Raviraja *et al.*, 2006
RTW	70%, 1 min	5.3%, 5 min[a]	75%, 0.5 min	ni	Several days, 27 °C	Agusta *et al.*, 2005
RTW [d]	70%, 1 min	6%, 5 min[a]	ni	twice, 1 min	30 °C	Chomcheon *et al.*, 2006
ni	95%	Ni	ni	ni	28 °C	Amna *et al.*, 2006
ni[f]	Ni	2,5% 20-30 min[a]	ni	once	7-14, 28 °C	Mirlohi *et al.*, 2006
RTW	95%, 1 min	6%, 5 min[a]	95%, 0.5 min	three times	4-5, 25 °C	Seena and Sridhar 2004
ni[f]	Ni	2%, 1 min	ni	twice, 1 min	2-15, 25 °C	Medina *et al.*, 2006
Water and detergent	70%, 1 min	3%, 4 min[a]	70%, 0.5 min	once	3-8, 18 °C	Souza *et al.*, (2004)
RTW [f]	75%, 1 min	34%, 10 min	75%, 0.5 min	ni	22 °C	Bayman *et al.*, 1998
Water and detergent	Not used, leaf contain latex	0.5%, 20 min	Not used, leaf contain latex	once	22 °C	Bayman *et al.*, 1998

[a] sodium hypochloride solution; [b] hydrogen peroxide solution; [c] the material was dried with sterile filter paper; [d] air-dried; [e] cultivated on banana leaf agar the fungi developed conidia, which permitted their identification; ni: not informed; [f] technique used for seeds

group includes drying, silica gel, freeze drying or lyophilization, liquid nitrogen (Onions, 1983) and prepared cryogenic freezer beads (Microbank) at -70°C (Baker and Jeffries, 2006). Numerous of these methods have been evaluated, and many have been found to be suitable for high quality storage (Borman *et al.*, 2006a; Ryan *et al.*, 2001; Smith, 1993). Cryopreservation was considered the ultimate method available for the long-term storage of microbial cultures due to the stability of secondary metabolite production and the minimized genetic alterations, though certain fungi have exhibited significant degrees of polymorphism after revival (Ryan *et al.*, 2001; Borman *et al.*, 2006a). However, such method is expensive and particularly suitable to the banking of valuable stocks such as type and patent application isolates (Onions, 1983). In addition, there are some factors that affect the recovery of the fungus cell as the cryoprotectant, the cooling rate, the storage temperature and the thawing rate (Smith, 1993). Lyophilization is more convenient and favored by large collections (Onions, 1983). Extend to 20-40 years of storage; however there are some restrictions such as the recovery time on resuscitation of the preserved isolate to obtain an optimum viability without sub-lethal damage, requiring a carefully design and controlled protocols. Besides, with this technique only the fungal conidia survive (Borman *et al.*, 2006a; Smith, 1993). Conceivable, the Castellani's method, besides being simple and economically feasible for small laboratories, yields satisfactory results for preservation of most species of fungi (Rodrigues *et al.*, 1992; Hartung de Capriles *et al.*, 1989; Onions, 1983), even if viability seems to be linked to the particular organisms involved, thus it is suggested that a second storage method should be provided in order to retain both viability and morphological stability over long periods (Borman *et al.*, 2006a). Storage in silica gel is cheap and simple and has been demonstrated to be trustworthy for the withholding of genetic markers. But fungi must be sporulating and usually belonging to the Zygomycetes, Ascomycetes and Deuteromycetes groups. In addition, the time of storage varies between 2-4 years (Smith, 1993; Onions, 1983). Although repeated passage in agar is a time consuming method, susceptible to contamination, and subculturing on synthetic media can lead to the loss of the ability of fungi sporulation, many collections have been maintained by this way (Smith, 1993; Amna *et al.*, 2006). Recently, this method has been evaluated by Mendonza *et al.* (2005) for isolates of *Sporothrix schenckii.* They concluded that the typical morphological feature of that fungus was preserved, but some physiological characteristics as capacity to hydrolyze starch and sensibility to potassium iodide were changed.

In conclusion, there is not an ideal method, but several of them depending on both the microorganism and the available resources. Also, there is a lack of studies taking into account expressly preservation of endophytic fungi.

5. FUNGAL TAXONOMY

Taxonomy is a dynamic and progressive discipline that frequently introduces changes in the classification of living species. Even the conception of what species is, the basic position in taxonomy, can diverge widely among different mycologists. Thus, a number of diverse concepts have been used to define the fungal species. There are several ways of approaching the concept based on morphological characters, combination of characters and adaptation to a particular ambient. However, some of them cannot be applied to any sexual or asexual fungi. Advancement on phylogenetic approaches and integration of molecular biological techniques, in particular the analysis of DNA nucleotide sequences, have deeply influenced and derived considerable changes into traditional systematic concepts. The novel phylogenetic concept has been found principally suitable for fungal groups in which no sexual reproduction has been detected (Guarro *et al.*, 1999).

Classification systems of fungi have been historically supported on readily observable morphological features and their comparison. Consequently, fungal culture is a requisite to enable the growth of fundamental structures applied in the identification. The most important sets of characteristics to be observed are the conidia and the process involved in their formation. Additionally, the pigmentation and shape of hyphae, the presence or absence of septa, the occurrence of sclerotia, chlamidospores or any other particular hyphal element may be very helpful to assist in classification of both anamorph (asexual state) or teleomorph (sexual state) phases (Bononi and Grandi, 1998). Yet, this method requires a taxonomic expertise taking into consideration its complexity.

Many attempts have been realized to classify fungi according to their secondary metabolite production pattern. However, the use of this method in fungal taxonomy has been queried since the production of compounds can be affected by environmental and culture conditions (Zain, 2004). A full metabolic picture requires the combination of several different analysis techniques. Hence hyphenated approaches using morphological, genetic and chemical traits, including mycotoxins, as well as metabolite profiling and metabolomics can highlight the utility of chemotaxonomic characters for fungal taxonomy overall (Russel and Paterson, 2006; Stadler *et al.*, 2005; Smedsgaard and Nielsen, 2005). In a similar way, the employment of cellular fatty acid profiles was more or less specific depending on the fungal genus and temperature variation in which fungus was cultured. Some groups can not be consistently distinguished such as ascomycetes from basidiomycetes. Nevertheless, the acquired information is particularly valuable at the intraspecific level, where disparities are limited, and for fungi like mycelia sterilia and vesicular-arbuscular mycorrhizal endophytes, in which morphological differentiation is minimal (Stahl and Klug, 1996; Frisvad *et al.*, 1998).

Cell wall polysaccharides have also performed a role as traits for fungal taxonomy and evolution. Subtle modifications in the basic polysaccharide structure produce new compounds constructive in distinguishing close genera (*Eupenicillium, Talaromyces, Neosartorya, Chaetosartorya* etc.) or groups of species belonging to the same genus, such as *Penicillium, Paecilomyces,* and *Fusarium.* Besides, different sugar sequences can delimitate higher taxa, for example the orders Onygenales and Eurotiales display sequences with a single sugar whereas Hypocreales shows glucuronic acid on its composition (Leal *et al.,* 2001).

Isoenzyme analysis, which is carried out by eletrophoretic methods, is other really powerful and adaptable technique that can be used to resolve many problems on fungal genetics, population biology and taxonomy; specifically it has also determined generic relationship and differentiated species. This technology is economically advantageous, faster and simpler in contrast with alternative ones. However, the shortcoming of isoenzyme analysis is a poorer level of polymorphism compared to DNA-based markers, but this regularly can be overcome by rising the amount of enzyme systems assayed (Goodwin, 2004).

Molecular identification and phylogenetic studies rely on a large extent on ribosomal DNA (rDNA) sequence polymorphism. The main reasons for the popularity of rDNA are that it is a multiple-copy, non-protein-coding gene but almost always treated as a single-locus gene. Additionally, ribosomes are present in all organisms displaying a common evolutionary origin (Guarro *et al.,* 1999). In order to classify fungi by this method, it is necessary to follow a procedure that includes DNA extraction, PCR amplification, DNA sequencing, sequencing alignment and phylogenetic analysis. DNA extraction of the fungal cultures is a laborious step taking over one hour to be completed. Nonetheless, a faster and recovered technique was developed derived from the employment of filter cards that expended around fifteen minutes in extraction (Borman *et al.,* 2006b). Regarding PCR amplification, regions of the molecule are transcribed, generally the 5.8S, 18S or 25S, along with internal and external transcribed spacers (ITS and ETS). Every spacer is joined out of the transcript. Between each cluster there is a nontranscribed or intergenic spacer (NTS or IGS) which provides separation among repetition on the chromosome. Afterward sequencing reactions are performed and sequences are aligned. Phylogenetic analysis can be conducted by methods like parsimony, neighbor-joining or Bayesian method of phylogenetic inference using a Metropolis-coupled Markov chain Monte Carlo (MCMC) approach and implemented in the computer program MRBAYES (Gao *et al.*, 2005; Pitt *et al.*, 2004). Nuclear rDNA sequences are very informative and allow comparison from genus to intra and interspecific levels (Hambleton *et al.*, 1998). Despite the use of huge amounts of rDNA data available, some species within a given genus remain indistinguishable (Abbasi *et al.*, 2005). For that reason, new target sequences have been selected and validated. For example, in order to distinguish eleven *Fusarium* species, in addition to the rDNA fragment encompassing the internal transcribed spacers ITS1 and ITS2 and the 5.8 S sequence, cellobiohydrolase-C and the topoisomerase II genes, which were found to be single copy genes, were used as new molecular markers enhancing the accuracy of the analysis (Hatsch *et al.*, 2004). Furthermore, large subunit gene (LSU) of rDNA was a useful molecular marker for clarifying taxonomic and phylogenetic relationships in arbuscular micorrhiza fungi (Silva *et al.*, 2006a). Yet aiming better identification based on sequence analysis, programs as the oligonucleotide barcode TrichOKEY, and TrichoBLAST, a multilocus database of vouchered sequences, have been developed. In conjunction, these progresses enabled 100 *Trichoderma* species classification (Druzhinina *et al.*, 2006).

The simultaneous use of highly conserved LSU rRNA-coding sequence and variable non-coding ITS1 sequence permitted the connection of genetically indistinguishable species *Polycytella hominis*, isolated from an Indian male mycetoma, and *Scedosporium apiospermum*, the anamorph of *Pseudallescheria boydii* which also causes mycetoma (Borman *et al.*, 2006c). Similar cospecificity was acquired between *Arthroderma olidum*, a new ascomycete taxon, and *Trichophyton eboreum*, a recently described isolate from human skin (Campbell *et al.*, 2006).

Frequently various microorganisms can be overlooked by culturing, during the usual isolation procedure, since the first fungus to grow of a fragment may inhibit growth of other fungi. Additionally, many microorganisms do not develop in artificial culture media and are difficult to identify in culture. Occasionally, herbarium species have been surveyed to evaluate the taxonomic implications, phylogenetic relationships and distribution of a complex fungus species (Abbasi *et al.*, 2005). In all these cases, culture-independent molecular techniques that rely on the rDNA and rRNA genes may detect a number of microorganisms without isolation of them. To amplify fungal rDNA fragments within the plant tissues, all epiphytic organisms attached on the surface are also removed by surface sterilization and the segments are dried for DNA extraction using standardizing protocols (Abbasi *et al.*, 2005; Gao *et al.*, 2005).

By employing molecular biology methods such as RFPL analysis and sequencing of rDNA ITS library or 5.8S region, the diversity of endophytic fungi communities that colonizes a tissue plant can be revealed, mainly for genus and species level identification.

6. CULTURE CONDITIONS

Ultimately, when an endophyte is acquired in pure culture, its ability to be grown is investigated in a numerous of different media and growth conditions. Subsequently, better medium and growth conditions are established, the microbe is fermented and extracted and the extract is submitted to several chromatographic procedures in order to yield the product of interest.

In most cases, the temperature, pH, composition of the medium, the length of growth period of culture and the degree of aeration are some of the factors that can affect the amount and kinds of compounds that are produced by a particular fungus and can be manipulated to improve yields and assortment of substances of bioactive significance (Elias *et al.*, 2006; Strobel *et al.*, 2004; Stahl and Klug, 1996). There are several media that can be used for the cultivation of filamentous fungi (Paterson and Bridge, 1994) and they can be selected according to the purpose and the species to be studied. To ascertain, for example, which medium supports the maximum production of a determined secondary metabolite, Stinson *et al.* (2003), after serial tests on potato dextrose agar (PDA), corn meal agar (CMA), oatmeal agar (OMA) and natural PDA (NPDA), discovered that *Gliocadium* sp., an endophyte isolated from *Eucryphia cordifolia*, produced a more efficient mixture of volatile antimicrobial compounds either on PDA or OMA media. Usually liquid media are chosen for physiological studies. However, for rapid screening of plentiful strains to afford taxonomic characters, agar media has attested to be more convenient (Paterson and Bridge, 1994; Santos *et al.*, 2003). In order to analyze the production of ochratoxin A, *Aspergillus ochraceus* and black aspergilli, endophytes from grapes, were cultured on Czapek Yeast Extract agar (CYA) during 7 days at 25 °C (Bellí *et al.*, 2006). On the contrary, *A. flavus* and *A. parasiticus*, isolated from barley seeds, were tested for identical aflatoxin production using Aflatoxin Production Ability (APA) broth whereas *A. ochraceus* and *A. niger* were cultured on Yeast Extract Sucrose (YES) broth supplemented with 5% (w/v) of bee pollen. In the same research, species of *Alternaria* were grown using Malt Broth (MB) while *Fusarium* isolates were cultured on maize kernels or rice seeds to test mycotoxins production (Medina *et al.*, 2006). In an attempt to stimulate production of conidia, Potato Sucrose Agar (PSA), fresh glucose-peptone agar, lactrimel medium and CMA were tested in strains that had been stored as hyphal and conidial suspensions in sterile water (Borman *et al.*, 2006a). Schulz *et al.* (2002) isolated about 6500 endophytic fungi which were cultured on semi-solid biomalt and malt extract soya media to be screened for their biological activities and chemical profiles. Besides, they compared the growth of *Fusarium* isolates in four different liquid media and concluded that the endophyte grew better in apoplastic washing fluid (AWF) of bean than in the other assessed synthetic media. Similarly, 500 strains of endophytic fungi were examined for their antimicrobial, cytotoxic, phytotoxic and nematicidal activities. Following optimization of culture conditions, the highest nematicidal activity was obtained in YMG medium in a stirred aerated fermentation (Schwarz *et al.*, 2004). Secondary metabolites produced on CYA and YES media by several *Phoma* species, isolated from diverse plants, were detected by thin layer chromatography as an approach to be used in the systematics of the genus (Montel *et al.*, 1991). In order to extract fungal DNA for subsequently molecular identification, isolated fungi are inoculated onto Sabouraud's (SAC) medium (glucose-peptone agar containing 0.05 mg/mL chloramphenicol) and incubated at 30 °C (Borman *et al.*, 2006b).

The cultivation of endophytic fungi for ensuing screening for extra- and intracellular enzymes and secondary metabolites are commonly performed in liquid culture, which is recommended as a more homogenous system than a solid medium. It is important to ensure satisfactory aeration to provide good growth yields without premature autolysis and to guarantee the maintenance of oxidative biochemical pathways to a viable extent. In methods employing stationary cultures it is usual to inoculate small volumes of the medium in bottles that are then incubated slanted. Shaken liquid cultures utilize large conical flasks containing approximately 1/5 of their volumes with culture medium to promote gas exchange, and shaken on an orbital shaker. Usually, the growth on shaken liquid media is faster than on solid media though morphological differences can be noted as the aptitude of sporulation, the length of mycelia and the like (Paterson and Bridge, 1994).

The achievability, age and amount of inoculum employed are all important in the determination of biochemical activities and attributes for filamentous fungi. It is essential that supplementary nutrients are not transferred to the test media from the inoculum. One way to avoid transfer is to use an inoculum with few storage products that has been produced on a nutritionally poor medium. This can be achieved by agar plugs of mycelium cut from first growing cultures on PDA agar for 3-7 days that will contain few nutrients, and the mycelium will also depleted storage products. These plugs are commonly employed as solid media inoculum but they can also be used for shaken liquids, however under this circumstance better growth is obtained by cutting the plugs to produce a particulate inoculum (Paterson and Bridge, 1994). A conidial or hyphal suspension can also be used as an inoculum of liquid medium. In both cases, the suspension is inoculated on a pre-fermentative medium and incubated with agitation of 120 rpm, at 30 °C for 24 or 48 h to stimulate growth. Than the resulting mycelia are harvested by vacuum filtration, rinsed with sterile water and transferred to the selected liquid medium and submitted to chosen conditions (Elias *et al.*, 2006).

7. STRAINS FREQUENTLY ISOLATED AS ENDOPHYTES

Literature review covering the past ten years demonstrated that about 770 fungal species have been isolated as endophytic. Nevertheless, an impressive number of those species carry on unidentified. *Alternaria* sp.; *A. infectoria*, *Aspergillus* sp.; *Colletrotrichum* sp.; *C. gloeosporioides*; *C. musae*; *Cordana musae*; *Guignardia* sp.; *Nigrospora oryzae*; *N. sphaerica*; *Penicillium* sp.; *Phomopsis* sp.; *Rhizoctonia* sp.; *Xylaria* sp. appear among the foremost identified species. Moreover, some new species have been found like *Phialocephala sphaeroides* B. J. Wilson, a dark septate root endophyte from boreal wetland plants of Canada (Wilson *et al.*, 2004). *P. compacta* and *P. scopiformis*, both identified by Kowalski and Kehr (1995), were isolated from the living branches of German forest trees. Some ascomycetes isolated from the wild ginger *Amomum siamense* named as *Gaeumannomyces amomi*, *Leiosphaerella amomi*, and *Pyricularia* sp. (Bussaban *et al.*, 2001), a pyrenomycete *Monosporascus ibericus* sp. isolated from plants occurring on Spanish saline soils (Collado *et al.*, 2002), *Gonatobotryum* sp., a conidial fungus from the Indian plant *Carissa carandas* L. (Jacob and Bhat, 2000), and *Mycena anoectochila* from the Chinese Orchidaceae *Anoectochilus roxburghii* (Guo *et al.*, 1997).

8. TROPICAL ENDOPHYTIC FUNGI

The majority of the researches concerning endophytic fungi have been focused on endophytes from temperate plants, although the number of tropical studies has increased (Rodrigues and Petrini, 1997). The research on tropical endophytes has been stimulated by the role played by these microorganisms on both global fungal diversity estimates (Fröhlich and Hyde, 1999; Hawksworth, 2001) and plant community dynamics (Arnold, 2001) as sources of novel bioactive compounds (Strobel and Long, 1998) and biological control agents for use in tropical agroforestry (Arnold *et al.*, 2001).

In order to enlarge knowledge on tropical fungal diversity, several tropical endophytes from Guyana (Cannon and Simmons, 2002), Panama (Arnold and Herre, 2003; Suryanarayanan *et al.*, 2003), India (Nair and Bhat, 2002;

Suryanarayanan *et al.*, 2003), Puerto Rico (Lodge, 1997; Otero *et al.*, 2002; Gamboa *et al.*, 2002), Brazil (Borges and Pupo, 2006; Campos *et al.*, 2005; Cafêu *et al.*, 2005; Souza *et al.*, 2004, Azevedo *et al.*, 2000), Australia (Parungao *et al.*, 2002; Fröhlich and Hyde, 1999), and Ecuador (Fisher *et al.*, 1995) have been studied, becoming a crucial issue in the accurate determination of global species amount and plant-fungus relationship. In this regard, some authors have found evidences for host preference, specialization on under-studied substrate (i.e., a tropical monocotyledonous host family) and spatial variability among morphospecies, suggesting that patterns of richness and preference may be consistent across host species and tropical sites (Arnold *et al.*, 2000; Fröhlich and Hyde, 1999). Quite the opposite, May (1991) hypothesized that tropical fungi tend to be less host specific than their temperate counterparts for the reason that highly diverse tropical forests could act against host specificity due to the disjunction in the distribution of hosts (May, 1990).

Tropical endophytes have produced bioactive secondary metabolites. For instance, 3-hydroxypropionic acid, from both *Phomopsis phaseoli*, which has been associated with a broad-leaved tree from French Guyana, and *Melanconium betulinum*, recovered from *Betula pendula* and *B. pubescens*, displayed nematicidal activity against the plant-parasite *Meloidogyne incognita*, presenting an LD_{50} between 12.5-15.0 mg/mL (Schwarz *et al.*, 2004). *Muscodor albus*, an endophyte from the small unidentified vine growing in the Tessa Nile area of Sumatra, synthesizes antimicrobial volatile antibiotic organic compounds (Atmosukarto *et al.*, 2005). Potential cholinesterase enzymes inhibitor extracts from *Pestalotiopsis guepinii*, *Phomopsis* sp., *Guignardia mangiferae* and *Physalospora* sp. have been observed (Rodrigues *et al.*, 2005). New and highly functionalized compounds were identified from tropical endophytes such as ambuic acid, an antifungal compound isolated from both *Pestalotiopsis* sp. and *Monochaetia* sp. (Li *et al.*, 2001).

Recent scattered but promising reports on the endophytic fungal natural products in Brazil offer the hope that continual investigations in tropical endophytes will reveal these fungi as a source of important bioactive compounds and novel structures (Rodrigues-Heerklotz *et al.*, 2001; Santos and Rodrigues-Filho, 2002; Corrado and Rodrigues, 2004). The endophytic fungi *Penicillium* sp. and *P. janthinellum*, both isolated from *Melia azedarach*, have been found as good producers of new chemical diversity. *Penicillium* sp. produced five new meroterpenes (7-11) (Santos and Rodrigues-Filho, 2002; Santos and Rodrigues-Filho, 2003), while the new steroid neocyclocitrinol (12) (Marinho *et. al.*, 2005a) and a novel modified anthraquinone, janthinone (13) (Marinho *et. al.*, 2005b), were isolated from *P. janthinellum*. Two novel anthraquinone derivatives, dendryol E (14) and dendryol F (15), have also been isolated from *Phoma sorghina*, endophyte found in association with the medicinal plant *Tithonia diversifolia* (Borges and Pupo, 2006). *Pestalotiopsis* sp., endophytic fungus associated with *Pinus taeda*, biosynthesized three new highly oxygenated caryophyllene sesquiterpenes (16-18) (Magnani *et al.*, 2003). *Phomopsis cassiae*, endophyte isolated from *Cassia spectabilis*, produced seven new compounds: ethyl 2,4-dihydroxy-5,6-dimethylbenzoate (19) and phomopsilactone (20), both with active against phytopatogenic fungi, in addition to five sesquiterpenes (21-25) (Silva *et al.*, 2005; Silva *et al.*, 2006b). Compounds 20 and 24 also presented cytotoxicity against HeLa cells. Two novel benzopyrans (26 and 27) have been isolated from *Curvalaria* sp., an endophytic fungus from *Ocotea corymbosa* (Teles *et al.*, 2006). Figure 2 shows the new chemical diversity produced by the Brazilian endophytic fungi.

9. NOVEL SECONDARY METABOLITES FROM ENDOPHYTES

Endophytes have the ability to produce a huge chemical diversity, including alkaloids, peptides, steroids, terpenoids, isocoumarins, quinones, phenylpropanoids and lignans, phenols and phenolic acids, aliphatic compounds, lactones, and others. Among these compounds, several have interesting biological activity. Comprehensive reviews regarding endophytic chemical diversity reported up to May 2006, and also biological activities, have been published and confirmed endophytes as an outstanding source of natural products (Tan and Zou, 2001; Gunatilaka, 2006; Zhang *et al.*, 2006). The continuous growing interest in the chemistry of endophytes can be illustrated by a considerable number of additional small molecules produced by endophytes reported after May 2006. Some of these novel structures recently reported are presented below (Fig. 3). Chaetominine (28), an alkaloid with a new framework, has been produced by *Chaetomium* sp. IFB-E015, an endophytic fungus from *Adenophora axiliflora*. Chaetominine showed cytotoxicity against the human leukemia K562 and colon cancer SW1116 cell lines higher than the drug 5-fluorouracil (Jiao *et al.*, 2006). Phaeosphoramides A (29) and B (30), two new carbon skeleton derivatives, were isolated from the endophytic fungus *Phaeosphaeria avenaria* (Maloney *et al.*, 2006). Phaeosphoramide A (29) was found to be an inhibitor of the signal transducer and activator of transcription (STAT)-3, which plays a vital role in regulating cell growth and survival, constituting a target for anticancer therapy (Maloney *et al.*, 2006). Two new 12-membered ring lactones (31 and 32) were isolated from the mycelial extracts of a brown alga endophytic fungus (Yang *et al.*, 2006). Additional poliketides 12-membered macrolides (33-35) have been produced by the endophytic fungal strain *Cladosporium tenuissimum* LR463 of *Maytenus hookeri* (Dai *et al.*, 2006a). *Nodulisporium* sp., endophytic fungus associated with *Juniperus cedre*, produced seven new metabolites (36-41).

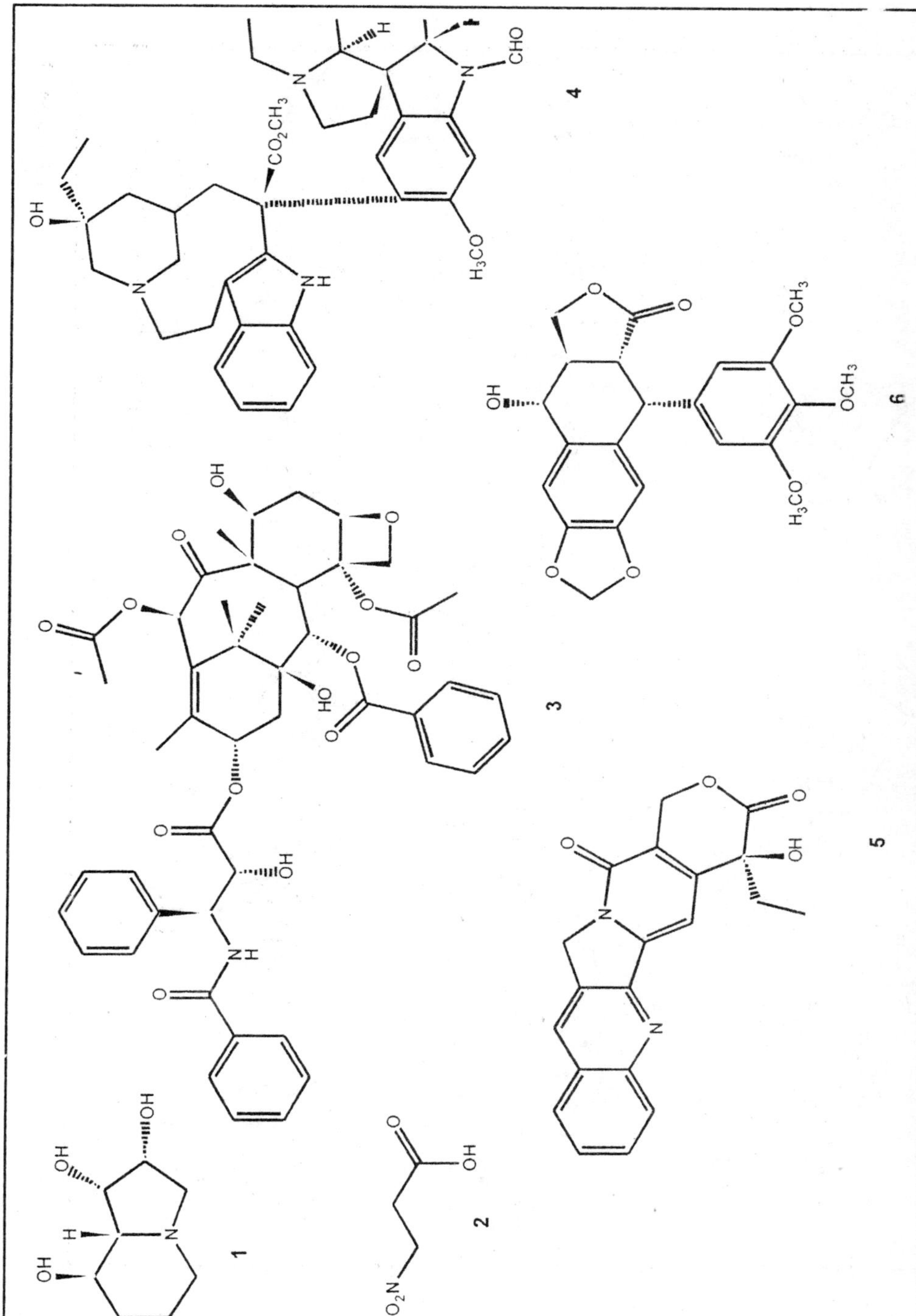

FIG. 1. Natural products produced by endophytic fungi and also by their host plants.

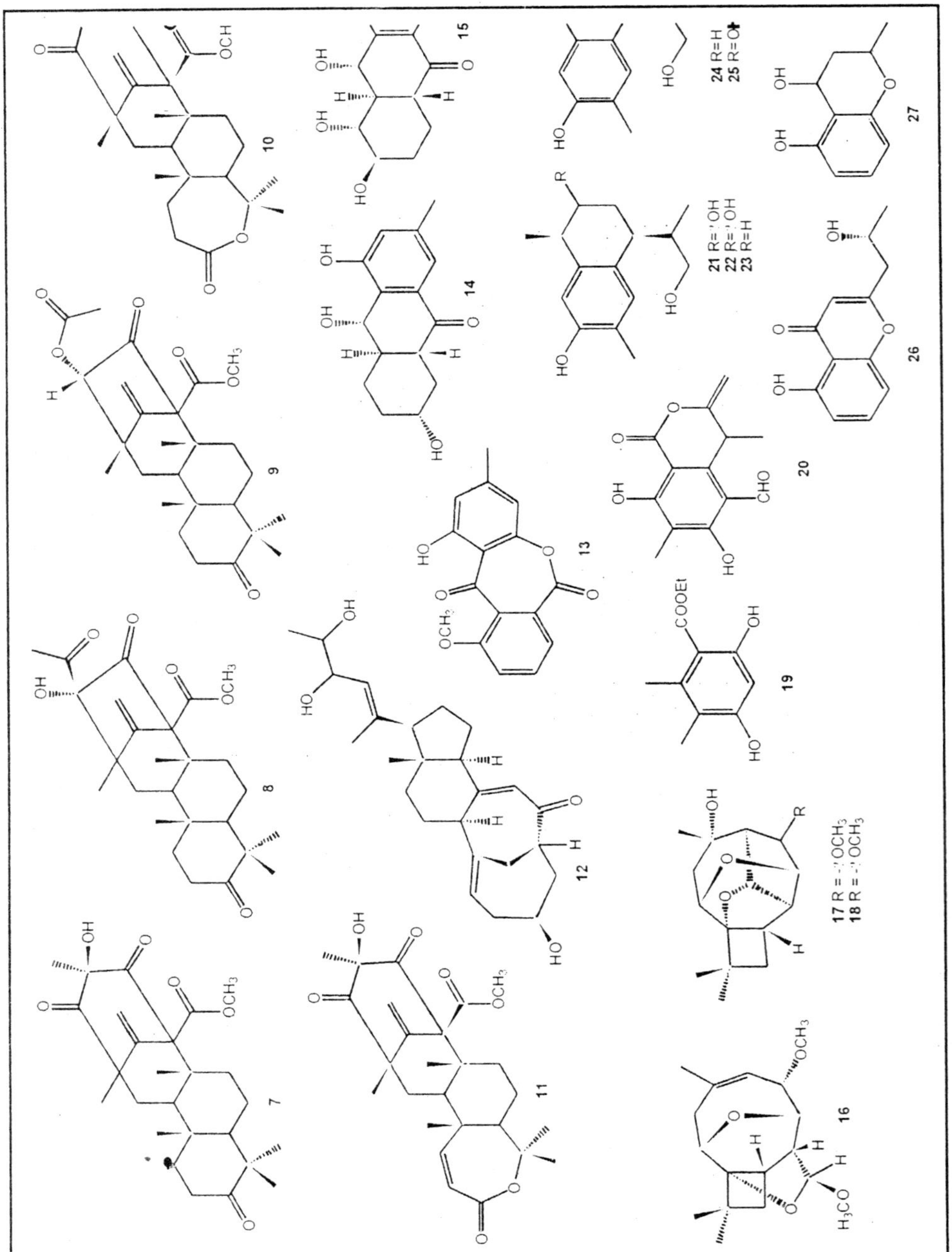

figure contd...

Figure 2 contd...

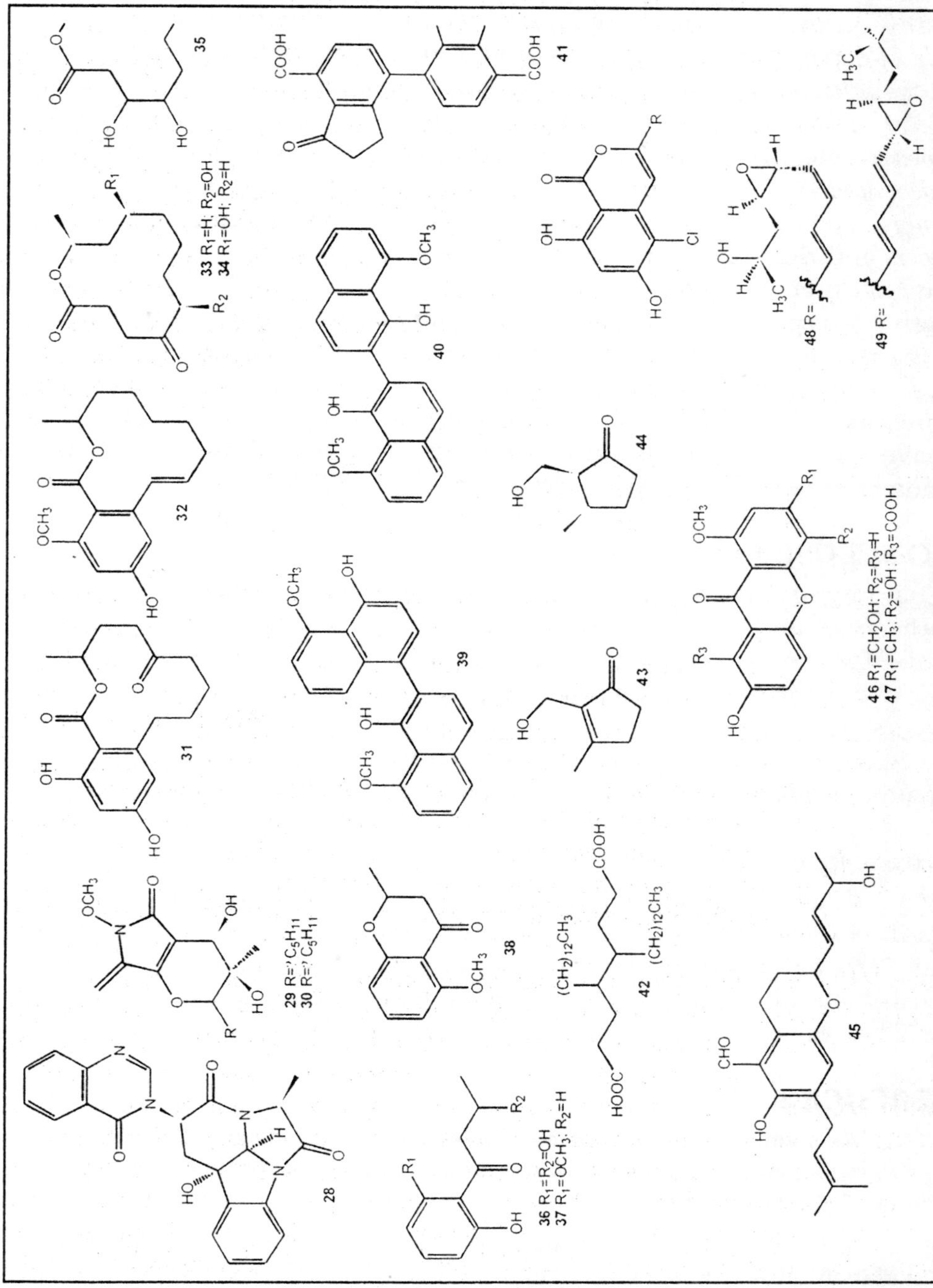

FIG. 2. New natural products produced by Brazilian endophytic fungi.

Compound 37 presented antimicrobial activity and compound 41 is the first naturally occurring dimeric indanone (Dai *et al.*, 2006b). The novel 4,5-ditridecyl-octanedioic acid (42) was produced together with known antifugal macrocyclic trichothecenes by the endophyte *Myrothecium* sp. Z16, isolated from *Argyrosomus argentatus* (Liu *et al.*, 2006). Two new natural products cyclopentenones (43 and 44) have been isolated from the endophyte mitosporic *Dothideomycete* sp. strain LRUB20, associated with the Thai medicinal plant *Leea rubra*. The cyclopentenones could be useful building blocks for organic synthesis and they can be obtained in good yields from the culture medium (Chomcheon *et al.*, 2006). Chaetopyranin (45), a novel benzaldehyde derivative, was produced by *Chaetomium globosum*, endophytic fungus derived from the red alga *Polysiphonia urceolata* (Wang *et al.*, 2006). The endophytic fungus *Xylaria* sp. FRR 5657 produced two new xanthones (46 and 47) (Davis and Pierens, 2006). Two new isocoumarins, chaetochiversins A (48) and B (49) have been isolated from the *Chaetomium chiversii*, endophytic fungus in Mormon tea (*Ephedra fasciculate*) (Wijeratne *et al.*, 2006).

10. CONCLUSIONS

Endophytic microorganisms have been proved as an outstanding source of both novel and bioactive natural products, which have an enormous potential for the development of new drugs and agricultural products. In spite of the increased number of reports about their natural products in the last years, endophytes are still a relatively poorly investigated group of microorganisms. Therefore, the research focusing on endophytes is a promising field in the chemistry and biological properties of natural products.

Acknowledgments

M. T. P. acknowledges Fundação de Amparo à Pesquisa do Estado de São Paulo (FAPESP) for the financial support for her researches on endophytes (grants 03/07535-5 and 04/07935-6). M. C. B. G. acknowledges FAPESP for her post-doctoral fellowship (05/56259-6). D.O.G and L. S. M. also thanks FAPESP for their Ph.D. fellowships (05/03791-2 and 04/06399-3).

REFERENCES

Abbasi, M., Goodwin, S. B. and Scholler, M. (2005). Taxonomy, phylogeny, and distribution of *Puccinia graminis*, the black stem rust: new insights based on rDNA sequence data. *Mycoscience* 46:241-247.

Agusta, A., Maehara, S., Ohashi, K., Simanjuntak, P. and Shibuya, H. (2005). Stereoselective oxidation at C-4 of flavans by endophytic fungus *Diaporthe* sp. isolated from a tea plant. *Chem. Pharm. Bull.* 53:1565-1569.

Amna, T., Puri, S. C., Verma, V., Sharma, J. P., Khajuria, R. K., Musarrat, J., Spiteller, M. and Qazi, G. N. (2006). Bioreactor studies on the endophytic fungus *Entrophospora infrequens* for the production of an anticancer alkaloid camptothecin. *Can. J. Microbiol.* 52:189-196.

Arnold, A. E. (2001). Fungal endophytes in neotropical trees: abundance, diversity, and ecological interactions. In: Ganeshaiah, K. N., Uma Shaanker, R. and Bawa, K. S. Eds. Tropical ecosystems: structure, diversity, and human welfare. Oxford and IBH Publishing Co. Pvt. Ltd., New Delhi, India, pp. 739-743.

Arnold, A. E. and Herre, E. A. (2003). Canopy cover and leaf age affect colonization by tropical fungal endophytes: Ecological pattern and process in *Theobroma cacao* (Malvaceae). *Mycologia* 95:388-398.

Arnold, A. E., Maynard, Z. and Gilbert, G. S. (2001). Fungal endophytes in dicotyledonous neotropical trees: patterns of abundance and diversity. *Mycol. Res.* 105:1502-1507.

Arnold, A. E., Maynard, Z., Gilbert, G. S., Coley, P. D. and Kursar, T. A. (2000). Are tropical fungal endophytes hyperdiverse? *Ecol. Lett.* 3:267-274.

Atmosukarto, I., Castillo, U., Hess, W. M., Sears, J. and Strobel, G. A. (2005). Isolation and characterization of *Muscodor albus* I-41.3s, a volatile antibiotic producing fungus. *Plant. Sci.* 169: 854-861.

Azevedo, J. L., Maccheroni, W. J., Pereira, J. O. and De Araújo, W. L. (2000). Endophytic microorganisms: a review on insect control and recents advances on tropical plants. *Electron. J. Biotechnol.* 3:40-65.

Baker, M. and Jeffries, P. (2006). Use of commercially available cryogenic vials for long-term preservation of dermatophyte fungi. *J. Clin. Microbiol.* 44:617-618.

Bayman, P., Angulo-Sandoval, P., Báez-Ortiz, Z. and Lodge, J. (1998). Distribution and dispersal of *Xylaria* endophytes in two tree species in Puerto Rico. *Mycol. Res.* 102:944-948.

Bellí, N., Bau, M., Marín, S., Abarca, M. L., Ramos, A. J. and Bragulat, M. R. (2006). Mycobiota and ochratoxin A producing fungi from Spanish wine grapes. *Int. J. Food Microbiol.* 111:S40-S45.

Bills, G. F. (1996). Isolation and analysis of endophytic fungal communities from woody plants. In: Redlin, S. C. and Carris, L. M. Eds. Endophytic Fungi in Grasses and Woody Plants, APS Press, St. Paul, pp.31-65.

Bononi, V. L. R. and Grandi, R. A. P. Eds. (1998). Zigomicetos, Basidiomicetos, Deuteromicetos: Noções Básicas de Taxonomia e Aplicações Biotecnológicas. Instituto de Botânica, Secretaria de Estado do Meio Ambiente, São Paulo, pp. 184.

Borges, W. S. and Pupo, M. T. (2006). Novel Anthraquinone Derivatives Produced by *Phoma sorghina*, an Endophyte Found in Association with the Medicinal Plant *Tithonia diversifolia* (Asteraceae). *J. Braz. Chem. Soc.* 17:929-934.

Borman, A. M., Szekely, A., Campbell, C. K. and Johnson, E. M. (2006a). Evaluation of the viability of pathogenic filamentous fungi after prolonged storage in sterile water and review of recent published studies on storage methods. *Mycopathologia* 161:361-368.

Borman, A. M., Linton, C. J., Miles, S., Campbell, C. K. and Johnson, E. M. (2006b). Ultra-rapid preparation of total genomic DNA from isolates of yeast and mould using Whatman FTA filter paper technology - a reusable DNA archiving system. *Med. Mycol.* 44:389-398.

Borman, A. M., Campbell, C. K., Linton, C. J., Bridge, P. D. and Johnson, E. M. (2006c). *Polycytella hominis* is a mutated form of *Scedoporium apiospermum*. *Med. Mycol.* 44:33-39.

Braun, K., Romero, J., Lidell, C. and Creamer, R. (2003). Production of swainsonine by fungal endophytes of locoweed. *Mycol. Res.* 107:980-988.

Bussaban, B., Lumyong, S., Lumyong, P., McKenzie, E. H. C. and Hyde, K. D. (2001). Endophytic fungi from *Amomum siamense*. *Can. J. Microbiol.* 47:943-948.

Cafêu, M. C., Silva, G. H., Teles, H. L., Bolzani, V. S., Araújo, A. A., Young, M. C. M. and Pfenning, L. H. (2005). Substâncias antifúngicas de *Xylaria* sp., um fungo endofítico isolado de *Palicourea marcgravii* (Rubiaceae). *Quim. Nova* 28:991-995.

Campbell, C. K., Borman, A. M., Linton, C. J., Bridge, P. D. and Johnson, E. M. (2006). *Arthroderma olidum*, sp. nov. A new addition to the *Trichophyton terrestre* complex. *Med. Mycol.* 44:451-459.

Campos, F. R., Barison, A., Daolio, C., Ferreira, A. G. and Rodrigues Fo, E. (2005). Complete ^{1}H and ^{13}C NMR assignments of aurasperone A and fonsecinone A, two bis-naphtopyrones produced by *Aspergillus aculeatus*. *Magn. Reson. Chem.* 43:962-965.

Cannon, P. F. and Simmons, C. M. (2002). Diversity and host preference of leaf endophytic fungi in the Iwokrama Forest Reserve, Guyana. *Mycologia* 94:210-220.

Cao, L., Qiu, Z., You, J., Tan, H. and Zhou, S. (2004). Isolation and characterization of endophytic Streptomyces strains surface-sterilized tomato (*Lycopersicon esculentum*) roots. *Lett. Appl. Microbiol.* 39:425-430.

Castellani, A. (1963). Further researches on the long viability and growth of many pathogenic fungi and some bacteria in sterile distilled water. *Mycopathol. Mycol. Appl.* 20:1-6.

Chomcheon, P., Sriubolmas, N., Wiyakrutta, S., Ngamrojanavanich, N., Chaichit, N, Mahidol, C., Ruchirawat, S. and Kittakoop, P. (2006). Cyclopentenones, scaffolds for organic syntheses produced by the endophytic fungus mitosporic Dothideomycete sp. LRUB20. *J. Nat. Prod.* 69: 1351-1353.

Chomcheon, P., Wiyakrutta, S., Sriubolmas, N., Ngamrojanavanich, N., Isarangkul, D. and Kittakoop, P. (2005). 3-Nitropropionic acid (3-NPA), a potent antimycobacterial agent from endophytic fungi: is 3-NPA in some plants produced by endophytes? *J. Nat. Prod.* 68:1103-1105.

Clardy, J. and Walsh, C. (2004). Lessons from natural molecules. *Nature* 432:829-837.

Clay, K. and Holah, J. (1999). Plant diversity in successional fields. *Science* 285:1742-1744.

Collado, J., Gonzalez, A., Platas, G., Stchigel, A. M., Guarro, J. and Pelaez, F. (2002). *Monosporascus ibericus* sp. nov., an endophytic ascomycete from plants on saline soils, with observations on the position of the genus based on sequence analysis of the 18S rDNA. *Mycol. Res.* 106:118-127.

Corrado, M. and Rodrigues, K. F. (2004). Antimicrobial evaluation of fungal extracts produced by endophytic strains of *Phomopsis* sp. *J. Basic Microb.* 44:157-160.

Dai, H. Q., Kang, Q. J., Li, G. H. and Shen, Y. M. (2006a). Three new polyketide metabolites from the endophytic fungal strain *Cladosporium tenuissimum* LR463 of *Maytenus hookeri*. *Helv. Chim. Acta.* 89:527-531.

Dai, J., Krohn, K., Flörke, U., Draeger, S., Schulz, B., Kiss-Szikszai, A., Antus, S., Kurtán, T. and van Ree, T. (2006b). Metabolites from the endophytic fungus *Nodulisporium* sp. from *Juniperus cedre*. *Eur. J. Org. Chem.* 3498-3506.

Davis, R. A. and Pierens, G. K. (2006). ^{1}H and ^{13}C NMR assignments for two new xanthones from the endophytic fungus *Xylaria* sp. FRR 5657. *Magn. Res. Chem.* 44:966-968.

Douds Jr., D. D. and Schenck, N. C. (1991). Germination and hyphal growth of fungi during and after storage in soil at five matrix potentials. *Soil Biol. Biochem.* 23:177-183.

Druzhinina, I. S., Kopchinskiy, A. G. and Kubicek, C. P. (2006). The first 100 *Trichoderma* species characterized by molecular data. *Mycoscience* 47:55-64.

Elias, B. C., Said, S., Albuquerque, S. and Pupo, M. T. (2006). The influence of culture conditions on the biosynthesis of secondary metabolites by *Penicillium verrucosum* Dierck. *Microbiol. Res.* 161: 273-280.

Eyberger, A. L., Dondapati, R. and Porter, J. R. (2006). Endophyte fungal isolates from *Podophyllum peltatum* produce podophyllotoxin. *J. Nat. Prod.* 69:1121-1124.

Fisher, P. J., Petrini, L. E., Sutton, B. C. and Petrini O. (1995). A study of fungal endophytes in leaves, stems and roots of *Gynoxis-oleifolia* Muchler (Compositae) from Ecuador. *Nova Hedwigia* 60: 589-594.

Frisvad, J. C., Bridge, P. D. and Arora, D. K. Eds. (1998). Chemical Fungal Taxonomy. Marcel Dekker, New York, pp. 398.

Fröhlich, J. and Hyde, K. D. (1999). Biodiversity of palm fungi in the tropics: are global fungal diversity estimates realistic? *Biodivers. Conserv.* 8:977-1004.

Gamboa, M. A., Laureano, S. and Bayman, P. (2002). Measuring diversity of endophytic fungi in leaf fragments: does size matter? *Mycopathology* 156:41-45.

Gao, X., Zhou, H., Yu, C., Chen, Y. and Qu, L. (2005). High diversity of endophytic fungi from the pharmaceutical plant *Heterosmilax japonica* Kunth revealed by cultivation-independent approach. *FEMS Microbiol. Lett.* 249:256-266.

Goodwin, S. B. (2004). Isozyme analysis in fungal taxonomy, genetics, and population biology. *Mycol. Series* 21(Fungal Biotechnology in Agricultural, Food, and Environmental Applications): 37-47.

Grayer, R. J. and Kokubun, T. (2001). Plant-fungal interactions: the search for phytoalexins and other antifungal compounds from higher plants. *Phytochemistry* 56:253-263.

Guarro, J., Gené, J. and Stchigel, A. M. (1999). Developments in fungal taxonomy. *Clin. Microbiol. Rev.* 12:454-500.

Gunatilaka, A. A. L. (2006). Natural products from plant-associated microorganisms: distribution, structural diversity, bioactivity and implications of their occurrence. *J. Nat. Prod.* 69:509-526.

Guo, S. X., Fan, L., Cao, W. Q., Xu, J. T. and Xiao, P. G. (1997). *Anoectochilus roxburghii* from Xishuangbanna, China. *Mycologia* 89:952-954.

Hahn, M. and Mendgen, K. (2001). Signal and nutrient exchange at biotrophic plant-fungus interfaces. *Curr. Opin. Plant Biol.* 4:322-327.

Hambleton, S., Egger, K. N. and Currah, R. S. (1998). The genus *Oidiodendron*: species delimitation and phylogenetic relationships based on nuclear ribosomal DNA analysis. *Mycologia* 90:854-869.

Hartung de Capriles, C., Mata, S. and Middelveen, M. (1989). Preservation of fungi in water (Castellani): 20 years. *Mycopathologia* 106:73-79.

Hatsch, D., Phalip, V. and Jeltsch, J. (2004). Use of genes encoding cellobiohydrolase-C and topoisomerase II as targets for phylogenetic analysis and identification of *Fusarium*. *Res. Microbiol.* 155:290-296.

Hawksworth, D. L. (2001). The magnitude of fungal diversity: the 1.5 million species estimated revised. *Mycol. Res.* 105:1422-1432.

Jacob, M. and Bhat, D. J. (2000). Two new endophytic conidial fungi from India. *Cryptogamie. Mycologia* 21:81-88.

Jiao, R. H., Xu, S., Liu, J. Y., Ge, H. M., Ding, H., Xu, C., Zhu, H. L. and Tan, R. X. (2006) Chaetominine, a cytotoxic alkaloid produced by endophytic *Chaetomium* sp. IFB-E015. *Org. Lett.* 8:5709-5712.

Kogel, K., Franken, P. and Hückelhoven, R. (2006). Endophyte or parasite - What decides? *Curr. Opin. Plant Biol.* 9:358-363.

Kowalski, T. and Kehr, R. D. (1995). Two new species of *Phialocephala* occurring on Picea and Alnus. *Can. J. Bot.* 73:26-32.

Leal, J., Prieto, A., Ahrazem, O., Pereyra, M. T. and Bernabe, M. (2001). Cell wall polysaccharides: characters for fungal taxonomy and evolution. *Rec. Res. Develop. Microbiol.* 5:235-248.

Lehtonen, P., Helander, M., Wink, M., Sporerzand, F. and Saikkonen, K. (2005). Transfer of endophyte-origin defensive alkaloids from a grass to a hemiparasitic plant. *Ecol. Lett.* 8: 1256-1263.

Lemons, A., Clay, K. and Rudgers, J. A. (2005). Connecting plant-microbial interactions above and belowground: a fungal endophyte affects decomposition. *Oecologia* 145:595-604.

Li, J. Y., Harper, J. K., Grant, D. M., Tombe, B. O., Bashyal, B., Hess, W. M. and Strobel, G. A. (2001). Ambuic acid, a highly functionalized cyclonexenone with antifungal activity from *Pestalotiopsis* spp. and *Monochaetia* sp. *Phytochemistry* 56:463-468.

Liu, J. Y., Huang, L. L., Ye, Y. H., Zou, W. X., Guo, Z. J. and Tan, R. X. (2006). Antifungal and new metabolites of *Myrothecium* sp. Z16, a fungus associated with white croaker *Argyrosomus argentatus*. *J. Appl. Microbiol.* 100:195-202.

Lodge, D. J. (1997). Factors related to diversity of decomposer fungi in tropical forests. *Biodivers. Conserv.* 6: 681-688.

Magnani, R. F., Rodrigues-Fo., E., Daolio, C., Ferreira, A. G. and Souza, A. Q. L. (2003). Three highly oxygenated caryophyllene sesquiterpenes from *Pestalotiopsis* sp., a fungus isolated from bark of *Pinus taeda*. *Z. Naturforsch. C* 58:319-324.

Maloney, K. N., Hao, W., Xu, J., Gibbons, J., Hucul, J., Roll, D., Brady, S. F., Schroeder, F. C. and Clardy, J. (2006). Phaeosphaeride A, an inhibitor of STAT3-dependent signaling isolated from an endophytic fungus. *Org. Lett.* 8:4067-4070.

Marinho, A. M. R., Rodrigues-Filho, E., Ferreira, A. G. and Santos, L. S. (2005a) C25 Steroid epimers produced by *Penicillium janthinellum*, a fungus isolated from fruits *Melia azedarach*. *J. Braz. Chem. Soc.* 16:1342-1346.

Marinho, A. M. R., Rodrigues-Filho, E., Moitinho, M. L. R. and Santos, L. S. (2005b) Biologically active polyketides produced by *Penicillium janthinellum* isolated as an endophytic fungus from fruits of *Melia azedarach*. *J. Braz. Chem. Soc.* 16:280-283.

Mauseth, J. D. (2003). Botany: An Introduction to Plant Biology, 3rd Ed. James and Bartlett Publishers, Boston. pp. 582-610.

May, R. M. (1990). How many species? *Philos. Trans. R. Soc. London Ser. B* 330:293-304.

May, R. M. (1991). A fondness for fungi. *Nature* 352:475-476.

Medina, A., Valle-Algarra, F. M., Mateo, R., Gimeno-Adelantado, J. V., Mateo, F. and Jiménez, M. (2006). Survey of the mycobiota of Spanish malting barley and evaluation of the mycotoxin producing potential of species of *Alternaria, Aspergillus* and *Fusarium*. *Int. J. Food Microbiol.* 108:196-203.

Mendonza, M., Alvarado, P., Diaz de Torres, E., Lucena, L. and Albornoz, M. C. (2005). Physiological comportment and in vivo sensitivity of *Sporothrix schenckii* isolates maintained for 18 years by two preservation methods. *Rev. Iberoamer. Micol.* 22:151-156.

Mirlohi, A., Sabzalian, M. R., Sharifnabi, B. and Nekoui, M. K. (2006). Widespread occurrence of *Neotyphodium*-like endophyte in populations of *Bromus tomentellus* Boiss. in Iran. *FEMS Microbiol. Lett.* 256:126-131.

Montel, E., Bridge, P. D. and Sutton, B. C. (1991). An integrated approach to *Phoma* systematics. *Mycopathology* 115:89-103.

Müller, C. B. and Krauss, J. (2005). Symbiosis between grasses and asexual fungal endophytes. *Curr. Opin. Plant Biol.* 8:450-458.

Nair, S. K. and Bhat, D. J. (2002). *Kumbhamaya jalapriya,* a new endophytic hyphomycete from India. *Mycotaxon* 84:65-68.

Onions, A. H. S. (1983). Preservation of Fungi. In: Smith, J. E., Berry, D. R. and Kristiansen, B. Eds. The Filamentous Fungi, vol. 4: Fungal Technology. Edward Arnold Publishers, Great Britain, pp. 373-390.

Otero, J. T., Ackerman, J. D. and Bayman, P. (2002). Diversity and host specificity of endophytic *Rhizoctonia*-like fungi from tropical orchids. *Am. J. Bot.* 89:1852-1858.

Parungao, M. M., Fryar, S. C. and Hyde, K. D. (2002). Diversity of fungi on rainforest litter in North Queensland, Australia. *Biodivers. Conserv.* 11:1185-1194.

Paterson, R. R. M. and Bridge, P. D. (1994). Biochemical Techniques for Filamentous Fungi. International Mycological Institute, CAB International, Bakehan Lane, Egham, U.K., pp. 125.

Petrini, O. (1991). Fungal endophytes of tree leaves. In: Microbial Ecology of Leaves. Andrews, J. H. and Hirano, S. S. Eds. Spring-Verlag, pp. 179-197.

Pitt, W. M., Goodwin, S. B., Ash, G. J., Cother, N. J. and Eric, J. (2004). *Plectosporium alismatis* comb. nov. a new placement for the Alismataceae pathogen *Rhynchosporium alismatis. Mycol. Res.* 108:775-780.

Pupo, M. T., Guimarães, D. O., Furtado, N. A. J. C. and Borges, W. S. (2006) Microbial natural products: a promising source of bioactive compounds. In: Modern Biotechnology in Medicinal Chemistry and Industry. Taft, C. A. Ed. Research Signpost, Kerala. pp. 51-78.

Puri, S. C., Nazir, A., Chawla, R., Arora, R., Riyaz-ul-Hasan, S., Amna, T., Ahmed, B., Verma, V., Singh, S., Sagar, R., Sharma, A., Kumar, R., Sharma, R. K. and Qazi, G. N. (2006). The endophytic fungus *Trametes hirsuta* as a novel alternative source of podophyllotoxin and related aryl tetralin lignans. *J. Biotechnol.* 122:494-510.

Puri, S.C., Verma, V., Amna, T., Qazi, G. N. and Spiteller, M. (2005). An endophytic fungus from *Nothapodytes foetida* that produces camptothecin. *J. Nat. Prod.* 68: 1717-1719.

Pyrozynski, K. A. and Hawksworth, D. L. (1998). Introduction and overview. In: Coevolution of Fungi with Plants and Animals. Pyrozynski, K. A. and Hawksworth, D. L. Eds. Academic Press, London. pp. 1-29.

Raviraja, N. S., Maria, G. L. and Sridhar, K. R. (2006). Antimicrobial Evaluation of Endophytic fungi inhabiting medicinal plants of the Western Ghats of India. *Eng. Life Sci.* 6:515-520.

Redman, R. S., Ranson, J. C. and Rodriguez, R. J. (1999). Conversion of the pathogenic fungus *Colletotrichum magna* to a nonpathogenic, endophytic mutualist by gene disruption. *Mol. Plant Microbe Interact.* 11:969-975.

Rodrigues, E. G., Lírio, V. S. and Lacaz, C. S. (1992). Preservation of fungi and actinomycetes of medical importance in distilled water. *Rev. Inst. Med. Trop.* 34:159-165.

Rodrigues, K. F. and Petrini, O. (1997). Biodiversity of endophytic fungi in tropical regions. In: Hyde, K. D. Ed. Biodiversity of Tropical Microfungi. Hong Kong, University of Hong Kong Press, pp. 57-69.

Rodrigues, K. F., Costa, G. L., Carvalho, M. P. and Epifanio, R. A. (2005). Evaluation of extracts produced by some tropical fungi as potential cholinesterase inhibitors. *World J. Microbiol. Biotechnol.* 21:1617-1621.

Rodrigues-Heerklotz, K. F., Drandarov, K., Heerklotz, J., Hesse, M. and Werner. C. (2001). Guignardic acid, a novel type of secondary metabolite produced by endophytic fungus *Guignardia* sp.: isolation, structure elucidation and asymmetric synthesis. *Helv. Chim. Acta* 84:3766-3771.

Rubini, M. R., Silva-Ribeiro, R. T., Pomella, A. W. V., Maki, C. S., Araújo, W. L., Santos, D. R. and Azevedo, J. L. (2005). Diversity of endophytic fungal community of cacao (*Teobroma cacao* L.) and biological control of *Crinipellis perniciosa*, causal agent of witches´ broom disease. *Int. J. Biol. Sci.* 1:24-33.

Russel, R. and Paterson, M. (2006). Identification and quantification of mycotoxigenic fungi by PCR. *Process Biochem.* 41:1467-1474.

Ryan, M. J., Jeffries, P., Bridge, P. D. and Smith, D. (2001). Developing cryopreservation protocols to secure fungal gene function. *Cryo Lett.* 22:115-124.

Saikkonen, K., Wäli, P., Helander, M. and Faeth, S. H. (2004). Evolution of endophyte-plant symbioses. *TRENDS Plant Sci.* 9:275-280.

Salminen, S. O., Richmond, D. S., Grewal, S. K., Grewal, P. S. and Parwinder, S. (2005). Influence of temperature on alkaloids levels and fall armyworm performance in endophytic tall fescue and perennial ryegrass. *Entomol. Experim. Appl.* 115:417-426.

Santos, R. M. G and Rodrigues-Filho, E. (2002). Meroterpenes from *Penicilliun* sp. found in association with *Melia azedarach. Phytochemistry* 61:907-912.

Santos, R. M. G and Rodrigues-Filho, E. (2003). Further meroterpenes produced by *Penicillium* sp., an endophyte obtained from *Melia azedarach*. *Z. Naturforsch. C* 58:663-669.

Santos, R. M. G., Rodrigues-Filho, E., Rocha, W. C. and Teixeira, M. F. S. (2003). Endophytic fungi from *Melia azedarach*. *World J. Microbiol. Biotechnol.* 19:767-770.

Schardl, C. L., Leuchtmann, A. L. and Spiering, M. J. (2004). Symbioses of grasses with seedborne fungal endophytes. *Annu. Rev. Plant Biol.* 55:315-340.

Schulz, B. and Boyle, C. (2005). The endophytic continuum. *Mycol. Res.* 109:661-686.

Schulz, B., Boyle, C., Draeger, S., Römmert, A. and Krohn, K. (2002). Endophytic fungi: a source of novel biologically active secondary metabolites. *Mycol. Res.* 106:996-1004.

Schwarz, M., Köpcke, B., Weber, R. W. S, Sterner, O. and Anke, H. (2004). 3-Hydroxypropionic acid as a nematicidal principle in endophytic fungi. *Phytochemistry* 65: 2239-2245.

Seena, S. and Sridhar, K. R. (2004). Endophytic fungal diversity of 2 sand dune wild legumes from southwest coast of India. *Can. J. Microbiol.* 50:1015-1021.

Shen, L., Jiao, R. H., Ye,Y. H., Wang, X. T., Xu, C., Song, Y. C., Zhu, H. L., and Tan, R. X. (2006). Absolute configuration of new cytotoxic and other bioactive trichothecene macrolides. Phylogenetic analysis of Glomeromycota by partial LSU rDNA sequences. *Mycorrhiza* 16:183-189.

Silva, G. H., Teles, H. L., Zanardi, L. M., Young, M. C. M., Eberlin, M. N., Haddad, R., Pfenning, L. H., Costa-Neto, C. M., Castro-Gamboa, I., Bolzani, V. S. and Araújo, A. R. (2006b). Cadinane sesquiterpenoids of *Phomopsis cassiae*, an endophytic fungus associated with *Cassia spectabilis* (Leguminosae). *Phytochemistry* 67:1964-1969.

Silva, G. H., Teles, H. L., Trevisan, H. C., Bolzani, V. S., Young, M. C. M., Pfenning, L. H., Eberlin, M. N., Haddad, R., Costa-Neto, C. M. and Araújo, A. R. (2005). New bioactive metabolites produced by *Phomopsis cassiae*, an endophytic fungus in *Cassia spectabilis*. *J. Braz. Chem. Soc.* 16:1463-1466.

Smedsgaard, J. and Nielsen, J. (2005). Metabolite profiling of fungi and yeast: from phenotype to metabolome by MS and informatics. *J. Exper. Bot.* 56:273-286.

Smith, D. (1993). Long-term preservation of test strains (fungus). *Int. Biodet. Biodeg.* 31:227-230.

Souza, A. Q. L., Souza, A. D. L., Astolfi Filho, S., Pinheiro, M. L. B., Sarquis, M. I. M. and Pereira, J. O. (2004). Antimicrobial activity of endophytic fungi isolated from amazonian toxic plants: *Palicourea longiflora* (aubl.) rich and *Strychnos cogens* bentham. *Acta Amaz.* 34:185-195.

Stadler, M., Laessoe, T. and Vasilyeva, L. (2005). The genus *Pyrenomyxa* and its affinities to other cleistocarpous Hypoxyloideae as inferred from morphological and chemical traits. *Mycologia* 97: 1129-1139.

Stahl, P. and Klug, M. J. (1996). Characterization and differentiation of filamentous fungi based on fatty acid composition. *Appl. Environ. Microbiol.* 62:4136-4146.

Stierle, A., Strobel, G. and Stierle, D. (1993). Taxol and taxane production by *Taxomyces andreanae*, an endophytic fungus of Pacific yew. *Science* 260:214-216.

Stinson, M., Ezra, D., Hess, W. M., Sears, J. and Strobel, G. (2003). An endophytic *Gliocadium* sp. of *Eucryphia cordifolia* producing selective volatile antimicrobial compounds. *Plant Sci.* 165:913-922.

Strobel, G. A. (2006). Harnessing endophytes for industrial microbiology. *Curr. Opin. Microbiol.* 9: 240-244.

Strobel, G., Daisy, B., Castillo, U. and Harper, J. (2004). Natural products from endophytic microorganisms. *J. Nat. Prod.* 67:257-268.

Strobel, G. A. (2002). Rainforest endophytes and bioactive products. *Crit. Rev. Biotechnol.* 22:315-333.

Strobel, G. A., Ford, E., Li, J., Sears, J., Sidhu, R. S. and Hess, W. M. (1999). *Seimatoantlerium tepuiense* gen. nov., a unique epiphytic fungus producing taxol from the Venezuelan Guyana. *System. Appl. Microbiol.* 22:426-433.

Strobel, G. A. and Long, D. M. (1998). Endophytic microbes embody pharmaceutical potential. *Am. Soc. Microbiologists News* 64:263-268.

Suryanarayanan, T. S., Venkatesan, G. and Murali, T. S. (2003). Endophytic fungal communities in leaves of tropical forest trees: Diversity and distribution patterns. *Curr. Sci.* 85:489-493.

Tan, R. X. and Zou, W. X. (2001). Endophytes: a rich source of functional metabolites. *Nat. Prod. Rep.* 18:448-459.

Teles, H. L., Silva, G. H., Castro-Gamboa, I., Bolzani, V. S., Pereira, J. O., Costa-Neto, C. M., Haddad, R., Eberlin, M. N., Young, M. C. M., Araújo, A. R. (2006). Benzopyrans from *Curvularia* sp., an endophytic fungus associated with *Ocotea corymbosa* (Lauraceae). *Phytochemistry* 66:2363-2367.

Tejesvi, M. V., Mahesh, B., Nalini, M. S., Prakash, H. S., Kini, K. R., Subbiah, V. and Shetty, H. S. (2006). Fungal endophyte assemblages from ethnopharmaceutically important medicinal trees. *Can. J. Microbiol.* 52:427-435.

Voll, E., Franchini, J. C., Da Cruz, R. T., Gazziero, D. L. P., Brighenti, A. M. and Adegas, F. S. (2003). Chemical Interactions of *Brachiaria plantaginea* with *Commelina benghalensis* and *Acanthospermum hispidum* in soybean cropping systems. *J. Chem. Ecol.* 30:1467-1475.

Voll, E., Voll, C. E. and Victória Filho, R. (2005). Allelopathic effects of aconitic acid in wild poinsettia (*Euphorbia heterophylla*) and morningglory (*Ipomoea grandifolia*). *J. Environm. Sci. Health* B40:69-75.

Wang, S., Li, X. M., Teuscher, F., Li, D. L., Diesel, A., Ebel, R., Proksch, P. and Wang B. G. (2006). Chaetopyranin, a benzaldehyde derivative, and other related metabolites from *Chaetomium globosum*, an endophytic fungus derived from the marine red alga *Polysiphonia urceolata*. *J. Nat. Prod.* 69:1622-1625.

Wijeratne, E. M. K., Paranagama P. A. and Gunatilaka, A. A. L. (2006). Five new isocoumarins from Sonoran desert plant-associated fungal strains *Paraphaeosphaeria quadriseptata* and *Chaetomium chiversii*. *Tetrahedron* 62:8439–8446.

Wilson, B. J., Addy, H. D., Tsuneda, A., Hambleton, S. and Currah, R. S. (2004). *Phialocephala sphaeroides* sp. nov., a new species among the dark septate endophytes from a boreal wetland in Canada. *Can. J. Bot.* 82:607-617.

Yang, X., Zhang, L., Guo, B. and Guo, S. (2004). Preliminary study of vincristine-endophytic fungus isolated from leaves of *Catharanthus roseus*. *Zhongcaoyao* 35:79-81.

Yang, R. Y., Li, C. Y., Lin, Y. C., Peng, G. T., She, Z. G. and Zhou, S. N. (2006) Lactones from a brown alga endophytic fungus (No. ZZF36) from the south China sea and their antimicrobial activities. *Bioorg. Med. Chem. Lett.* 16:4205-4208.

Zain, M. E. (2004). Secondary metabolites as taxonomic markers in fungal taxonomy. *African J. Mycol. Biotechnol.* 12:45-54.

Zhang, H. W., Song, Y. C. and Tan, R. X. (2006). Biology and chemistry of endophytes. *Nat. Prod. Rep.* 23:753-771.

❑❑❑

CHAPTER **8**

Marine Microbial Diversity and its Application in Biotechnology

Imelda-Joseph

Central Marine Fisheries Research Institute, Post Box 1603
Ernakulam North P.O., Cochin 682 018, Kerala, India

ABSTRACT

The world ocean with a coastline of approximately 3,50,000 km is the largest ecosystem on earth. Marine organisms are the major, sustaining components of ecosystem processes and are responsible for biogeochemical reactions that drive our climate changes. Despite this, many marine organisms are poorly described and little is known of broad spatial and temporal scale trends in their abundance and distribution. With new molecular and analytical techniques we can advance our knowledge of marine biodiversity at the species level to understand how marine biodiversity supports ecosystem structure, dynamics and resilience. We can then interpret environmental, ecological and evolutionary processes controlling and structuring marine ecosystem biodiversity. In biotechnology applications marine microbes represent a potential source for numerous commercially important biomolecules and their bioremediation capabilities are also remarkable. They also play a crucial role in decomposition of organic matter and cycling of nutrients. Microbes also serve as food for some bottom-living organisms. Understanding and preserving biodiversity was one of the most important global challenges for the past 20 years and will continue to be an important scientific issue into the new millennium. Our knowledge of marine microbial diversity has, however, been severely limited by relying on microorganisms that have been cultured which are less than 1%.

Key words: marine microbial diversity, mangrove ecosystem, metagenomics

1. INTRODUCTION

The world's oceans cover 70 percent of the Earth's surface and these areas are dominated numerically by microscopic protists and prokaryotes. They are the major, sustaining components of ecosystem processes and are

Email: imeldajoseph@rediffmail.com

responsible for biogeochemical reactions that drive our climate changes. With new molecular and analytical techniques we can advance our knowledge of marine biodiversity at the species level to understand how marine biodiversity supports ecosystem structure, dynamics and resilience. We can then interpret environmental, ecological and evolutionary processes controlling and structuring marine ecosystem biodiversity (Sweijd *et al.*, 2004). More than two billion years ago, the generation of oxygen by photosynthetic marine microorganisms helped to shape the chemical environment in which plants, animals, and all other life forms have evolved. Primary production is typically carried out by microorganisms (phytoplankton) whose biodiversity and activity will ultimately influence fish stocks. The smallest microscopic photosynthetic organisms in the world (0.6 µm *Prochlorococcus* and *Synechococus cyanobacteria*) occur in the ocean and they contribute 10-80% of the primary production in over about 75% of the sea (Fuhrman and Campbell, 1998). A great deal of research on the biogeography of marine microorganisms has been carried out, but many unknowns persist, and more work is needed to elucidate and understand their complexity. The habitats of marine microorganisms are diverse and include open water, sediment, bodies of marine macro- and microorganisms, estuaries, and hydrothermal vents. Symbiosis and pathogenesis are the two important direct interactions of marine microorganisms with larger organisms. Beneficial microbial symbioses have enabled many invertebrate species to take advantage of habitats that would otherwise be unavailable to them. Invertebrates in these relationships may also enjoy the benefits of bioactive compounds of microbes that may produce prevention of bio-fouling or ward off predators. The metabolic diversity of marine microorganisms allows them to assume many roles in the biogeochemical cycles that other organisms can not complete. Marine microbes are also able to adapt to the many extreme environments in the oceans. Potential applications for marine microorganisms in ameliorating environmental degradation also exist (Hunter-Cevera *et al.*, 2005). As humans continue to alter the environment, climate change will inevitably impact marine microbial communities and the biogeochemical cycles in which they participate, but the exact nature of these impacts cannot yet be predicted. Human health relies on a number of critical equilibriums that marine microorganisms mediate, including the balance between viruses and their hosts in the oceans, the balances that keep harmful algal blooms in check, the processes that control nutrient concentrations in marine waters, and others. The metabolic capabilities of marine microbes can be put to work in any number of biotechnology applications, including the manufacture of industrial products and energy production. Marine microbes are sources of novel bioactive compounds that may have application as pharmaceuticals (Costello, 2000).

2. MARINE MICROBIAL HABITATS

2.1 Mangrove Ecosystem

Mangrove forests generate considerable amount of detritus such as leaf litter, woody debris and inflorescence and hence constitute an ideal environment for many detritus-dependent microbes. There are different groups of bacteria which get nourished by detritus and in turn help the mangrove ecosystem in different ways. The mangrove environment is a potent source for the isolation of antibiotic-producing actinomycetes (Ratnakala and Chandrika, 1995). It was reported that bacterial isolates from mangrove swamps can be effectively used for biodegradation of phenol at high concentrations (Imelda-Joseph and Chandrika, 1999). Imelda-Joseph and Chandrika (2000) have also reported the aerobic heterotrophs in mangrove swamps.

2.2 Coral Reefs

Reef sediments are one of the main sinks for nutrients, especially phosphorus and nitrogen which are the result of bacterial action. Coral produces mucus which plays a major role in reef metabolism as an important source of organic material and supports high bacterial activity (Richman, *et al.*, 1975). Coral mucus consists mainly of polysaccharides and protein (Meikle *et al.*, 1988), and the mucus released from the corals serves as good growth substrates for bacteria.

2.3 Oceanic Ecosystem

Microbial habitats in the oceans are influenced by an almost innumerable array of forces and factors, including salinity, currents, terrestrial inputs, and climate. Salinity is relatively constant in the open ocean, but is less stable in coastal areas. Ocean and sea floor currents have been shown to behave in ways other than previously thought, greatly affecting our understanding of transport processes in the deep sea. Terrestrial inputs create gradients of nutrients, pollutants, and other matter that affect habitats. Climate effects represent the largest scale of influence on microbial habitats. Temperature, precipitation, and wind (including windborne particulate matter) can each impact marine communities in a number of ways. Importantly, marine microbes themselves exert influence on their habitats by consuming, producing, and sequestering a variety of compounds. In the oceans, gradients of materials important to micro- and macro organisms are often controlled by processes carried out by microbes itself. In the marine environment and elsewhere, interfaces tend to be hotspots of diversity and biological activity. Marine microbial habitats at interfaces include the air-water, water-sediment, water-ice, and host organism-water interfaces. The sub-millimeter scale of physical and chemical variability in these habitats poses a serious challenge to studying interface habitats in detail.

The abundance of prokaryotes in the open ocean, in soil, and oceanic and terrestrial sub-surfaces is 1.2×10^{29}, 2.6×10^{29}, 3.5×10^{30} and $0.25 - 2.5 \times 10^{30}$, respectively (Whitman *et al.*, 1998). Prokaryotes represent likely the largest living surface on earth and this enormous interface between the living and the abiotic world is one of the reasons for their significance in transforming organic matter. It has been estimated that there are ca. 7.5×10^{29} viruses in the world ocean (Weinbauer and Wenderoth, 2002). Recent evidence suggests that the domains *Archaea* and *Bacteria*, which together are also known as prokaryotes, are equally abundant in the deep sea and that about one third of the prokaryotes in the ocean are *Archaea* (Karner *et al.*, 2001). The remarkable specialisation of *Archaea* include: growing at temperature conditions of 113 °C, pH near zero, anaerobic and high salt concentrations and combinations of these conditions (Jarrell *et al.*, 1999).

An important mechanism by which microbes can interact in the environment is gene transfer (Lorenz and Wackernagel, 1994). Studies on the process of conjugation, transduction and transformation provided direct evidence for the exchange of genetic information between prokaryotes. The ability of microorganisms to exchange DNA or release DNA, which transforms or transduces other bacterial cells enable populations to adapt or evolve. Prokaryotes have several possibilities to transfer genes including a mechanism called transduction, where genes are transferred by the activity of viruses. This result in a horizontal spread of genes (as opposed to a 'vertical' spread by cell division) within a community and may contribute to diversity. Although there is considerable interest in horizontal gene transfer, it has not been often investigated in the environment due to methodological problems. One of the concerns to study gene transfer is the potential of unwanted gene transfer such as the spread of genes such as e.g. for antibiotics or frost resistance. The huge number of prokaryotes (and viruses) on earth certainly provides for an enormous potential of gene transfer. An evaluation of whole genome sequencing also suggests that lateral gene transfer occurs frequently (Doolittle, 1999). Recent evidence suggests that antibiotic resistance of bacteria due to fish farming can be spread worldwide within years (Sørum, 2002). It is not known how this spread occurs, but transport in ship ballast or atmospheric transmission is possible. Although it is not clear, how this resistance was transmitted, it is clear that such events can occur in natural communities on a global scale with a previously unimagined speed. This also points to the need of a better risk assessment of biotechnological applications.

3. MARINE MICROBIAL DIVERSITY

The study of marine bacterial diversity is important in order to understand the community structure and pattern of distribution. In the marine environment, 90% of bacteria are gram-negative with different characteristics (Zobell, 1946) and the gram-negative cell wall is better adapted for survival in the marine environment.

3.1 Role of Bacteria in the Marine Environment

The marine environment is characterized by high pressure, salinity, low temperature, absence of light etc. The marine heterotrophic bacteria which have adapted themselves to survive in this environment require Na+ for growth because it is essential to maintain the osmotic environment for protection of cellular integrity. Oligotrophy is also another adaptation because of the limited availability of nutrients. However, heterotrophic bacterial action promotes organic degradation, decomposition and mineralization processes in sediments and in the overlying water, and releases dissolved organic and inorganic substances (Purushothaman, 1998). The mineralization of organic matter, which is derived from primary producers, results in it being recycled, so that these substances are again available for primary producers. Heterotrophic microorganisms are the major agents shaping the organic composition of the ocean. These heterotrophic bacteria comprise the bulk of microbial populations inhabiting the water column of oceans and are responsible for much of the biological transformation of organic matter and production of carbon dioxide (Sherr and Sherr, 1996). Distribution of bacteria depends on changes in water temperature, salinity and other physicochemical parameters (Alavandi, 1990). Bacteria also serve as an important source of food for a variety of marine organisms. Thus, bacteria not only maintain the pristine nature of the environment, but also serve as biological mediators through their involvement in the biogeochemical processes (Das *et al.*, 2006).

4. APPLYING MARINE MICROBES IN BIOTECHNOLOGY

The great metabolic diversity of marine microbes makes the oceans a rich source of biological materials for biotechnology applications. A variety of secondary metabolites (Hans-Peter *et al.*, 2005; Jensen *et al.*, 2007), enzymes (Imada, 2005), polymers, and metabolic processes can be found only in the microbes that inhabit the oceans. Marine microorganisms have been recognized as a productive source of novel secondary metabolites. To date, the majority of natural products of marine bacterial origin have arisen from a small number of taxonomic groups that include *Streptomyces, Alteromonas, Pseudomonas, Vibrio, Agrobacterium*, and the cyanobacteria (Wagner-Dobler *et al.*, 2002; Burja *et al.*, 2001) Monks *et al.* (2002) found that extracts of eight out of ten different Brazilian sponge species exhibited anti-bacterial, anti-tumor, or anti-chemotactic activities. According to chemical ecologists, secondary metabolites are produced as part of a chemical arsenal designed to deter grazers by imparting toxicity or low palatability to the metazoan host. These marine invertebrate-microbial assemblages also produce toxic compounds to prevent colonization by other non-beneficial microbial species. Interestingly, some of the small molecules isolated from marine metazoans display a striking resemblance to prokaryotic-borne metabolites; it has thus been suggested that the source of many of these bioactives is in fact the symbiotic microorganisms (Piel *et al.*, 2004). One well-studied example is the synthesis of bryostatin-1 by the marine bryozoan, *Bugula neritina*, which has been directly linked to the presence of the uncultured bacterium *Candidatus Endobugula sertula* (Davidson *et al.*, 2001). Similarly, *Bugula simplex* hosts the symbionts *Candidatus Endobugula glebosa*, which are linked to bryostatin production (Lim and Haygood, 2004). The biotechnology potential of these organisms is a major driving force behind the push to characterize marine microbial diversity. The technical applications of bacteriorhodopsin comprise holography, spatial light modulators, optical computing, and optical memories. Compatible solutes are useful as stabilizers of biomolecules and whole cells, salt antagonists, or stress-protective agents (Hampp *et al.*, 1990; Hampp, 2000; Fischer and Hampp, 2005). Biopolymers, such as biosurfactants and exopolysaccharides (EPS), produced by bacterial strains like *Bacillus licheniformis* are of interest for microbially enhanced oil recovery (MEOR). The ability to grow on hydrocarbons is widespread for bacteria such as *Pseudomonas, Alcaligenes, Micrococcus, Norcadia, Bacillus, Rhodococcus* and *Proteus* (Rosli *et al.*, 2001). Other useful biosubstances are enzymes, such as new isomerases and hydrolases that are active and stable at high salt contents. Halotolerant microorganisms play an essential role in food biotechnology for the production of fermented food and food supplements (Margesin and Schinner, 2004). The degradation or transformation of a range of organic pollutants and the production of alternative energy are other fields of applications of microorganisms like cyanophyceae (Kuritz and Wolk, 1995). Different marine habitats will yield microbes suited to different applications. Exploring hot environments, for example, will unearth thermophilic bacteria that produce thermostable polymerases that could be used in industrial applications. Collections of libraries and culture collections of marine microorganisms would be very useful for harvesting biotechnological products from marine microbes. Basic research needs to provide the fundamental libraries, tools, and metadata that can be linked and passed on to biotechnological and biomedical interests for devising applications for marine microbes and their products. Innovative ways of linking research to real-world applications are needed.

4.1 Applications in Health, Engineering, Agriculture and Allied Sectors

Potential applications of marine microorganisms are many and varied, including:

- The interface between microbiology, nanotechnology, and engineering may yield new nanostructures or electronic components produced with the help of marine microbes.
- Magnetic beads from magnetotactic marine bacteria could have applications in optics and electronics.
- Certain marine organisms, including *Prochlorococcus* and *Pelagibacter*, have streamlined genomes that may be useful as metabolic engines for manufacturing industrial products.
- Developing novel drugs for treating diseases
- Development of sophisticated biosensors and diagnostic devices for medicine, aquaculture and environmental biomonitoring.
- Thermostable enzymes for industrial applications and research.
- Cold-active or cold adapted enzymes for use in cleaning agents, leather processing, degradation of xenobiotics in cold climate, food processing etc.
- Biopolymers and non-toxic biosurfactants for environmental waste management and treatment.

4.2. Energy Production

Humankind faces vast energy needs, but with the help of marine microorganisms, biofuels, cleaner energy production, and alternative fuel technologies may all be within reach. New processes that could build sustainable sources of energy may arise from a better understanding the diversity of processes available in marine microorganisms. Understanding the diversity of microbial processes will provide the evolutionary, structural, and functional information that are needed for process organization and optimization. Blue-green algae is a potential hydrogen producer with production rate of one cc per ml algal suspension per day in a nitrogen-free medium under an argon atmosphere, make it an attractive source for hydrogen fuel. Many blue-green algae fix nitrogen and may serve as fertilizers. In addition, the soft cell walls of blue-green algae may make them suitable for processing as nutrients. Methane production aided by algal species also receives attention (Mitsui, 1978).

4.3. Medical Applications

Enzymes and bioactive compounds from marine microbes are very useful in medical applications. Today, polysaccharides are being used in many new products, and several promising marine compounds are in the drug development pipeline. Promising bioactive compounds initially discovered in marine invertebrates may be produced by symbionts rather than by the invertebrates themselves. It has been shown that bacterial species that produce antimicrobial compounds are intimately associated with sponges. Cross sections of sponges from Caribbean reef revealed that 50% of the sponge mass is comprised of prokaryotic structures which are thought to be bacteria (Colwell, 1997). Although most of these bacterial species have not yet been cultured, several of those that have been cultured have been found to produce bioactive compounds with potential pharmacological value. Indeed, it is now thought that many of the bioactive compounds synthesized by sponges are actually produced by the diverse bacteria living within their tissues (Colwell, 1997). Successful culture of these symbionts can ensure an economic, sustainable supply of these compounds, improving the odds of their commercial development as drugs. Marine microbes may also be used to manufacture compounds called "nutriceuticals," natural products used as diet supplements to promote health. Docosahexaenoic acid, a fatty acid necessary for proper brain function, for example, could be harvested from cultures of marine dinoflagellates and processed for sale as a nutriceutical (Beach *et al.*, 1974; Kyle *et al.*, 1996). Taxonomy has some predictive value in linking physiological potentials to specific taxa. For example, some groups, like actinomycetes and cyanobacteria, are reliable producers of bioactive compounds. Similarly, thermostable enzymes are likely to be found in taxa that thrive in hot environments, like *Thermocales*. However, researchers should take caution in following these types of generalities so as to avoid the self-fulfilling prophesy: finding bioactive compounds only where they were pursued. Currently, the allocation of resources for taxonomic research is disproportionately allocated towards macroorganisms; more resources are needed in taxonomic research on microbial species.

4.4 Using Marine Microbes to Ameliorate Environmental Deterioration

The metabolic diversity of marine microorganisms not only makes them useful in biotechnology applications, it makes them versatile tools for addressing environmental problems. One prime environmental application for these organisms is bioremediation- the treatment of chemical contamination using microorganisms. Microbially mediated removal of contaminants from natural systems over time can occur entirely in the absence of human intervention. The clean up of hydrocarbons, specifically petroleum products, is an especially pressing matter in the oceans, where accidents aboard oil tankers can release thousands of gallons of oil in a single incident (Atlas, 1981). In some cases, fertilizing the indigenous microbial communities on the affected beaches with nitrogen and phosphorus can speed the degradation of the spilled oil, but bioremediation options in open water are limited because of difficulties in delivering sufficient nutrients to sustain biodegradation. It is now known that the actions of an entire microbial community are necessary to break down complex organic matter, including petroleum (A super-organism that could carry out the entire process was sought, but was never isolated or designed. A cluster of genes that carry out the degradation of chlorinated biphenyls, called the BPH cluster, have been isolated and are now being applied in treating dredged contaminants from the Hudson River Estuary (Hunter-Cevera *et al.*, 2005). Today, most research in bioremediation in the marine environment is focused on the organisms already present in the affected areas. For a number of reasons, few efforts are underway to develop engineered microorganisms to address problems of chemical contamination. There are particular problems with respect to releasing microbes for bioremediation into aquatic environments; confining the organisms to the site of concern, for example, would be difficult. To maximize the potential benefits of microbial community in combating pollution problems, it is vital to have fundamental understanding of a microbe's degradative potential under various conditions, its biochemical systems and its molecular biology (Jain *et al.*, 2005).

5. GENOMICS AND MARINE MICROBES

Genomics, the study of all or part of the genome of an organism to answer questions about that organism, describes a relatively new and powerful set of research techniques that can be applied in marine microbiology. Genomics will allow researchers to uncover previously unknown abilities and metabolisms and define the interrelationships that hold marine microbial communities together, revolutionizing our understanding of the field in the coming years. A genome comprises the full set of genes that the organism can draw on and it defines not only what genes the organism has but what genes it does not and the genome size is not merely dependent on the microbial size, but also on species (Das *et al.*, 2006). Despite their power, genomics approaches cannot be the only tools used to address marine microbial biology. Traditional laboratory approaches in genetics, biochemistry, physiology, and ecology are still necessary to validate the biology. Functional studies will often be necessary to follow up on genomics studies that uncover unknown genes, for example. Conserved hypothetical genes pose another problem; it is not often possible to discern protein function from the sequence of a hypothetical gene alone. Better motif recognition tools could help improve the ability to discern function from sequence. Microarray studies could also help by identifying where and under what conditions the gene of interest is expressed. The integration of genomics, transcriptomics and proteomics with ecological, biophysical, and chemical techniques will be necessary to delineate fundamental physiological processes, understand their regulation, and determine how they relate to biogeochemical cycles (Hunter-Cevera *et al.*, 2005).

5.1 Metagenomics

Genomic sequencing of single species has made great contributions to our understanding of marine microbial systems, and metagenomics (in which the genomes of a group of organisms are lumped and studied together) approaches will extend that knowledge and add to our understanding of marine microbial processes. Metagenomics confers the potential to map the metabolisms of microbes in space and time. It may even be possible to identify differences in metabolic potential between different ocean regions using metagenomics approaches (Hunter-Cevera *et al.*, 2005). Archiving of genomic and metagenomic samples will be necessary to provide baseline data against which microbial community structures can be assessed over time. Moreover, genomic and metagenomic information provide the fundamental for microarray and proteomics- based technologies that have shown the promise for unveiling complex microbial functions. Community-based proteomics have been applied to investigate the major expressed proteins from estuarine bacterioplankton. To make the best progress in the field, it will be critical to tie metagenomics studies to the existing knowledgebase in biological oceanography. It will be important to make metagenomic data and other genomics data sets accessible to other researchers and as easy to use as possible (Handelsman, 2004).

5.2 16S Dependent Techniques rRNA

The level of resolution of techniques that rely on the 16S ribosomal RNA (16S rRNA) gene is another pressing issue in research on marine microbes. Ribosomal RNA gene sequences are often used to identify microorganisms and 16S group probes are used assess the diversity of groups of microbes. New information has brought up questions about whether 16S genes offer sufficient resolution to inform researchers about the distribution of different microbes. It may be that 16S genes are simply too highly conserved across related strains of bacteria and archaea to be useful in many analyses and that studies that rely on denaturing gradient gel electrophoresis (DGGE) or other 16S-dependent methods are seriously underestimating microbial diversity. Greater resolution of microbial species can now be achieved using other tools, including the internal transcribed spacer (ITS) region between the 16S and 23S genes, and even measures such as genome size. Studies using these methods have shown that 16S genes can miss out on a lot of genetic diversity, confirming that the definition of a microbial species is indeed a complicated issue.

A combination of fluorescent in situ hybridization (FISH) with microautoradiography (MICRO-FISH, STAR-FISH) is a means to link diversity and function for in situ studies (Ouverney and Fuhrman, 1999; Cottrell and Kirchman, 2000). This technique uses radiolabeled substrates to determine the uptake of specific substrates by cells and specific taxonomic probes to assess, which species are taking up such substrates. Genus- to species-specific FISH probes, which are available for some taxa, can be used in the future. With such and other methods, we might learn more about the role of as-of-yet uncultured prokaryotic species, and this may allow us to get a handle on their culturability. Moreover, by using molecular tools it has been recently shown that unicellular cyanobacteria in the ocean might fix molecular nitrogen (Zehr *et al.*, 2001) and that aerobic anoxygenic phototrophs are abundant and diverse in the ocean (Beja *et al.*, 2000).

5.3 What Questions can be Answered Using Molecular Biology Techniques?

Molecular tools in general offer the possibility to estimate biodiversity at all levels, e.g., kingdom/class/family/species level, in a comparatively small environmental sample. In some cases even a few milliliters of seawater may be enough. Moreover, some of the techniques are very sensitive, e.g., offer the possibility to detect single cells in a sample. Depending on the question(s) being asked the molecular tools to answer them differ greatly. One may wish to detect as many species as possible in a given sample. In this case the establishment of an rRNA clone library with subsequent sequencing of as many clones as possible can uncover the biodiversity in the sample in great detail. General assessment of comparative biodiversity in a larger number of samples can be achieved with fingerprinting methods based on restriction fragment length polymorphisms (RFLPs), denaturing or temperature gradient gel electrophoresis (DGGE, TGGE) or single strand conformation polymorphisms (SSCP). Presence or absence of a known species can be monitored with species-specific probes using chemiluminescent detection with dot blot techniques or, more sophisticated, with fluorescent *in-situ* hybridization (FISH). Distinction of individuals at the family or even species level can be obtained using highly variable molecular markers such as ITS sequences or microsatellites. Finally per se non-molecular techniques like flow cytometry that have already been used in the "premolecular age" can be combined with DNA techniques (staining of the nucleus, hybridization with fluorescence-labeled specific oligonucleotide probes) to distinguish and quantify species in environmental samples.

In general molecular techniques have some significant advantages over traditional methods like-only very small samples (in the range of milliliters up to a liter) are required for most analyses; sensitivity of many methods is very high, e.g., enabling the researcher to detect even single specific cells among thousands of others; dead or non-culturable cells can be analyzed; species-specific data (such as sequences) can be obtained without the need to culture or even isolate a species.

5.4 Molecular Probes for Identification and Characterization of Marine Microorganisms

Quite often morphological features as seen by light microscopy are not sufficient to distinguish clearly between species or groups of marine bacteria. Therefore, more sophisticated methods like electron-microscopy or analysis of specific chemical components by HPLC are needed to identify a species for confirmation, but these are laborious and time-consuming. An alternative approach is the development of specific molecular probes (Hunter-Cevera *et al.*, 2005). These probes are short oligonucleotides of normally 16-24bp length that are hundred percent homologous only to a complementary sequence in a gene of the species of interest and differ by at least one position to all other organisms. In hybridization experiments these probes can therefore be used to identify species of interest by binding to the target's sequence and later detection by a probe-attached label, e.g., Digoxigenin (DIG) or a fluorochrome like fluorescein (Hopman *et al.*, 1998). The application range of these probes extends from answering ecological questions like species composition and its change through space and time to the development of an early warning system for harmful algal blooms using probes for toxic species.

6. IMPLICATIONS FOR BIOTECHNOLOGY IN DEVELOPING COUNTRIES

A broad definition of biotechnology is the (industrial) use of living organisms to produce food, drugs or other products. Two types of implications emerge from our knowledge on microorganisms in the environment. One type of implications is that the diversity and functions of the unseen majority are also the unmined riches (Wilson, 1992) the source of new biotechnological applications. The other type of implication is that biotechnological applications such as all types of agriculture and fisheries or the release of genetically engineered microorganisms (GEMs) might threaten microbial biodiversity and the environment.

7. CONCLUSION

Study of marine microbial biodiversity is of vital importance to the understanding of the different processes of the ocean, which may present potent novel microorganisms for screening of bioactive compounds. Marine microbial diversity can be best studied by a combination of techniques of both conventional and modern approaches for better understanding. Molecular techniques can enhance our understanding of microbial diversity in an environment as vast as the world's oceans. Phylogenetic diversity can be recovered without dependence on more traditional, often biased, preservation or culturing methods. Molecular techniques can reconstruct the phylogenetic history of a group and can document the spatial and temporal structuring of genetic diversity, i.e., biodiversity below the species level. A variety of molecular tools may need to be invoked in order to find the resolution needed to separates species, populations or individuals in the marine environment.

REFERENCES

Alavandi, S. V. (1990). Relationship between heterotrophic bacteria and suspended particulate matter in the Arabian Sea (Cochin). *Indian J. Mar. Sci.*, 30:89-92.

Atlas, R.M. (1981). Microbial Degradation of Petroleum Hydrocarbons: an Environmental Perspective. *Microbiol.Revi.*, 45:180-209.

Azam, F. and Worden, A. Z. (2004). Microbes,molecules and marine ecosystems. *Science*, 303:1622-1624.

Beach, D. H, Harrington G. W., Gellerman, J. L., Schlenk, H. and Holz, GG Jr. (1974). Biosynthesis of oleic acid and docosahexaenoic acid by a heterotrophic marine dinoflagellate *Crypthecodinium cohnii. Biochim Biophys Acta.*, 369:16-24.

Beja, O., Aravind, L., Koonin, E.V., Suzuki, M.T., Hadd, A., Nguyen, L.P., Jovanovich, S.B., Gates, C.M., Feldmann, R.A., Spudich, J.L., Spudich, E.N. and Delong, E.F. (2000.) Bacterial rhodopsin: evidence for a new type of phototrophy in the sea. *Science*, 289:1902-1906.

Burja, A.M., Banaigs, B., Abou-Mansour, E., Burgess, J.G. and Wright, P.C. (2001).Marine cyanobacteria-a prolific source of natural products. *Tetrahedron*, 57:347–9,377.

Colwell, R.R. (1997). Microbial biodiversity and biotechnology. In: Biodiversity II: Understanding and protecting our biological resources. Reaka-Kudla, M.L., Wilson, D.E. & Wilson, E.O. (eds). Joseph Henry Press, University of Washington, D.C.

Costello, M.J. (2000). A Framework for an Action Plan on Marine Biodiversity in Ireland, Prepared for The Marine Institute, Ireland. 51p http://www.marine.ie/NR/rdonlyres/553C3EF1-FAD9-4601-BD95-EBF05A06F300/0/r14.pdf

Cottrell, M. and Kirchman, D. (2000). Natural assemblages of marine proteobacteria and members of the *Cytophaga-Flavobacter* cluster consuming low- and high-molecular-weight dissolved organic matter. *Appl. Environ. Microbiol.* 66:1692-1697.

Das, S., Lyla P. S. and Khan S. A. (2006). Marine microbial diversity and ecology: importance and future perspectives. *Curr. Sci*, 90:1325-1335.

Davidson, S.K., Allen, S.W., Lim, G.E., Anderson, C.M. and Haygood, M.G. (2001). Evidence for the biosynthesis of bryostatins by the bacterial symbiont "*Candidatus Endobugula sertula*" of the bryozoan *Bugulaneritina*. *Appl. Environ. Microbiol.* 67:531-4,537.

De Long, E.F. (2001). Microbial seascapes revisited. *Curr. Opin. Microbiol.* 4:290-295.

Doolittle, W.F. (1999).Phylogenetic classification and the universal tree. *Science*, 284:2124-2129.

Fuhrman, J.A. and Campbell, L. (1998). Microbial microdiversity. *Nature*, 393:410-411.

Handelsman, J. (2004). Metagenomics: Application of Genomics to Uncultured Microorganisms. *Microbiol. Mol. Biol. Rev.* 68(4):669–685.

Hampp, N., Bräuchle, C. and Oesterhelt D. (1990). Bacteriorhodopsin wild type and variant aspartate-96 ? asparagine as reversible holographic media. *Biophys. J.* 58: 83-93.

Hampp, N. (2000). Bacteriorhodopsin as a photochromic retinal protein for optical memories. *Chemical Revi.*, 100:1755-1776.

Fischer, T. and Hampp, N. (2005). Two-photon-absorption of bacteriorhodopsin: Formation of a red-shifted thermally stable photoproduct F620. *Biophys. J.* 89:1175-1182.

Hans-Peter F, Bruntner, C., Bull, A.T., Ward A.C., Goodfellow, M., Potterat, O., Puder C. and Mihm G. (2005). Marine actinomycetes as a source of novel secondary metabolites. *Antonie van Leeuwenhoek*, 87(1):37-42

Holguin, G., Vazquez, P. and Bashan, Y. (2001). The role of sediment microorganisms in the productivity, conservation, and rehabilitation of mangrove ecosystems: an overview. *Biol. Fertil. Soils*, 33:265–278.

Hopman, A.H.N., Ramaekers, F.C.S. and Speel, E. J.M. (1998). Rapid synthesis of biotin, digoxigenin-, trinitrophenyl-, and fuorochrome-labeled tramides and their aplication for in situ hybridization Using CARD amplification. *Jour. Histochem. Cytochem.* 46: 771–777.

Hunter-Cevera, J., Karl, D. and Merril, B. (2005). Marine microbial diversity: the key to earth's habitability. A report from the American Academy of Microbiology, American Academy of Microbiology, 1752 N Street, N.W., Washington, DC 20036. 28p.

Imada, C. (2005). Enzyme inhibitors and other bioactive compounds from marine actinomycetes *Antonie van Leeuwenhoek*, 87: 59-63.

Imelda-Joseph and Chandrika, V. (1999). Biodegradation of phenol using bacteria from different brackishwater habitats. *Indian J. Mar. Sci.*, 28:438-442.

Imelda Joseph and Chandrika, V. (2000). Seasonal variations of sediment phenolics and aerobic heterotrophs in mangrove swamps. *Indian J. Mar. Sci.*, 29:52-56.

Jain, R.K., Kapur, M., Labana1, S., Lal, B., Sarma, P. M., Bhattacharya, D. and Thakur, I. S. (2005). Microbial diversity: Application of microorganisms for the biodegradation of xenobiotics. *Curr. Sci.* 89 (1):101-112.

Jarrell, K.F., Bayley, D.P., Correia, J.D. and Thomas, N. A. (1999). Recent excitement about the Archaea. The Archaea are valuable for studying basic biological questions and have novel biotechnology application. *Bioscience*, 490: 530-541.

Jensen, P. R., Williams, P. G., Dong-Chan O., Zeigler, L. and Fenical, W. (2007). Species-Specific Secondary Metabolite Production in Marine Actinomycetes of the Genus *Salinispora. Appl. Environ. Microbiol.* 73 (4):1146-1152.

Karner, M.B., Delong, E. F. and Karl, D.M. (2001). Archaeal dominance in the mesopelagic zone of the Pacific Ocean. *Nature*, 409:507-10.

Kuritz, T. and Wolk, P.C. (1995). Use of filamentous cyanobacteria for biodegradation of organic pollutants. *Appl. Environ. Microbiol.* 61 (1):234–238

Kyle D.J., Reeb S.E., and Sicotte V.J. (1996). Single-cell docosahexaenoic acid from dinoflagellates. *Trends Food Sci. Techn.*, 7 (4):143-143.

Lim, G.E. and Haygood, M.G. (2004). "*Candidatus* Endobugula glebosa," a specifi c bacterial symbiont of the marine bryozoan *Bugula simplex. Appl. Environ. Microbiol.* 70:921–4,929.

Lorenz, M.G. and Wackernagl, W. (1994). Bacterial gene transfer by natural genetic transformation in the environment. *Microbiol. Rev.* 58:563-602.

Margesin, R. and Schinner, F. (2004). Potential of halotolerant and halophilic microorganisms for biotechnology, *Extremophiles*, 5 (2):73-83.

Meikle, P., Richards, G. N. and Yellowless, D. (1988).Structural investigation on the mucus from six species of corals. *Mar. Biol.*, 99:187–193.

Mitsui, A. (1978). Alternative energy sources; Proceedings of the Miami International Conference, Miami Beach, Fla., December 5-7, 1977. Volume 2. (A79-34002 13-44) Washington, D.C., Hemisphere Publishing Corp., 1978, Pp. 1013-1017.

Monks, N.R., Lerner, C., Henriques, A.T., Farias, F.M., Schapoval, D.D.S., Suyenaga, E.S., Da Rocha, A.B., Schwarsmann, G. and Mothes, B. (2002). Anticancer, antichemotactic and antimicrobial activities of marine sponges collected off the coast of Santa Catarina, southern Brazil. *J. Exp. Mar. Biol. Eoc/.* 281:1-12.

Ouverney, C.C. and Fuhrman, J.A. (1999). Combined microautoradiography-16S rRNA probe technique for determination of radioisotope uptake by specific microbial cell types in situ. *Appl. Environ. Microbiol.* 65:1746-1752.

Piel, J., Hui, D., Wen, G., Butzke, D., Platzer, M., Fusetani, N., and Matsunaga, S. (2004). Antitumor polyketide biosynthesis by an uncultivated bacterial symbiont of the marine sponge *Theonella swinhoei. Proceedings of the National Academy of Sciences* [USA] 101: 222- 16,227.

Purushothaman, A. (1998). Microbial diversity. In *Proceedings of the Technical Workshop on Biodiversity of Gulf of Mannar Marine Biosphere Reserve*, M. S. Swaminathan Research Foundation, Chennai, pp. 86-91.

Rathna Kala, R. and Chandrika, V. (1995). Microbial production of antibiotics from mangrove ecosystem. *CMFRI Spl. Publ.*, 61:117-122.

Richman, S., Loya, Y. and Slobodkin, L. B. (1975). The rate of mucus production by corals and its assimilation by the coral reef copepod *Acartia negligens. Limnol. Oceanogr.* 20:918-923.

Rosli, MD.I, Wei, O. S., Kamal, A. and Aizan, W. (2001). Isolation and characterization of halotolerant aerobic bacteria from oil reservoir. *Jurnal Teknologi*, 35:1-10.

Sherr, E. B. and Sherr, B. F. (1996). Temporal offset in oceanic production and respiration process implied by seasonal changes in atmospheric oxygen: The role of heterotrophic microbes. *Aquat. Microb. Ecol.* 1:91-100.

Sørum, H. (2002). *Antibiotics in aquaculture - a dead end.* Eau-water-acqua-wasser, 61st Annual Meeting of the Swiss Society for Microbiology. Luzern, Pp.29.

Sridhar, K. R. (2004). Mangrove fungi in India. *Curr. Sci.*, 86:1586-1587.

Sweijd, N. A., Bowie, R. C. K., Evans, B. S. and Lopata, A. L (2004). Molecular genetics and the management and conservation of marine organisms. *Hydrobiologia*, 420:153-164.

Wagner-Dobler, I., Beil, W., Lang, S., Meiners, M. and Laatsch, H. (2002). Integrated approach to explore the potential of marine microorganisms for the production of bioactive metabolites. *Adv. Biochem. Eng. Biotechnol.* 74:207-238.

Weinbauer M.G. and Wenderoth D.F. (2002). Microbial Diversity and Ecosystem Functions - the Unmined Riches EJB Electronic Journal of Biotechnology ISSN: 0717-3458, *Biotechnology issues for developing countries by Universidad Católica de Valparaíso – Chile*

Whitman, W. B., Coleman, D. C. and Wiebe, W. J. (1998) Prokaryotes: the unseen majority. *Proc. Natl. Acad. Sci.* USA, 95:6578–6583.

Wilson, E. (1992). *The diversity of life*. New York, W.W. Norton and Company, pp 424

Zehr, J.P., Waterbury, J.B., Turner, P.J., Montoya, J.P., Omoregie, E., Steward, G.F., Hansen, A. and Karl, D.M. (2001).Unicellular cyanobacteria fix N2 in the subtropical North Pacific Ocean. *Nature*, 412:635-8.

Zobell, C. E. (1946). *Marine Microbiology*, Chronica Botanica Co, Waltham, Mass., USA, pp. 240.

□□□

CHAPTER **9**

The Importance of Extracellular Metal Immobilization for *Rhizobium* Persistence in Heavy Metal Contaminated Soils

Sofia Isabel Almeida Pereira and Etelvina Maria de Almeida Paula Figueira*

Centro de Biologia Celular, Departamento de Biologia, Campus de Santiago, Universidade de Aveiro, 3810-193 Aveiro, Portugal

ABSTRACT

Industrial and agricultural activities and the disposal of waste products have resulted in the contamination of many terrestrial environments. Industrial effluents often contain considerable amount of potentially toxic compounds such as Cd, As, Pb and Hg. Due to their small size, which provides a large contact area that can interact with the surrounding environment, microorganisms are the first biota showing the impact of toxic compounds. Microorganisms being in intimate contact with the soil environment are considered to be the best indicators of soil pollution. In general, they are very sensitive to low concentrations of contaminants and provide a rapid response to soil perturbation. *Rhizobium* spp. are ubiquitous gram-negative soil bacteria that have a profound scientific and agronomic significance due to their ability to establish nitrogen-fixing symbiosis with legumes, which is of major importance to the maintenance of soil fertility and to feed livestock and populations in many developing countries. Therefore, the present work aimed to evaluate the impact of heavy metals on *Rhizobium* populations isolated from different locations contaminated with heavy metals. In order to reach this goal some physicochemical parameters that influence metal bioavailability were determined, as well as their concentrations in soil. Cadmium tolerance of *Rhizobium* isolates was screened in artificial media and the highest cadmium concentration allowing growth was defined as the maximum resistance level (MRL). The heavy metal tolerance variability found among *Rhizobium* isolates, suggests that beyond basic metabolic machinery also exist variations that allow isolates to display distinct tolerance responses to Cd stress. To better understand *Rhizobium* cadmium tolerance, wall-bound and intracellular Cd was determined. Sensitive and moderately tolerant isolates showed low intracellular Cd amounts

*Email: efigueira@bio.ua.pt

when compared to extremely tolerant isolates, which were able to withstand intracellular Cd concentrations 40 times higher. Lipopolysaccharides (LPS) bind cations unspecifically, so they constitute the initial defence mechanism to heavy metals since they sequestrate cations and therefore prevent their entry into the cell. LPS quantity varied significantly between isolates. LPS increases were observed in most sensitive, moderately tolerant and tolerant isolates, evidencing an attempt of cells to adjust to cadmium entry. These results may be of great importance for understanding how *Rhizobium* persists in contaminated soils. They also provide important data to the development of bioremediation processes and to the improvement of soil productivity.

Key words: cadmium, lipopolysaccharides, soil contamination, tolerance, *Rhizobium*

1. INTRODUCTION

Industrialization has long been accepted as a hallmark of civilization. However, the fact remains that many agricultural and industrial practices have been adversely affecting the environment, introducing several toxic compounds, such as heavy metals (Malik, 2004; Trajanovska *et al.*, 1997). In the last decades an increase of heavy metal contamination in water and soils, has been observed (Alloway, 1995a). These elements are ubiquitous and persistent pollutants that are introduced into the environment through anthropogenic activities, such as smelters, mining, power station industry and the application of metal-containing pesticides, fertilizers and sludges (Alloway, 1995a; Carrasco *et al.*, 2005; Giller *et al.*, 1989; Malik, 2004; McGrath *et al.*, 1995; Robinson *et al.* 2001). Agricultural soils often present deficiencies in nutrients that affect plant growth and development, being necessary the addition of fertilizers and sludges (Alloway, 1995a). According to Alloway (1995c), phosphatic fertilizers are widely regarded as being the most ubiquitous source of agricultural soils contamination with Cd (< 500 mg/kg), since relatively high concentrations of this metal can be found in phosphorites used for the manufacture of these fertilizers. Sewage sludge application to land is also very common, resulting in an improvement of the soil physical characteristics (Abd-Alla *et al.*, 1999; Giller *et al.*, 1989; Obbard, 2001). Furthermore, sludges are a rich source of inorganic nutrients, such as N, P, Ca and Mg (Chander and Brookes, 1993). However, there has been concern over the use of sludge contaminated with heavy metals and other potential toxic compounds from both industrial and domestic sources (Horswell *et al.*, 2003; Obbard, 2001; Purchase and Miles, 2001), since they can contribute to increase soil contamination.

The soil is a dynamic system that is subjected to short-term fluctuations, such as variations in moisture, pH and redox conditions. These changes in soil properties affect the form and bioavailability of metals. So, the amounts of metals accumulated in soils are dependent on the emission levels, the transport of the metal from the source to the accumulation site and the retention of the metal once it has reached the soil (Alloway, 1995b).

Heavy metal contamination has been subject of increasing concern to the scientific community because they become irreversibly immobilised in soil components being a cause of toxicity to microorganisms, plants, animals and in particular to humans (McGrath and Lane, 1989).

Since microorganisms are the most abundant and most genetically diverse living organisms adapted to almost all environments that exist on the Earth, they have been object of study in numerous works on environmental stress.

Several studies have shown that metals adversely influence microorganisms (Shi *et al.*, 2002), affecting their growth, morphology and activities (Baath *et al.*, 1998; Lakzian *et al.*, 2002), including symbiotic N_2 fixation (Chaudhary *et al.*, 2004). It is well known that metals are direct and/ or indirectly involved in all aspects of microbial growth, metabolism and differentiation. Some metals such as Zn, Cu, Ni and Cr are essential or beneficial micronutrients for plants, animals and microorganisms, whereas others such as Cd, Hg and Pb have no known biological and/or physiological functions being toxic even at very low concentrations (Gadd, 1992).

Bacterial exposure to environmental stresses results in a high number of physiological changes (Saxena *et al.*, 1996). Some metals are indispensable for enzymatic functions and bacterial growth, so uptake mechanisms exist that allow for the entrance of these ions into the cell (Nies, 1999). When organisms face high metal concentrations these ions are accumulated by unspecific systems, constitutively expressed (Nies and Silver, 1995).

Metals can generally exert an inhibitory action on microorganisms by blocking essential functional groups, displacing essential metal ions or modifying the active conformations of proteins and nucleic acids (Doelman *et al.*, 1994). However, the ability to grow at high metal concentration it is found in many microorganisms and may be the result of intrinsic or induced mechanisms (Pereira *et al*, 2006). Some investigations have indicated that bacteria, which survive and indeed flourish in such environments, have developed or acquired systems that counteract the effects of high metal ions concentrations (Bruin *et al.*, 2000). These mechanisms are diverse and may involve energy-dependent efflux of metals (Ji and Silver, 1995; Nies, 1999; Nies and Silver, 1995), precipitation as insoluble salts (Blake *et al.*, 1993), alteration in membrane permeability (Bruins et *al.*, 2000), metal immobilization in the cell wall (Beveridge and Murray, 1976, 1980; Mullen *et al.*, 1989; Scott and Palmer, 1990), production of chelating agents (Figueira *et al.*, 2005; Lima *et al.*, 2006; Silver and Phung, 1996) or biochemical transformation of metal ions (Bruins *et al.*, 2000).

The cell wall of bacteria represents the first contact between the cell and the external environment. Bacteria synthesise a wide variety of complex molecules on their surfaces, which contain charged chemical groups that usually give the cell surface a net anionic charge density (Langley and Beveridge, 1999). Since the cell surface is in direct contact with the environment, the charged groups within the surface are able to interact with ions. As a result, metals can become electrostatically bound to the cell surface (Beveridge and Murray, 1976; 1980; Mullen *et al.*, 1989). Lipopolysaccharides comprise a family of complex macromolecules functioning as integral membrane components in the cell envelope of Gram-negative bacteria (Sutherland *et al.*, 1985). These molecules are relevant both for outer membrane stability and for the interaction of the bacteria with the environment, due to their ability to bind different cations including heavy metals (Valverde *et al.*, 1997). Different genotypes often display differences in LPS, hence obtaining a natural advantage to cope with the metal presence by retaining cations extracellularly. These differences could be an important factor for bacterial survival in metal contaminated sites.

The soil microbial ecosystem is functionally complex. It contains key groups of microorganisms, which have an integral role in maintaining soil fertility, via general decomposition activity or via symbiotic relationships (Obbard *et al.* 1993). The most important symbiotic relationship occurring in agriculture is that between the genus *Rhizobium* and their leguminous host plants (Obbard *et al.*, 1993; Somasegaran and Hoben, 1994). *Rhizobium* are ubiquitous bacteria that have a profound agronomic importance, since the symbiotic association between legumes and rhizobia is by far the most important contributor to the world's supply of biologically fixed N_2. Rhizobial symbioses with more than 100 agriculturally important legumes contribute nearly half the annual quantity of biological nitrogen fixation entering soil ecosystems (Somasegaran and Hoben, 1994). For this reason and taking into consideration the importance of legumes in animal and human consumption, some attention has been given to the effects that these elements exert on *Rhizobium* (Ibekwe *et al.*, 1995). Several authors reported that leguminous species grown on contaminated soils, exhibited reduced yields and N content (Chaudhary *et al.*, 2004), since rhizobia isolated from host plants root nodules were completely ineffective to fix N_2 (Giller *et al.*, 1989; Hernandez *et al.*, 2002). In others studies (Figueira, 2000; Pereira et al., 2006) *Rhizobium* has shown to be a genus very sensitive to environmental stresses, being a good indicator of soil contamination.

Metals exert a selective pressure on organisms, resulting in microbial populations with higher tolerance, but with lower diversity, when compared to unpolluted neighbouring areas (Baath, 1992; Baath *et al.*, 1998). The soil microbial population presents variability, which in a non-stressed environment would not bring any advantage. However, when the conditions are changed and this population is submitted to a selective pressure, the genotypes presenting tolerance ability can survive, hence yielding a tolerant population, with lower genetic variability. The phenomenon of bacterial resistance is of fundamental importance and is particularly relevant to microbial ecology, especially in connection with the roles of microbes in polluted ecosystems and in the reclamation of metal-contaminated habitats (Gadd and Griffiths, 1978; McGrath *et al.*, 1995).

Most studies on stress tolerance, particularly those concerning heavy metals, usually focus on the more tolerant genotypes, as an attempt to understand the reason behind their higher tolerance, hence neglecting the more sensitive genotypes. Nevertheless, the analysis of strains with different degrees of tolerance can be very useful for pointing out the specific mechanisms underlying metal coping, as well as their relative importance for the different degrees of resistance and persistence observed in a given population. The analysis of sensitive genotypes is of extreme importance since their disappearance depletes the population's variability, reducing its ability to persist if other types of stress are imposed. Under this context, in this chapter we will proceed to the analysis of different strains, ranging from sensitive to extremely tolerant strains to evaluate the effects that increasing levels of heavy metals imposed on indigenous *Rhizobium* populations, isolated from soils subjected to different anthropogenic activities, such as mining, metalomechanics industry and sludge application. A relationship between cadmium tolerance levels and the natural conditions experienced by the populations in their place of origin may be established. The determination of cadmium distribution in the cell can reveal the relative importance of intracellular and extracellular mechanisms that help to cope with metal stress. Since plasmids code for genes that are important for bacterial adaptation and survival to environmental constraints, plasmid profile analysis can be a basic genetic tool to elucidate if different populations have the same or different genetic features to cope with heavy metal stress. One of the characteristics encoded by plasmids that can interfere with heavy metal tolerance is the LPS production. LPS quantity analysis can provide evidences about the reason behind different degrees of tolerances observed between isolates, and how these biomolecules can act as a barrier to the entry of metals into the cell.

2. CADMIUM TOLERANCE OF *RHIZOBIUM* ISOLATED FROM HEAVY METAL CONTAMINATED AREAS

Seven composite soils were collected from arable field in the centre of Portugal (Aveiro region). A non-heavy metal contaminated agricultural soil was used as control (C). The other soils have been exposed to different sources of heavy metals: MIA (metalomechanics industry A) and MIB (metalomechanics industry B) soils were collected from an area in close proximity (1 and 2 Km respectively) to metalomechanics industries that have been in operation for nearly 30 years; MA and MB soils were placed in a river margin directly influenced by lead mining for 100 years, however, this activity ceased 30 years ago; SA and SB soils have suffered the application of sewage sludge for 12 and 4 years, respectively.

Soil is a dynamic, multi-component system, whose properties are continually modified by microbial, chemical, geological and anthropogenic processes (Saeki *et al.*, 2002). Some environmental factors such as pH and organic matter can influence metal bioavailability. In addition, soil pH value and organic matter content *per se* may influence the presence of rhizobia in soil (Smith, 1997). According to Ibekwe *et al.* (1997, 1998), soil pH has a significant effect on rhizobia survival. These authors reported that in the more acid soils clover nodules were ineffective for N_2 fixation irrespective of soil metal content. Under lower pH conditions, rhizobia numbers and nodule number *per* plant were both significantly reduced. The pH values reported in Table 1 varied significantly ($P< 0.05$), being in most soils below the optimal pH (6.8) for rhizobia growth. MA and MB soils were quite acid (4.4 and 4.9 respectively), while the other soils have pH values around 6.0.

Total heavy metal concentrations (Zn, Pb, Co, Cd, Ni and Cr) varied significantly between locations ($P< 0.05$) as shown in Figure 1. Generally, contaminated soils presented higher concentrations of metals than the control soil (C), with the exception of SA and SB soils, where metal concentrations were similar or lower than control. MIA and MIB soils showed the highest metal concentrations, being therefore the most polluted soils, especially for zinc whose levels (464,97 and 299,49 mg kg^{-1}, respectively) were near or exceed EC limits (Commission European Communities, 1986). All soils presented Cd concentrations below the EC limits (Fig.1). Despite its concentration in contaminated soils is generally low, cadmium is a well-studied element due to be one of the most toxic heavy metals that accumulates in agricultural soils and biomagnifies in food chain.

Eighty-five isolates of *R. leguminosarum* were obtained from contaminated soils (MIA, MIB, MA, MB, SA and SB) and a non-contaminated soil (C).

According to Hirsch *et al.* (1993) and Ibekwe *et al.* (1997) the distribution of *Rhizobium* strains in agricultural soils is affected both by current and previous cultivation: the presence of the host plant generally leads to a population increase which persists for some years. Rhizobia were isolated from soils where no host plants were grown on the last five years, which lead us to conclude that free-living was more vulnerable to the direct influence of heavy metal contamination.

TABLE.1. Physicochemical analysis of soils from different locations. C - control isolates; MIA and MIB, metalomechanics industry isolates; MA and MB, mining isolates; SA and SB, sewage sludge isolates. Data are the mean ± SE from three replicate measurements. Different superscripts letters (a-f) represent significant differences ($P < 0.05$) between values at same column.

Soil	Physicochemical analysis			
	Water (%)	**pH**	**Redoxpotential (Eh)**	**Organic matter (%)**
C	19.1±0.67^{a}	5.5±0.07^{a}	376.3±42.25^{a,b}	5.87±0.06^{a}
MIA	21.3±0.90^{a,b}	6.1±0.08^{b,c}	376.3±28.75^{a,b}	6.24±0.04^{a,b,c}
MIB	25.1±0.19^{b,c}	5.8±0.06^{a,b}	392.7±3.77^{a,b,c}	8.39±0.25^{d,e}
MA	27.3±0.09^{d}	4.4±0.52^{d}	471.0±9.17^{c}	11.66±0.35^{f}
MB	22.3±0.80^{a}	4.9±0.62^{a,d}	444.7±44.46^{b,c}	9.44±0.80
SA	27.3±0.61^{c,d}	6.6±0.12^{c}	404.0±9.54^{a,b}	7.44±0.80^{c,d}
SB	18.8±0.67^{a}	5.7±0.12^{a,b}	349.3±11.15^{a}	5.73±0.24^{a,b}

In other reports (Chaudri *et al.*, 1993), although the majority of metal concentrations determined in soils were below the EC limits, they were enough to eliminate *Rhizobium* from soils. Nevertheless, in this work, we were able to isolate rhizobia from all soils, indicating that these bacteria were capable to survive under the metal concentrations determined. Isolates survival can be related to the existence of resistance mechanisms that bring advantage in hazardous environments. However, according to Giller *et al.* (1998) and Ibekwe *et al.* (1997) survival can also be related to the physical protection of clay minerals and organic matter or to the existence of microsites where metal contamination may be minimal. These "niches" may harbour rhizobia that are not resistant to heavy metals and may explain the presence of sensitive genotypes in contaminated soils.

Based on the highest heavy metal concentration allowing growth screening of *Rhizobium* cadmium tolerance originated four groups: sensitive (MRL ≤ 125 μM $CdCl_2$), moderately tolerant (125 ≤ MRL ≤ 210 □μM $CdCl_2$), tolerant ($250 \leq$ MRL ≤ 500 μM $CdCl_2$) and extremely tolerant (MRL ≥ 750 μM $CdCl_2$).

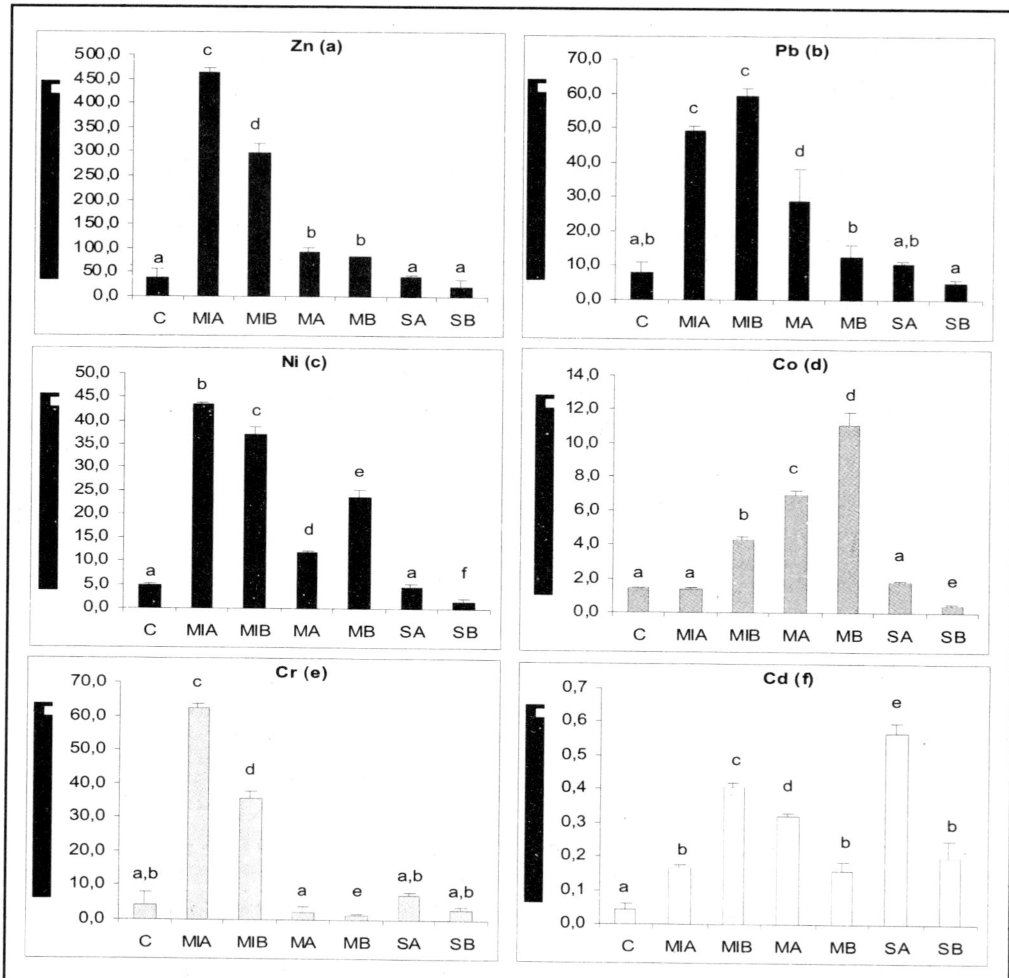

FIG. 1. Total heavy metal concentration (Zn, Pb, Ni, Co, Cr and Cd) in the different soils. Data are mean ± SE of three replicate measurements. Different letters indicate means that are significantly different ($P < 0.05$) from each other.

Considering the metal contamination levels in the different soils and the isolates ability to cope with Cd stress, it seems evident that the degree of exposure to heavy metals, i.e., the time and concentration of exposure, are the major modelling factors of metal resistance in the rhizobial populations. The most contaminated soils (MIA, MIB and MA) presented heavy metal concentrations enough high to originate a selective pressure leading to the appearance of genotypes with high cadmium tolerance (Fig. 2 (A-C)), fact already described by Baath (1992) and Baath *et al.* (1998), who considered the possibility of heavy metals to exert a selective pressure in bacteria of contaminated soils, resulting in the presence of more tolerant organisms in these locations.

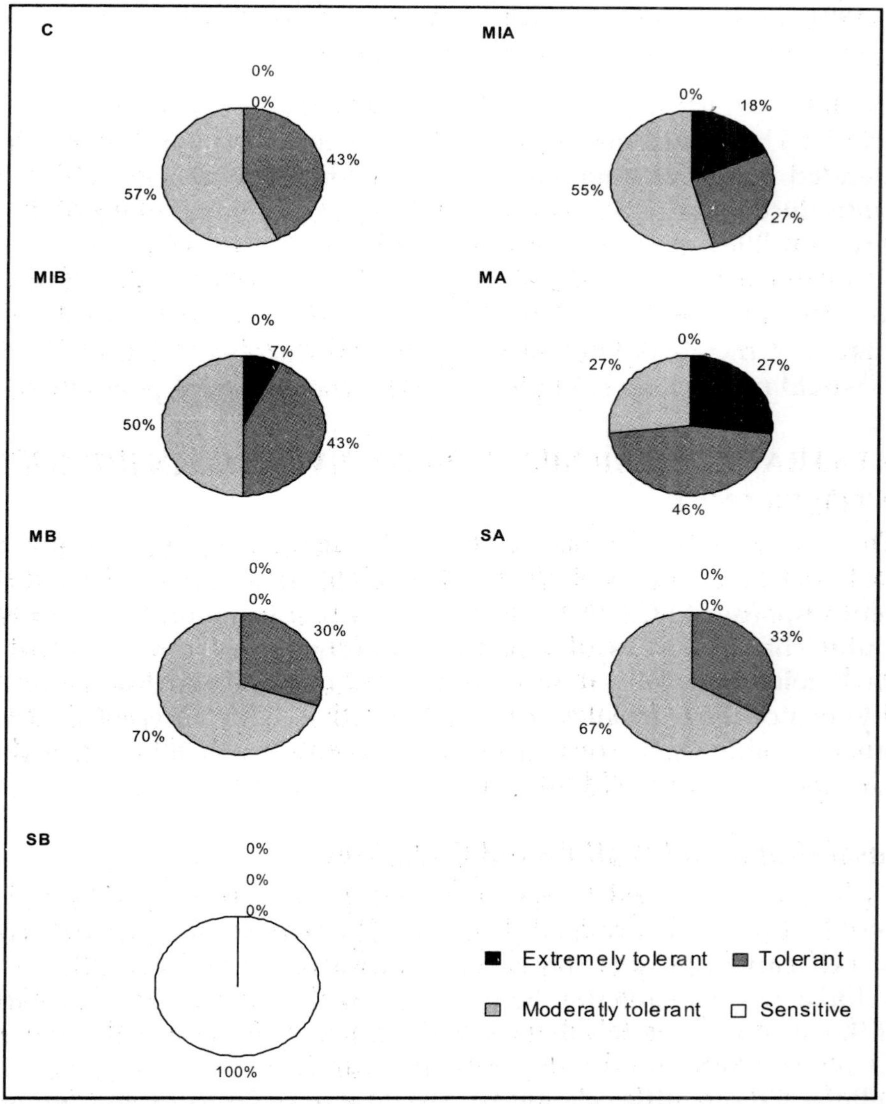

FIG. 2. Percentage of extremely tolerant, tolerant, moderately tolerant and sensitive *Rhizobium* strains isolated from each location.

Interestingly, results showed that *Rhizobium* populations with higher resistance to Cd are not related to soils with higher cadmium contaminations, but rather to soils presenting the higher overall total heavy metal contamination. This fact suggests that the selective pressure induced by these metals allowed the development of resistance mechanisms that bestow tolerance to different metal stresses, which is corroborated by Díaz-Raviña *et al.* (1994) and Nies (1992), who reported the existence of heavy metal multiple tolerance patterns in soil microorganisms.

In the agricultural soil C, the anthropogenic pressures mainly characterized by animal manures and chemical fertilizers applications during an extended period of time has yielded a cumulative toxicity which, with time, introduced a selective pressure in the microbial populations of the soil, being responsible for a more tolerant population that could be expected. On the other hand, in the SB soil *Rhizobium* population consisted mainly of sensitive isolates. This fact can be explained by the short period between the single application of metal-contaminated sludge and rhizobia isolation (4 years), which would not be long enough to evolve a more tolerant population.

3. AN EXTRACELLULAR MECHANISM INVOLVED IN *RHIZOBIUM* CD TOLERANCE

The variability of tolerances observed suggests that behind the basic metabolic machinery exists biochemical variability that allows isolates to show different responses to Cd stress. For this reason, it is important to study the molecular characteristics of sensitive, moderately tolerant, tolerant and extremely tolerant isolates in order to identify mechanisms that can provide metal tolerance. Four isolates from each location, with different degrees of tolerances (Table 2), were studied for screening possible extracellular mechanisms involved in Cd tolerance.

3.1 Intracellular and Wall Bound Cadmium

To better understand *Rhizobium* cadmium resistance, wall-bound and intracellular Cd was determined (Figure 3). The analysis of metal distribution in the cell can highlight the relative importance of intracellular and extracellular mechanisms that help to cope with metal stress. Cadmium is one of the most toxic metals that cause detrimental effects in cell metabolism and therefore injure the development and survival of microorganisms. So, intracellular Cd quantities should be very low in order to protect important metabolic pathways from the detrimental effects of this metal. The results presented in Figure 3 show that sensitive, moderately tolerant and tolerant

TABLE 2. Classification of *Rhizobium leguminosarum* isolates according maximum resistance level (MRL). MRL correspond to the maximum Cd concentration (μM) tolerated. Sensitive isolates (MRL ≤ 125 μM), moderately tolerant isolates (165 ≤MRL ≤210 μM), tolerant isolates (250 ≤MRL ≤500 μM), extremely tolerant isolates (MRL ≥750 μM).

Isolates	MRL	Classification
SB 3-1	65	Sensitive
SB 13-1	65	
SB 7-2	125	
SB 12-2	125	
C 2-6	210	Moderately Tolerant
C 2-8	210	
MIA 2-1	210	
MIB 10-1	210	
MB 1-2	210	
MB 7-1	210	
SA 5-2	210	
SA 10-1	210	
MIA 3-2	250	Tolerant
MIA 8-1	250	
MIB 12-2	250	
MA 18-1	250	
SA 2-2	250	
MIB 9-2	500	
MB 7-2	500	
MIB 15-1	750	Extremely Tolerant
MA 11-1	750	
MA 2-2	1000	
MIA 9-1	2000	

isolates evidenced wall-bound Cd quantities 100 fold higher than Cd intracellular levels, while in extremely tolerant isolates wall-bound Cd was only 15 fold higher than intracellular levels. This leads to conclude that cadmium entry into the cell was immediately driven out, supporting the putative existence of extracellular sequestration or exclusion mechanisms. In sensitive isolates intracellular Cd amounts were near 0.2 ng 10^7cells^{-1}, while extremely tolerant isolates were able to withstand intracellular Cd amounts 40 times higher (8 ng 10^7 cells^{-1}), indicating that in extremely tolerant isolates Cd resistance must be dependent on intracellular mechanisms. These findings are in agreement with Figueira *et al.* (2005), Purchase *et al.* (1997) and Purchase and Miles (2001), who demonstrated that *Rhizobium* tolerant isolates accumulated intracellularly more Cd than sensitive ones. However, in sensitive and moderately tolerant isolates extracellular sequestration seems to be the major mechanism allowing their survival in these environments. Thus, in less tolerant *Rhizobium* populations persistence in heavy metal contaminated soils seems to be supported, fundamentally, by the maintenance of low intracellular metal concentrations, which is dependent of the ability to accumulate these ions in the cell wall.

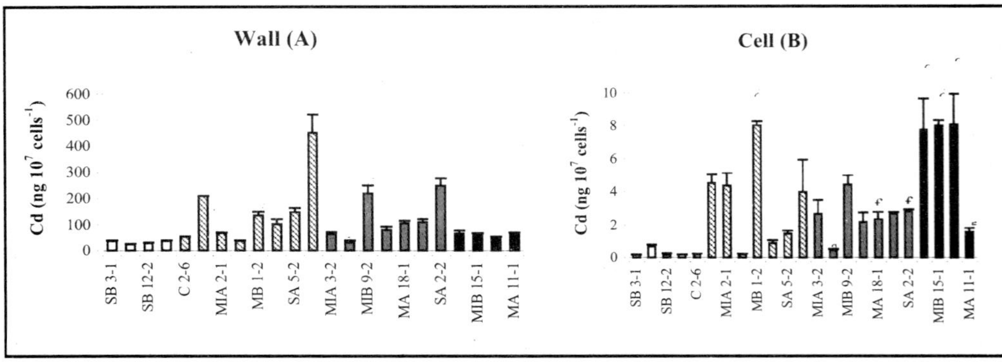

FIG. 3. Wall (A) and cell (B) cadmium concentrations (ng 10^7cells^{-1}) of *Rhizobium leguminosarum* isolates grown on YEM plates supplemented with different $CdCl_2$ concentrations: white - sensitive (65 μM); stripes - moderately tolerant (125 μM); grey - tolerant (210 μM); black - extremely tolerant (250 μM). Data are expressed as the mean ± SE from four replicate experiments. Different letters indicate means that are significantly different ($P< 0.05$) from each other. *In*: Pereira *et al*. (2006).

3.2 Possible Involvement of Plasmids in Resistance

Although chromosomal genes are important to bacterial adaptation and survival to environmental constraints, resistance genes are mainly located in plasmids. Their analysis can be a genetic tool to understand the way that bacteria tolerate stress.

Our results demonstrate striking differences on plasmid profiles of different *Rhizobium* isolates; in addition to variation in the number and in the molecular weight were observed. Lakzian *et al*. (2002) showed that *Rhizobium* isolates from contaminated soils, with 6-9 plasmids, were more tolerant than isolates from the same locations with only 3 plasmids, and they proposed that metal tolerance is related to the number of plasmids. In our study this relationship was not found. The maximum number of plasmids was four and they were distributed regularly by tolerant, moderately tolerant and sensitive isolates. Most isolates tolerating between 250 and 500 μM of $CdCl_2$ displayed two plasmids with molecular weights of 485 and 415 kb, which may code for cadmium tolerance. Plasmid-mediated metal resistance may be ecologically important since resistance can be rapidly transferred from resistant to sensitive bacteria (Lawlor *et al.*, 1999). Weaver and Holt (1990) showed that some rhizobial plasmids influenced survival under stress conditions. These results opened new perspectives for the improvement of the *Rhizobium*-legume symbiosis.

3.3 Lipopolysaccharides Analysis

LPS quantity varied significantly between isolates (Figure 4). This may be related with genetic variability of *Rhizobium* populations. Hynes and McGregor (1990) referred that genes carried on plasmids codify for the synthesis of lipopolysaccharides. Therefore, the absence of some plasmids can be enough to lower the amount of LPS in some isolates. However, no relation was observed between LPS levels and the presence of certain plasmids in rhizobia Portuguese isolates.

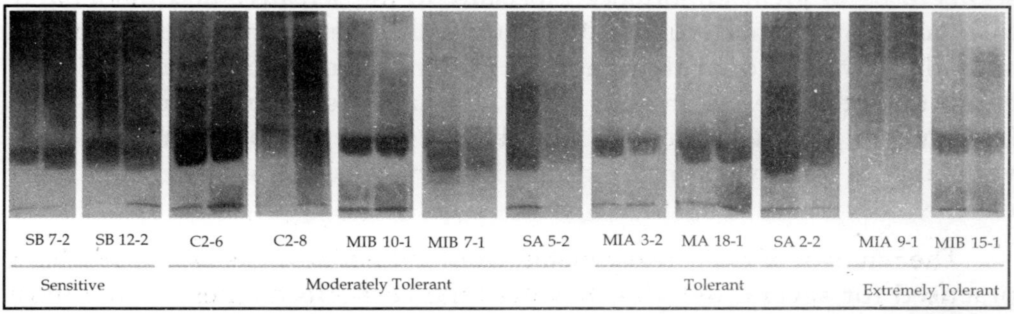

FIG. 4. LPS profiles of some isolates of *Rhizobium leguminosarum* with different cadmium tolerances. For each isolate are shown 2 lanes, the left lane corresponds to the LPS profile in the absence of cadmium, and the right lane to the LPS profile under cadmium stress. *In*: Pereira *et al.* (2006).

LPS bind cations unspecifically (Valverde *et al.*, 1997), so they constitute the initial defence mechanism to heavy metals since they sequestrate cations and therefore prevent their entry into the cell. LPS increases were observed in most of sensitive, moderately tolerant and tolerant isolates, evidencing an attempt of cells to adjust to the new conditions, i.e., to protect the cell against cadmium entry into the cell. This is corroborated by the fact that in moderately tolerant and tolerant isolates wall-bound Cd is near 100 fold higher than intracellular Cd. Thus, LPS seem to be an efficient barrier to the entry of Cd in the cell. Extremely tolerant isolates did not present an increase of LPS when grown under cadmium stress, which may explain why wall-bound cadmium values were lower than in most of moderately tolerant and tolerant isolates in spite of grown under higher Cd concentrations. These extremely tolerant genotypes may comprise other resistance mechanisms more efficient to cope with Cd stress.

4. CONCLUSIONS AND FUTURE PERSPECTIVES

The pressure that soil microflora experience under the direct effect of heavy metals is evident. Even though *Rhizobium* isolates had different origins and had been exposed to different sources of heavy metal contamination for long periods of time, resistance mechanisms do not seem to be influenced by location.

It was observed that there is a tolerance mechanism common to all isolates, which is characterized by the presence and/or induction of LPS quantity that sequestrate most of the metal extracellularly, acting as a first defence barrier against heavy metal stress. This extracellular immobilization of metals seems to be of great importance especially to the sensitive and moderately tolerant isolates, since it allows their survival in heavy metal contaminated soils. It is well known that bacteria undergo the influence of different environmental stresses in soils. Thus, the persistence of these isolates is fundamental for the ecological equilibrium, since a population with high variability of genotypes may increase its survival when experiencing other stresses.

The removal of toxic heavy metals from industrial waste waters has been practised for several decades, however the cost-effectiveness of the most common physico-chemical processes is limited. Strong and contaminating reagents are used for desorption, resulting in toxic sludge and secondary environmental pollution. Bioremediation seems to be a good solution for this problem, since several reports (Malik, 2004) show the efficiency of microorganisms in the recovery of contaminated soils.

The results showed that *Rhizobium* populations are well correlated with soil heavy metal contamination, and so may be used for the evaluation of agricultural soils pollution and for the establishment of new bioremediation procedures, which may be used in the improvement of soil characteristics and on the reclaim of contaminated soils for crop production. Furthermore, the knowledge about the effects of contaminants in the soils may provide important data to alert not only the public opinion to the detrimental and recalcitrant effects of heavy metals on soils, but also to alert governmental, municipal and industrial entities in order to improve or develop new treatments for the decontamination of polluted areas.

Acknowledgments

The authors wish to thank to Dr. Robert Grace for his revision of the English language. This work was supported by a grant from the Centre for Cell Biology.

REFERENCES

Abd-Alla, M.H., Yan, F. and Schubert, S. (1999). Effects of sewage sludge application on nodulation, nitrogen fixation, and plant growth of Faba, Bean, Soybean, and Lupin. *J. Appl. Bot.* 73: 69-75.

Alloway, B.J. (1995a). Introduction. In: Alloway, B.J. (eds). Heavy Metals in Soils. Blackie Academic and Professional. New York, pp. 3-9.

Alloway, B.J. (1995b). Soil processes and the behaviour of metals. In: Alloway, B.J. (ed). Heavy Metals in Soils. Blackie Academic and Professional. New York, pp. 11-37.

Alloway, B.J. (1995c). Cadmium. In: Alloway, B.J. (ed). Heavy Metals in Soils. Blackie Academic and Professional. New York, pp. 122-147.

Baath, E. (1992). Measurement of heavy metal tolerance of soil bacteria using thymidine incorporation into bacteria extracted after homogenization-centrifugation. *Soil Biol. Biochem.* 24: 1167-1172.

Baath, E., Díaz-Ravina, M., Frostegard, A. and Campbell, C.D. (1998). Effect of metal-rich sludge amendments on the soil microbial community. *Appl. Environ. Microbiol.* 64: 238-245.

Beveridge, T.J. and Murray, R.G.E. (1976). Uptake and retention of metals by cell walls of *Bacillus subtilis*. *J. Bacteriol.* 127: 1502-1518.

Beveridge, T.J. and Murray, R.G.E. (1980). Sites of metal deposition in the cell wall of *Bacillus subtilis*. *J. Bacteriol.* 141: 876-887.

Blake, R.C., Choate, D.M., Bardhan, S., Revis, N., Barton, L.L. and Zocco, T.G. (1993). Chemical transformation of toxic metals by a *Pseudomonas* strain from a toxic waste site. *Environ. Toxicol. Chem.* 12: 1365-1376.

Bruins, M.R., Kapil, S. and Oehme, F.W. (2000). Microbial resistance to metals in the environment. *Ecotoxicol. Environ. Safe.* 45: 198-207.

Carrasco, J.A., Armario, P., Pajuelo, E., Burgos, A. Caviedes, M.A., López, R., Chamber, M. A. and Palomares, A.J. (2005). Isolation and characterisation of symbiotically effective *Rhizobium* resistant to arsenic and heavy metals after the toxic spill at the Aznalcóllar pyrete mine. *Soil Biol. Biochem.* 37: 1131-1140.

Chander, K. and Brookes, P.C. (1993). Residual effects of zinc, copper and nickel in sewage sludge on microbial biomass in a sandy loam. *Soil Biol. Biochem.* 25: 1231-1239.

Chaudhary, P, Dudeja, S.S. and Kappor, K.K. (2004). Effectivity of host *Rhizobium leguminosarum* symbiosis in soils receiving sewage water containing heavy metals. *Microbiol. Res.* 159: 121-127.

Chaudri, A.M., McGrath, S.P., Giller, K.E., Rietz, E. and Sauerbeck, D.R. (1993). Enumeration of indigenous *Rhizobium leguminosarum* biovar *trifolii* in soils previously treated with metal-contaminated sewage sludge. *Soil Biol. Biochem.* 25: 301-309.

Commission of the European Communities. (1986). Council Directive on the protection of the environment, and in particular of the soil, when sewage sludge is used in agriculture. *Off. J. Eur. Com.* L181. Annex 1A, 10.

Díaz-Raviña, M., Baath, M. and Frostegård, A. (1994). Multiple heavy metal tolerance of soil bacterial communities and its measurement by thymidine incorporation technique. *Appl. Environ. Microbiol.* 60: 2238-2247.

Doelman, P., Jansen, E., Michaels, M. and van Til, M. (1994). Effects of heavy metals in soil on microbial diversity and activity as shown by the sensivity-resistance index, an ecologically relevant parameter. *Biol. Fertil. Soil.* 17: 177-184.

Figueira, E.M.A.P. (2000). Aspectos da tolerância salina em *Pisum sativum* L.: influência da nutrição azotada. PhD Thesis. Universidade de Aveiro.

Figueira, E.M.A.P., Lima, A.I.G. and Pereira, S.I.A. (2005). Monitoring glutathione levels as a marker for cadmium stress in *Rhizobium leguminosarum* biovar *viciae*. *Can. J. Microbiol.* 51: 7-14.

Gadd, G.M. (1992). Metals and microorganisms: A problem of definition. *FEMS Microbiol. Lett.* 100: 197-204.

Gadd, G.M. and Griffiths, A.J. (1978). Microorganisms and heavy metal toxicity. *Microb. Ecol.* 4: 303-317.

Giller, K.E., McGrath, S.P. and Hirsch, P.R. (1989). Absence of nitrogen fixation in clover grown on soil subject to long-term contamination with heavy metals is due to survival of only ineffective *Rhizobium*. *Soil Biol. Biochem.* 21: 841-848.

Giller, K.E., Witter, E. and McGrath, S.P. (1998). Toxicity of heavy metals to microorganisms and microbial process in agricultural soils: a review. *Soil Biol. Biochem.* 30: 1389-1414.

Hernandez, T., Moral, R., Perez-Espnosa, A., Moreno-Caselles, J., Perez-Murcia, M.D. and Garcia, C. (2002). Nitrogen mineralization potential in calcerous soils amended with sewage sludge. *Bioresour. Technol.* 83: 213-219.

Hirsch, P.R., Jones, M.J., McGrath, S.P. and Giller, K.E. (1993). Heavy metals from past applications of sewage sludge decrease the genetic diversity of *Rhizobium leguminosarum* biovar. *trifolii* populations. *Soil Biol. Biochem.* 25: 1485-1490.

Horswell, J., Speir, T.W. and van Schaik, A.P. (2003). Bio-indicators to assess impacts of heavy metals in land-applied sewage sludge. *Soil Biol. Biochem.* 35: 1501-1505.

Hynes, M.F. and McGregor, N.F. (1990). Two plasmids other than the nodulation plasmid are necessary for formation of nitrogen-fixing nodules by *Rhizobium leguminosarum*. *Mol. Microbiol.* 4: 567-574.

Ibekwe, A.M., Angle, J.S., Chaney, R.L. and van Berkum, P. (1995). Sewage sludge and heavy metal effects on nodulation and nitrogen fixation of legumes. *J. Environ. Qual.* 24: 1199-1204.

Ibekwe, A.M., Angle, J.S., Chaney, R.L. and van Berkum, P. (1997). Differentiation of clover *Rhizobium* isolated from biosolids-amended soils with varying pH. *Soil Sci. Soc. Am. J.* 61: 1679-1685.

Ibekwe, A.M., Angle, J.S., Chaney, R.L. and van Berkum, P. (1998). Zinc and cadmium effects on Rhizobia and white clover using chelator-buffered nutrient solution. *Soil Sci. Soc. Am. J.* 62: 204-211.

Ji, G. and Silver, S. (1995). Bacterial resistance mechanisms for heavy metals of environmental concern. *J. Ind. Microbiol.* 14: 61-75.

Lakzian, A., Murphy, P., Turner, A., Beynon, J.L. and Giller, K.E. (2002). *Rhizobium leguminosarum* bv. *viciae* populations in soils with increasing heavy metal contamination: abundance, plasmid profiles, diversity and metal tolerance. *Soil Biol. Biochem.* 34: 519-529.

Langley, S. and Beveridge, T.J. (1999). Effect of O-side-chain-lipopolysaccharide chemistry on metal binding. *Appl. Environ. Microbiol.* 65: 489-498.

Lawlor, K., Chaudri, A.M., McGrath, S.P. and Hirsch, P.R. (1999). Gene transfer in bacteria from soils contaminated with heavy metals. *Lett. Appl. Microbiol.* 28: 317-320.

Lima, A.I.G., Corticeiro, S.C. and Figueira, E.M.A.P. (2006). Glutathione-mediated cadmium sequestration in *Rhizobium leguminosarum*. *Enz. Microbial. Technol.* 39: 763-769.

Malik, A. (2004). Metal bioremediation through growing cells. *Environ. Inter.* 30: 261-278.

McGrath, S.P. and Lane, (1989). An explanation for the apparent losses of metals in a long-term field experiment with sewage sludge. *Environ. Pollu.* 60: 235-256.

McGrath, S.P., Chaudri, A.M. and Giller, K.E. (1995). Long-term effects of metals in sewage sludge on soils, microorganisms and plants. *J. Ind. Microbiol.* 14: 94-104.

Mullen, M.D., Wolf, D.C., Ferris, F.G., Beveridge, T.J., Flemming, C.A., Bailey, G.W. (1989). Bacterial sorption of heavy metals. *Appl. Environ. Microbiol.* 55: 3143-3149.

Nies, D.H. (1992). Resistance to cadmium, cobalt, zinc and nickel in microbes. *Plasmid.* 27: 17-28.

Nies, D.H. (1999). Microbial heavy-metal resistance. *Appl. Microbiol. Biotechnol.* 51: 730-750.

Nies, D.H. and Silver, S. (1995). Ion efflux systems involved in bacterial metal resistances. *J. Ind. Microbiol.* 14: 189-199.

Obbard, J.P. (2001). Ecotoxicological assessment of heavy metals in sewage sludge amended soils. *Appl. Geochem.* 16: 1405-1411.

Obbard, J.P., Sauerbeck, D.R. and Jones, K.C. (1993). *Rhizobium leguminosarum* bv. *trifolii* amended with heavy metal contaminated sewage sludges. *Soil Biol. Biochem.* 25: 227-231.

Pereira, S.I.A., Lima, A.I.G. and Figueira, E.M.A.P. (2006). Screening possible mechanisms mediating cadmium resistance in *Rhizobium leguminosarum* bv. *viciae* isolated from contaminated portuguese soils. *Microb. Ecol.* 52: 176-186.

Purchase, D., Miles, R.J. and Young, T.W.K. (1997). Cadmium uptake and nitrogen fixing ability in heavy-metal-resistant laboratory and field strains of *Rhizobium leguminosarum* biovar *trifolii*. *FEMS Microb. Ecol.* 22: 85-93.

Purchase, D. and Miles, R.J. (2001). Survival and nodulating ability of indigenous and inoculated *Rhizobium leguminosarum* biovar *trifolii* in sterilized and unsterilized soil treated with sewage sludge. *Curr. Microbiol.* 42: 59-64.

Robinson, B., Russell, C., Hedley, M. and Clothier, B. (2001). Cadmium adsorption by rhizobacteria: implications for New Zealand pastureland. *Agri. Eco. Environ.* 87: 315-321.

Saeki, K., Kunito, T., Oyaizu, H. and Matsumoto, S. (2002). Relationships between bacterial tolerance levels and forms of copper and zinc in soils. *J. Environ. Qual.* 31: 1570-1575.

Saxena, D., Amin, M. and Khanna, S. (1996). Modulation of protein profiles in *Rhizobium* sp. under salt stress. *Can. J. Microbiol.* 42: 617-620.

Scott, J.A. and Palmer, S.J. (1990). Sites of cadmium uptake in bacteria used for biosorption. *Appl. Environ. Microbiol.* 33: 221-225.

Shi, W., Bischoff, M., Turco, R. and Konopka, A. (2002). Long-term effects of chromium and lead upon the activity of soil microbial communities. *Appl. Soil Ecol.* 21: 169-177.

Silver, S. and Phung, L.T. (1996). Bacterial heavy metal resistance: new surprises. *Annu. Rev. Microbiol.* 50: 753-789.

Smith, S.R. (1997). *Rhizobium* in soils contaminated with copper and zinc following the long-term application of sewage sludge and other organic wastes. *Soil Biol. Biochem.* 29: 1475-1489.

Somasegaran, P. and Hoben, H.J. (1994). Handbook for Rhizobia, Springer-Verlag, Berlin.

Sutherland, I.W. (1985). Biosynthesis and composition of Gram-negative bacterial extracelular and wall polysaccharides. *Annu. Rev. Microbiol.* 39: 243-270.

Trajanovska, S., Britz, M.L. and Bhave, M. (1997). Detection of heavy metal ion resistance genes in Gram-positive and Gram-negative bacteria isolated from a lead-contaminate site. *Biodegrad.* 8: 113-124.

Valverde, C., Hozbor, D.F. and Lagares, A. (1997). Rapid preparation of affinity-purified lipopolysaccharide samples for electrophoretic analysis. *BioTechniques.* 22: 230-236.

Weaver, R.W. and Holt, E.C. (1990). Short-term survival of rhizobia on arrow leaf clover seed at different depths. *Plant Soil.* 122: 147-150.

□□□

CHAPTER **10**

Bioremediation : A Natural Approach for Heavy Metal Contaminated Sites

Anju Rani[1], Alok Kumar[2] and Reeta Goel[1*]

[1]Department of Microbiology, [2]Department of Molecular Biology and Biotechnology, G.B. Pant University of Agriculture and Technology, Pantnagar - 263 145, Uttarakhand

ABSTRACT

An understanding of the metabolism of heavy metals facilitates biotechnological use of these functions just recently, although heavy metals comprise the major part of the elements in the periodic table. Due to their ability to form complex compounds, some heavy metals are essential trace elements. In any case, essential or not, most heavy metals are toxic at higher concentration. Heavy metal pollution from last few decades has led to an increased environmental concern. Heavy metal like cadmium is a non-essential toxic heavy metal and is also being increased in soil during last few years posing a serious threat to environment and human health worldwide. Therefore, remediation strategies of the contaminated sites are great concerns and can be achieved by a number of different physico-chemical and biological means.

Key words: bioremediation, heavy metal, bioaccumulation, biosorption, extremophiles

1. INTRODUCTION

Microorganisms, a cosmopolitan conglomerate are a pioneer compatible organism of this world. Of all forms, bacteria are not only oldest, but they also inhabit the greatest diversity of habitats, form a major proportion of the earth global biomass. These abound in habitats with all sorts of varying conditions viz. high temperature (90-113^{0}C), low temperature (0-12^{0}C), pH (acidophiles 1-4; alkaliphiles 8.5-12), pressure (500-1000 atm.), salt concentration (15-32%), heavy metals (Cd, Pb and As) and other toxic compounds. Some microorganisms are known to resist the higher concentration of heavy metals and thus, help in removing these from contaminated sites (Roane and Pepper, 2000; Tripathi *et al.*, 2004). The known

*Email: rg55@rediffmail.com

mechanisms of heavy metal toxicity include induction of oxidative stress and interfering with protein folding and function (Nies, 1999). The ability of microorganisms to survive and reproduce in a metals habitat may depend on genetic as well as physiological adaptations. The former leads to the wide spread appearance of metal resistant organisms. The mechanism of heavy metal resistance and its genetic basis, however, varies with the microbe and the metal in question. Therefore, an ideal candidate should be considered for bioremediation in natural mean.

2. HEAVY METAL POLLUTION AND ENVIRONMENTAL CONCERNS

Increasing industrialization led to the disruption of biogeochemical cycles and has led to an increase in the concentration of variety of pollutants in the soil. Some organic compounds are degraded naturally, but pollutants like heavy metals can not be degraded easily, hence accumulate at different useful sites. Some metals are essential (like Cu, Co, Mg, Ni, Zn etc.), while other are non-essential (like Ag, Cd, Pb, As, Au etc.). Although, elevated levels of toxic heavy metals are present in some natural locations such as volcanic soil, hot springs and low deep sea vents. But increased industrial activity has disrupted natural biogeochemical cycles resulting in addition of more and more toxic metals into the environment through anthropogenic activities, such as mining, smelting, dispersal of sewage sludge, use of fertilizers as well as through industrial waste.

3. HEAVY METAL TOXICITY

Heavy metals are metals with density greater than 5 g/cm^3 so accumulated in soil contaminate groundwater (Table 1) and enter the food chain (Smith *et al.*, 1996; Herawati *et al.*, 2000). Heavy metals affect soil microbial population especially rhizospheric which are important for plant growth promotion and other activities. Almost every index of microbial metabolism and activity can be adversely affected by elevated concentrations of heavy metals. These includes primary conductivity, methanogenesis, nitrogen fixation, respiration, motility, biogeochemical cycling of C, N, S, P, and other elements, organic matter decomposition, enzymes synthesis and activity in soil sediments and water (Gadd, 1992). Also, heavy metal ions and metalloids are chemical species that can be very toxic to cells in a variety of ways, e.g. through binding to essential respiratory chain proteins, oxidation damage etc (Canovas *et al.*, 2003). In *Escherichia coli*, functional heterogeneity of RpoS (the stationary-phase sigma factor) in stress tolerance has been studied (Bhagwat *et al.*, 2006).

TABLE 1. Sources of heavy metals

Metal	Sources	Disease
Lead	Mining, coal, automobile, paper dyeing, petrochemicals	Learning disability, mental retardation
Chromium	Leather/tanner, thermal power plant, mining fertilizers, textile photography	Bronchial asthma, allergies
Cadmium	coal, nuclear and coal power plant, batteries , ceramics, toys	Itai Itai disease
Nickel	Mining, coal, power plant, phosphate fertilizers, chocolate, automobile electroplating	Dermatitis, Pneumonia
Mercury	Mining, paper and pulp, coal power plant, cement, electrical equipments, pesticides cosmetics	Minimata disease
Zinc	Phosphate fertilizers, distillery, pharmaceuticals	Fever

Bollag and Barabasz (1979) reported that heavy metal inhibits denitrification in soil, and since denitrification causes the reduction of nitrate to nitrite and nitrogenous gases, this investigation concludes the possible impact of heavy metals on nitrogen cycle. A summary of heavy metal toxicity and its mechanisms towards microbes (Tripathi, 2005) is illustrated in the figure below (Fig. 1).

3.1 How Heavy Metal Harm Cell?

Once inside the cell, especially heavy metal cations with high atomic numbers tend to bind to SH groups, e.g. Hg2+, Cd2+ and Ag+. The MIC (minimal inhibitory concentration) of these metal ions is a function of the complex dissociation constants of the respective sulfides. By binding to SH groups, the metals may inhibit the activity of sensitive enzymes. Other heavy metal cations may interact with physiological ions, Cd2+ with Zn2+ or Ca2+, Ni2+ and Co2+ with Fe2+, Zn2+ with Mg2+ thereby inhibiting the function of the respective physiological cation. Heavy metal cations may bind to glutathione, the resulting bis-glutathiono-complexes tend to react with molecular oxygen to oxidize bis-glutathione GS-SG (Kachur *et al.*, 1998), the metal cation and hydrogen peroxide (H_2O_2). Since the oxidized bis-glutathione has to be reduced again in a NADPH-dependent reaction and the metal cations immediately catches another two glutathione molecules, heavy metal cations cause a considerable oxidative stress. Finally, heavy metal oxyanions interfere with the metabolism of the structurally related non metal (chromate with sulfate, arsenate with phosphate) and reduction of heavy metal oxyanion leads to the production of radicals, e.g. in case of chromate.

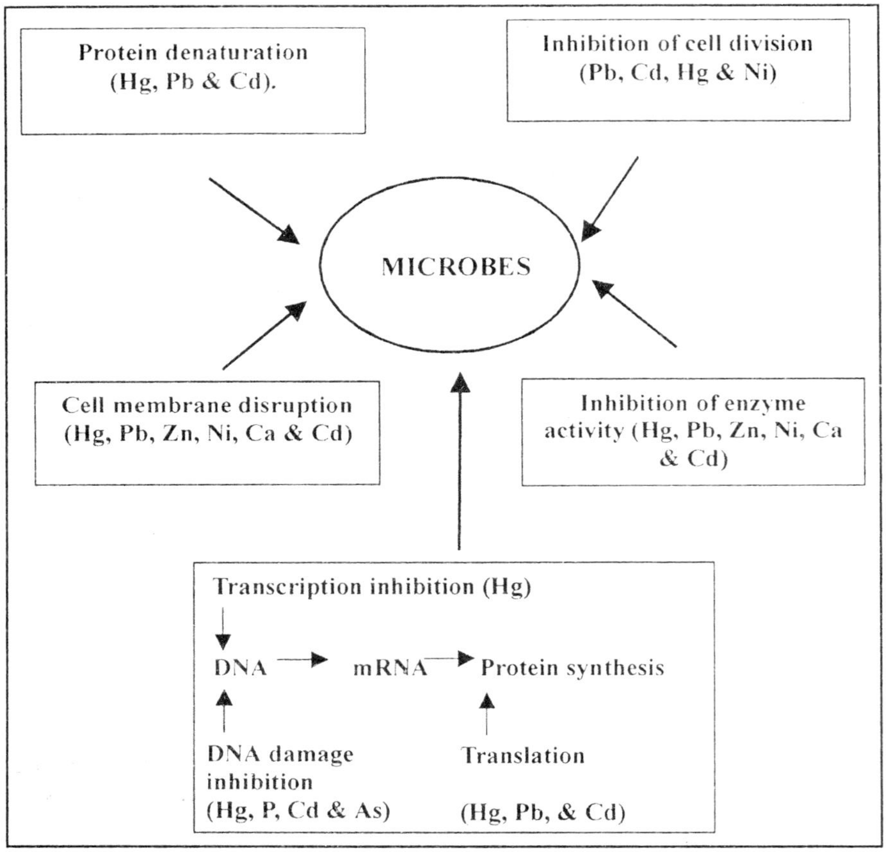

FIG. 1. Heavy metal toxicity mechanisms to microbes (*Courtesy:* Tripathi, 2005).

Meargeay *et al.* (1985) tested the MICs of several different metal ions for *Escherichia coli* on agar medium, and the most toxic metal, (with the lowest MIC) was mercury, whereas, the least toxic metal tested was manganese (Table 2); a) the MIC was determined on Tris-buffered minerals salts medium (Meargeay *et al.*, 1985), starting pH 7.0, containing 2 g sodium gluconate/l as carbon source, and 1 g yeast extract/l to complement for *E. coli* auxotrophies. The plates were incubated for 2 days at 30°C; b) weak acidification of the medium had to be allowed to keep the metal ion in solution; c) acidification of the medium had to be allowed to keep the metal ion in solution.

TABLE 2. Minimal inhibitory concentrations (MICs) of several heavy metals in *Escherichia coli*

MIC (mM)*a*	Heavy metal ions
0.01	Hg^{2+}
0.02	Ag^{+}, Au^{3+}
0.2	CrO_4^{2-}, Pd^{2+}
0.5	Pt^{4+}, Cd^{2+}
1.0	Co^{2+}, Ni^{2+}, Cu^{2+}, Zn^{2+}
2.0	Tl^{+}, UO^{2-}, (La^{3+}, Y^{3+} Sc^{3+})*b*, (Ru^{3+}, Al^{3+})*c*
5.0	Pb^{2+} (Ir^{3+}, Os^{3+}, Sb^{3+}, Sn^{2+}, In^{3+}, Rh^{2+}, Ga^{3+}, Cr^{3+}, V^{3+}, Ti^{3+},Be^{2-})*c*
10.0	(Cr^{2+})*c*
20.0	Mn^{2+}

Courtesy: Meargeay *et al.* (1985)

3.2 Metal Tolerance Mechanisms

In high concentrations, heavy metal ions react to form toxic compounds in cells (Nies, 1999). However, heavy metal ions must first enter the cell for this activity. Because some heavy metals are necessary for enzymatic functions and bacterial growth, uptake mechanisms exist that allow for the entrance of metal ions into the cell (Table 3). There are two general uptake systems (Anne, 2003).

a. An unspecific and quick, driven by a chemiosmotic gradient across the cell membrane and thus requiring no ATP.

b. Slower and more substrate-specific, driven by energy from ATP hydrolysis.

While the first mechanism is more energy efficient, it results in an influx of a wider variety of heavy metals, and when these metals are present in high concentrations, they are more likely to have toxic effects once inside the cell (Nies and Silver, 1995).

b"parts" are proteins or protein domains. *c*CPM = cytoplasmic membrane. *d*ATP-binding cassette (Fath and Kolter, 1993). *e*(Fagan and Saier, 1994). *f*(Saier, 1994). *g*Resistance-Nodulation-Cell division (Saier *et al.*, 1994; Saier, 1994). *h*Chromate-transport. *i*Metal inorganic transport (Paulsen *et al.*, 1998). *j*Cation diffusion facilitators (Nies and Silver 1995; Nies, 1995; Paulsen and Saier, 1997).

TABLE 3. Protein families important for heavy metal transport

Family	Direction of transport	Energy	Metal ions	Composition
ABCd	Uptake	ATP	Mn^{2+}, Zn^{2+}, Ni^{2+}, Fe^{2+}	2 membrane-integral parts b + 2 ATPase parts = ABC-core + periplasmic binding protein
	Efflux	ATP	-	ABC-core + membrane fusion protein and outer membrane factor
P-typee	Both	ATP	Mg^{2+}, Mn^{2+}, Ca^{2+}, K^{+}, Cu^{2+}, Zn^{2+}, Cd^{2+}, Pb^{2+}, Ag^{+}	one membrane-bound protein as core
A-typef	Efflux	ATP	arsenite	1 membrane-integral protein + a dimeric ATPase subunit
RNDg	Efflux	proton gradient	Co^{2+}, Zn^{2+}, Cd^{2+}, Ni^{2+}, Cu^{2+}?, Ag^{+}?	1 CPM c proton/cation antiporter + membrane fusion protein (dimer?) + outer membrane factor: CBA-transport systems
HoxN	Uptake	chemi-osmotic	Co^{2+}, Ni^{2+}	membrane-integral protein
CHRh	Antiport?	chemi-osmotic	chromate	membrane-integral protein (ChrA)
MITi	Uptake	chemi-osmotic	most cations	membrane-integral protein (CorA)
CDFj	Efflux	chemi-osmotic	Zn^{2+}, Cd^{2+}, Co^{2+}, Fe^{2+}?	membrane-integral protein (CzcD, ZRC1p, ZnT1)

Microorganisms which are indigenous to heavy metals containing environments have evolved into several distinct mechanisms of heavy metal tolerance as confined below.

I. Heavy metal can be sequestered by bioaccumulation and biosorption to cell wall (Scott and Palmer, 1990).

II. A common plasmid encoded mechanism employs heavy metal diffusion and transport which specially capture and ejects undesirable divalent cations through the cell membrane (Nies, 1992; Silver and Mishra, 1988; Silver and Phung, 1996).

III. By binding to detoxifying ligands, proteins or polymers (Aiking *et al.*, 1984; Gekeler, *et al.*, 1988; Inoube *et al.*, 1989; Kurek *et al.*, 1991; Nies, 1992; Volesky and Prasetyo, 1994).

IV. Microbial mediated precipitation of heavy metals as insoluble sulfides, carbonates, phosphates or hydroxides can also reduce the bioavailable concentration of the toxic ions (Aiking *et al.,* 1984; 1982; Tebo, 1995)

Enzymes can convert metal ions into organometalic compounds, thereby detoxifying or volatilizing the metals ion (Nies, 1992; Tebo, 1995).

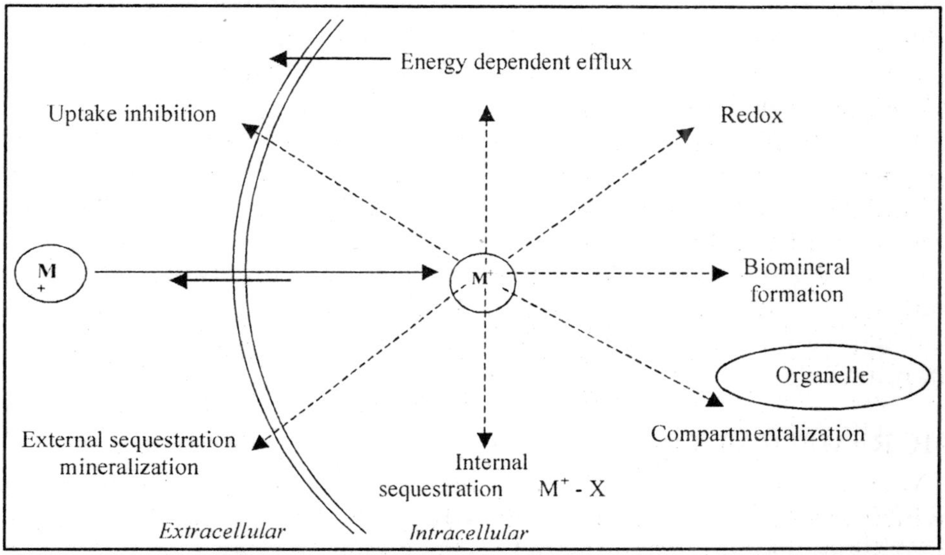

FIG. 2. General mechanisms for metal ion detoxification (Dameron and Harrison, 1998)

3.3 Heavy Metal Remediation

Over the past century, mining, manufacturing and urban activities have all contributed to extensive soil contamination. Various physical, chemical and biological processes are already being used to remediate contaminated soil. These processes either 'decontaminate' the soil, or 'stabilize' the pollutant within it. Decontamination reduces the amount of pollutants within the soil by removing them, stabilization does not reduce the quantity of pollutant at a site, but makes use of soil amendments to alter the soil chemistry and sequester or absorb the pollutant into the matrix so as to reduce or eliminate environmental risks (Cunningham *et al.,* 1995).

Heavy metals in soil are associated with various forms which related to their solubility, directly bear on the bioavailability (Xian, 1989). These forms are-(1) in soil solution as free metal ions and soluble metal complexes; (2) adsorbed to inorganic soil constituents at ion exchange sites; (3) bound to soil organic matter; (4) precipitated such as oxides, hydroxides, and carbonates; and (5) embedded in structure of the silicate minerals. A large proportion of many metals are adsorbed or occluded by carbonates, organic matters, Fe-Mn oxides and primary or secondary minerals (Loveley *et al.*, 1991; Dilek and Yetis, 1992; Garbisu and Alkorta, 2001). Influence of biofilm structure and antibiotic resistance mechanism also help in indirect pathogenecity and resistance (O'Connel *et al.*, 2006).

Once metals are introduced and contaminate the environment, they will remain in respective environment. Metals do not degrade like carbon-based (organic) molecules. The only exceptions are mercury and selenium, which can be transformed and volatilized by microorganisms. However, in general it is very difficult to eliminate metals from the environment. Traditional treatments for metal contamination in soils are expensive and cost prohibitive when large areas of soil are contaminated. Treatments can be done in situ (on-site), or ex situ (removed and treated off-site). Both are extremely expensive. Some treatments that are available viz. i. high temperature treatments (produce a vitrified, granular, non leachable material); ii. solidifying agents (produce cement-like material); iii. washing process (leaches out contaminants).

4. BIOREMEDIATION

Microorganism can interact with metals via different mechanisms, some of which may be used as the basis of potential bioremediation strategies (Fig. 3). Bioremediation of industrial wastes containing heavy metals has been demonstrated by several biotechnology companies employing bioaccumulation, biosorption, bioprecipitation, and uptake by purified biopolymers derived from microbial cells provide alternative and/or additive processes for conventional physical and chemical methods. Intact microbial cells live or dead and their products can be highly efficient bioaccumulators of both soluble and particulate forms of metals (Gupta *et al.*, 2004). The cell surfaces of all microorganisms are negatively charged owing to the presence of various anionic structures. This gives bacteria the ability to bind metal cations. Various microbial species, mainly *Pseudomonas*, have been shown to be relatively efficient in bioaccumulation of the different heavy metals from polluted effluents (Hussein *et al.*, 2003). Bioremediation was also studied by using sulfate reducing bacteria for cadmium accumulation through DNA homology (Naghma *et al.*, 2005).

4.1 Bioaccumulation and Biosorption

Microorganisms can physically remove heavy metals and radionuclides from solution through either bioaccumulation or biosorption. Bioaccumulation is the retention and concentration of a substance by an organism wherein,

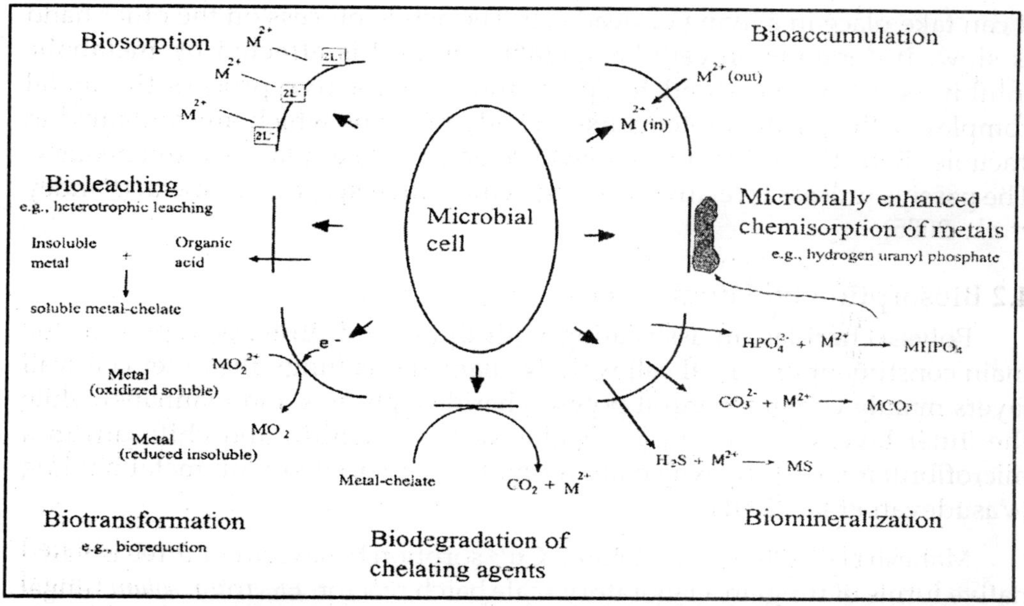

FIG. 3. Mechanisms of metal-microbe interaction (Atlas and Philip, 2005a)

metals are transported from the outside of the microbial cell, through the cellular membrane, and into the cell cytoplasm where the metal is sequestered and therefore immobile.

Biosorption does not consume cellular and positively charged metal ions which are sequestered primarily through the absorption of metal to the negative ionic groups on cell surfaces. The polysaccharide coating found on most form of bacteria, or other extracellular structures such as capsules or slime layer are generally instrumental for this binding. Common binding sites on microbial cell surface are carboxyl residues, phosphate residues, SH groups or hydroxyl group. The amount of metal biosorbed to the exterior of bacterial cells often exceeds the amount predicted using information about the charge density of the cell surface. It has been demonstrated that charged functional groups serve as nucleation sites for deposition of various metal bearing precipitates.

Vasudevan *et al.* (2001) defined biosorption as a property of certain inactive or dead microbial biomass to bind and concentrate heavy metals from even very dilute aqueous solutions. The uptake of metal could be active, passive or both in nature. Passive uptake is independent of cellular metabolisms and metal binds to polyionic cell walls through ion exchange as the process is not affected by physical conditions such as pH and ionic strength, and it is rapid taking only 5-10 minutes to complete. The process is also reversible and it can take place in a living or dead cell. The active process on the other hand is slow. It depends on cellular metabolism and is affected by metabolic inhibitors, uncouplers and temperature. In the active process the metal complex with specific proteins like metallothioneins which are contained in vacuole. Both the active and passive mode may take place simultaneously. The passive process is relatively non-specific with respect to the metal (Pandey *et al.*, 2005).

4.2 Biosorption by Fungi

Polysaccharides, in association with lipids and proteins, represent the main constituent of fungal cell wall. In filamentous fungi, the outer cell wall layers mainly contain, natural polysaccharides (glycans and mannans) while the inner layers contain more of glucosamines (chitin and chitosan) in a microfibrillar structure, which are found to be associated with metal binding (Vasudevan *et al.*, 2001).

Mahesh *et al.* (2001) investigated Cu^{+2} sorption behaviour of three isolated native forms of fungi in a laboratory scale batch reactor. *Fusarium solani* fungal species tend to remove 90 percent Cu^{+2} from solution at alkaline pH (8.0-10.0). Metal uptake capacities of some fungi are listed by Bagdwal *et al.* (2003).

4.3 Biosorption by Algae

Special polysaccharides are present in the algal cell wall. The number and nature of binding sites depends on the chemical composition of the cell wall. In pheophycean members, algin is present and contributes significantly to metal binding.

4.4 Biosorption by bacteria

The anionic nature of bacterial surface enables them to bind metal cations through electrostatic interactions. Because of their thickness and anionic character which is mainly due to peptidoglycan, teichoic acid and teichuronic acids, the cell walls of gram positive bacteria have high capacity for metal binding. It has been reported by Beveridge and Murray (1980) that peptidoglycan is the major cell wall component responsible for metal binding by *Bacillus subtilis*. In contrast metal binding by *B. licheniformis*, is predominantly due to teichuronic acid. Three strains of thermo tolerant polymer producing bacteria; *B. subtilis* WD90, *B. subtilis* SH29, and *Enterobacter agglomerans* 5M38 as well as their bioflocculants were capable of nickel and cadmium removal. However, the bioflocculant of the three isolates gave higher metal adsorption than the cells, showed extracellular polysaccharide and cell wall component play important role in metal uptake

(Kaewchai and Prasertsan, 2002). *Pseudomonas aeruginosa* CW-96-1 is able to grow in concentration of cadmium up to 5 mM. One hundred forty hour after inoculation strain CW-96-1 removed > 99% of the cadmium from solution. Electron micrographs and energy dispersive microanalysis indicated that cadmium was bound to the cell wall as a sulfur complex with a 1:1 stoichiometry (Wang *et al.*, 1997). *P. putida* actively accumulates cadmium from the medium and the resistance mechanisms involve both polyphosphate and a series of low molecular weight cysteine rich cadmium proteins that are induced during different growth phases. Nuclear magnetic resonance (NMR) study on the major, native cadmium protein ($CdBP_1$) establishes a definite relationship to cadmium metallothioneins (Higham *et al.*, 1984). Metal uptake capacities of some bacteria are listed by Bagdwal *et al.* (2003) which are further updated in Table 4.

TABLE 4. Metal uptake capacity of bacteria

Metal ion	Metal uptake capacity (mol/g)	Organisms
Cd	8.0	*B. cereus*
	9.5	*B. subtilis*
	10.8	*Bacillus* sp.
	11.8	*B. sphaericus*
Pb	0.7	*B. sphaericus*
	1.1	*B. cereus*
	1.4	*Bacillus* sp.
	1.8	*B. subtilis*
Zn	4.3	*B. sphaericus*
	4.6	*B. cereus*
	4.8	*Bacillus* sp.
Cu	5.6	*B. sphaericus*
	5.9	*B. cereus*
	5.9	*B. subtilis*

Assessment of morphological changes in response to heavy metals can be done by scanning electron microscopy (SEM) and energy dispersive X-ray spectrometer (EDX). SEM analysis has revealed increase in cell size (Tripathi, 2005) of bacteria *Proteus vulgaris* KNP3 and *P. putida* KNP9 in the presence of lead and cadmium and distribution of some round shiny uniformly distributed bodies over the surface of bacterial cell wall (Bagdwal, 2002) in the presence of copper. EDX analysis was used for confirming the accumulation of respective metals (Tripathi, 2005; Bagdwal, 2002).

5. GENETICALLY MODIFIED MICROORGANISMS

Sauge-Merle *et al.* (2003) expressed the *Arabidopsis thaliana* PC synthase gene (AtPCS) in *Escherichia coli*. They showed that expression of AtPCS allow PC synthesis resulting in a large increase in intracellular metal content (20, 6 and 50 fold for Cd, Cu and As, respectively) when bacterial cells are grown on metal enriched media. Similarly Ueki *et al.* (2003) constructed *E. coli* strains that expressed vanabin 1 or vanabin 2 fused to maltose binding protein (MBP) in the periplasmic space. They found that both strains accumulated about twenty times more copper ($^{+}2$) ions than the control BL21 strain. *P. fluorescens* strains PRS_9 and GRS_1 wild type were made mercury resistant PRS_9 Hg and GRS_1 Hg^r, respectively and their in situ colonization studies were carried out (Gupta *et al.*, 2005).

6. ENZYMATICALLY CATALYZED BIOTRANSFORMATIONS

Microorganisms can catalyze the transformation of toxic metals to less soluble or more volatile forms. For example, the microbial reduction of Cr (VI) to Cr (III), Se (VI) to Se (0) , V (V) to V (III), Au (III) to Au (0) results in metal precipitation under physiological conditions. In many cases the high valence metal can be used as an electron acceptor under anoxic conditions (Lloyd, 2003). Metal detoxification is also possible through biotransformation of toxic metals. The bioreduction of Hg (II) to relatively nontoxic Hg (0) is perhaps the best-studied example. Finally, in addition to redox biotransformation, biomethylation may also increase the volatility of metals with the methylation of mercury, cadmium lead, tin, and selenium (Ehrlich, 1996). These mechanisms could also offer potential use for *in situ* remediation of contaminated soil (Losi and Frankenberger, 1997).

7. BIOMINERALIZATION *via* MICROBIALLY GENERATED LIGANDS

Microorganisms are able to precipitate metals as carbonates and hydroxides via plasmid borne resistance mechanisms, whereby proton influx countercurrent (antiport) to metal efflux results in localized alkalinization at the cell surface (Diels *et al.*, 1995; Van Roy *et al.*, 1997). Alternatively, metals can precipitate with enzymatically generated ligands, e. g., sulfide or phosphate (Macaskie *et al.*, 1992). Most of the metal should be removed from solution if an excess of ligand is supplied. This is difficult to achieve when chemical precipitation methods and dilute solutions are used. The advantage of microbial ligand generation is that high concentrations of ligand are achieved in juxtaposition to the cell surface, which can also provide nucleation foci for the rapid onset of metal precipitation; effectively, the metals are concentrated "uphill" against concentration gradient.

Sulfide precipitation, catalyzed by a mixed culture of sulfate-reducing bacteria, has been utilized first to treat water contaminated by sulfate and zinc (Barnes *et al.*, 1991) and also soil leachate contaminated with sulfate alongside metal and radionuclides. The ubiquitous distribution of sulfate - reducing bacteria in acid, neutral, and alkali environments (Postgate, 1979) suggests that they have the potential to treat a variety of effluents, while the ability of organisms to metabolize a wide range of electron donors may also allow co-treatment of other organic contaminants.

8. MICROBIALLY ENHANCED CHEMISORPTION OF HEAVY METALS

Microbially enhanced chemisorption of heavy metals (MECHM) is a generic term to describe a class of reactions whereby microbial cells first precipitate a biomineral of one metal ("priming deposit"). The priming deposit then acts as a nu-cleation focus, or "host crystal," for the subsequent deposition of the metal of interest (target metal), acting to promote and accelerate target metal precipitation reactions (Macaskie *et al.*, 1994). The priming deposit is made initially by the sulfide or phosphate biomineralization routes.

9. BIOREMEDIATION BY USING EXTREMOPHILES

Extremophiles are microorganisms that thrive under conditions that from a human perspective are clearly extreme. Extreme environments include high temperature, pH, pressure, salt concentration and water availability and also conditions having high levels of radiation, harmful heavy metals and toxic compound. Extremophiles exposed to different physiological conditions alter the protein synthesis. A different band pattern was observed in alkalophilic, acidophilic and psychrophilic bacteria when grown in presence and absence of cadmium (Kumar, 2006). Change in banding pattern may help in conferring resistance in psychrophilic isolate which may be due to reduced metal bridges or proline content or reduced aromatic interaction or hydrophobic clustering (Khan *et al.*, 2003; Khan, 2005).Therefore, these isolate could help in cadmium accumulation and bioremediation of the contaminated sites at extreme conditions of pH and temperature, respectively.

10. MONITORING BIOREMEDIATION

Performance monitoring is a critical part of remediation effort. The needs for monitoring of bioremediation projects clearly start even before technology selection. First the nature of the contamination must be determined, in terms of the specific contaminants and their concentrations. Then the nature of the environmental matrix containing those contaminants must be considered. That will determine whether bio-remediation can be considered as a possible remediation strategy for future or not.

As in any remediation effort, one must establish the initial parameters and to show that in the end the target goal of reducing the contaminant to a safe level has been achieved. In some cases, only the starting and end points need to be measured to demonstrate success, but in other cases it is critical to follow the progress of the remediation effort and to monitor critical parameters, including those that may need to be modified to optimize the remediation process. In the case of bioremediation, monitoring may involve measuring not only concentrations of contaminants, but also the microbial populations involved in the degradation or transformation of those contaminants and the environmental parameters that influence rates of microbial metabolism.

11. MICROBIAL METHODS FOR MONITORING BIOREMEDIATION

For bioremediation to be considered as a remediation technology, it is critical to establish that there is an adequate active microbial population that is capable of attacking the specific contaminant(s). If the site has the requisite values for water content and pH, has porosity within the desirable range, and is contaminated with only petroleum hydrocarbons, then there is a very good chance that there will be an active population of hydrocarbon-oxidizing microorganisms in the soil and that bioremediation may be able to succeed (Atlas and Philip, 2005b).

Use of fundamental chemical analyses for pollutant identification and standard microbiological techniques for quantification of viable populations of microorganisms is the starting point for monitoring. If the contaminants are more recalcitrant, then the soil may not contain large numbers of microbes that can degrade them, and enumeration might not be possible in this circumstance. A decision whether bioaugmentation may be needed may then be made. Even when it is difficult to cultivate specific microbial populations, however, the technique of enrichment culture can still reveal the presence of the necessary degrading microbes and establish that they have the necessary potential to degrade the contaminant at an acceptable rate of performance.

12. MOLECULAR BIOLOGY TOOLS FOR BIOREMEDIATION MONITORING

In recent years, molecular methods have been used for the study of microbial community structure and function because of the realization that the majority of microbes in the environment are unculturable (Amann *et al.*, 1995). It is thought that the percentage of culturable organisms is something in the range of 1 to 10% of the total community; in fact, the figure for the number of bacteria described so far is thought to be approximately 1% (Cox and Moore, 2000).

The methods available now can be used in various ways to provide separate pieces of information. Some are used in order to give a profile of the total community present at a site such as PCR amplification of 16S rRNA gene, DGGE (Muyzer *et al.*, 1993), length heterogeneity PCR (LH-PCR)(Suzuki *et al.*,1998) and terminal restriction fragment length polymorphism (tRFLP) (Liu *et al.*, 1987), while other identify the organisms performing a specific function. Another set of methods focuses on the fluorescent probing includes FISH, DNA microarray (Small *et al.*, 2001). More recently, new technologies which are substrate based, labeling organisms according to their catabolic potential, have emerged.

Besides the above mentioned techniques flow cytometery can be used for analysis of membrane potential and membrane integrity in response to heavy metals. Toxic metal ions, including Co, Cu, Cd, Ni, As, Hg inhibits plasma membrane ATPas by means of various binding interaction (Ochiai, 1987). The above effect lead to increased permeability of the cell to external material i. e adverse effect on membrane integrity and a reduced ability to maintain electrochemical gradient or membrane potential. Membrane potential and integrity was recorded for strain AG27 and AGM13 (Gupta, 2006). Flow cytometery analysis for both *Bacillus cereus* AG27and strain AGM13 showed significant reduction in membrane potential after addition of sodium arsenate to cells growing in log phase.

13. CONCLUSION

Heavy metal like cadmium, lead are non-essential toxic heavy metals and are being increased in soil during last few years posing a serious threat to environment and human health worldwide. However, it is realized that large areas of land contaminated with metal cannot be economically remediated by conventional chemical approaches. Microbial bioremediation can prove to be the best alternative for remediation of heavy metal contaminated sites. Supported by genomics enabled studies ongoing in many laboratories worldwide, we can expect this research area to develop further in the near future, delivering more robust technologies for the bioremediation of metal contaminated waters and land. Bioremediation using microbes well adapted to diverse physiological conditions can be utilized for remediation of heavy metal contaminated sites. Use of extremophilic heavy metal resistant microorganisms give a good insight for remediation of heavy metal contaminated sites in the near future.

REFERENCES

Aiking, H., Kok, K., Heerikhuizen, H.V. and Riet, J.V.T. (1982). Adaptation to cadmium by *Klebsiella aerogenes* growing in continuous culture proceeds mainly via formation of cadmium sulfide. *Appl. Environ. Microbiol.* 44:938-944.

Aiking, H., Slijnman, H.A., Garderen, C.V., Heerikhuizen, H.V. and Riet, J.V.T. (1984). Inorganic phosphate accumulation and cadmium detoxification in *Klebsiella aerogens* NCTC418 growing in continuous culture. *Appl. Environ. Microbiol.* 47:374-377.

Amann, R. I., Ludwig, W. and Schleifer, K.H. (1995).Phylogenetic identification and *in situ* detection of individual microbial cells without cultivation. *FEMS Microbiol* Rev.59:143-169.

Anne, S. (2003). Implications of Microbial Heavy Metal Tolerance in the Environment. *Rev. Undergraduate Res.* 2:1-6.

Atlas, R.M. and Philp, J. (2005a).Bioremediation of metals and radionuclides In: Bioremediation: Applied microbial solutions for real - world environment clean up, ASM press Washington,D.C. Pp. 293- 317.

Atlas, R.M. and Philp, J. (2005b). Monitoring of bioremediation In: Bioremediation: Applied microbial solutions for real - world environment clean up, ASM press Washington, D.C. Pp. 237- 268.

Bagdwal, N. (2002). Biosorption studies of fluorescent Pseudomonads. M. Sc. Thesis, Deptt. of Microbiology, G.B.P.U.A.T. Pantnagar, India, Pp. 373-377.

Bagdwal, N., Gupta, A. and Goel,R. (2003). Metal resistant growth promontory fluorescent *Pseudomonads. In*: Microbial Biotechnology. Aavishkar Publishers and Distributors Jaipur, India. EdTrevide, P.C. Pp.126 - 161.

Barnes, L. J., Janssen, F.J., Sherren, J., Versteegh, J.H., Koch, R.O and Scheeren, P.J.H. (1991). A new process for the microbial removal of sulfate and heavy metal from contaminated water extracted by a geohydrological control system. *Chem. Eng. Res. Res. Des.* 69A:184-186.

Beveridge, T. J. and Murray, R. G. E. (1980). Sites of metal deposition in the cell wall of *Bacillus subtilis*. *J. Bacteriol.* 142 (2):876-887.

Bhagwat, A.A., Tan, J., Sharma, M., Kothory, M., Low, S., Tall, B.D. and Bhagwat, M. (2006). Functional Heterogeneity of RpoS in stress tolerance of Enterohomorrhagic *Escherichia coli* strains. *Appl. Environ. Microbiol.* 72(7):4978-4986.

Bollag, J. M. and Barabasz, W. (1979). Effects of heavy metals on the gentrification process in soil. *J. Environ. Qual.* 8:196-201.

Canovas, D., Cases, I. and DeLorenzo, V. (2003). Heavy metal tolerance and metal homeostasis in *Pseudomonas putida* as revealed by complete genome analysis. *Environ. Microbiol.* 5(12):1242-1256.

Cox, C.B. and Moore, P.D. (2000). Biogeography: *An Ecological and Evolutionary Approach,* 6th ed. Blackwell Science, Oxford, United Kingdom.

Cunningham, S.D., Berti, W.R. and Huang, J.W. (1995). Phytoremediation of contaminated soils. *Trends. Biotechnol.* 13:393-397.

Dameron, C.T. and Harrison, M.D. (1998). Mechanism for protection against copper toxicity. *Am. J. Clin. Nutr.* 67:1091-1097.

Diels, L., Dong, Q., Lelie, D. Van der, Baeyens. W. and Mergeay, M. (1995). The czc operon of *Alcaligenenes eutrophus* Ch34: from resistance mechanism to the removal of heavy metals. *Ind. .J. Microbiol* 14:142-153.

Dilek, F.B and Yetis, U. (1992). Effects of heavy metals on activated sludge process. *Water Sci. Technol.* 26:801-813.

Ehrlich, H.L. (1996). *Geomicrobiology,* 3rd ed.Marcel Dekker, Inc., New York, N.Y.

Fagan, M.J. and Saier, M.H.J. (1994). P-type ATPases of eukaryotes andbacteria: sequence comparisions and construction of phylogenetic trees. *J Mol Evol* 38: 57-99.

Fath, M. J. and Kolter, R. (1993).ABC transporters: the bacterial exporters. *Microbiol Rev.* 57: 995-1017.

Gadd, G. M. (1992). Heavy metal pollutants: Environmental and Biotechnological aspects. *Encyl. Microbiol.* 2:351-360.

Garbisu, C. and Alkorta, I. (2001). Phytoextraction: a cost-effective plant based technology for the removal of metals from the environment. *Bioresource Technol.* 77:229-236.

Gekeler, W.E., Grill, E., Winacker, and Zenk, M.H. (1988). Algae sequester heavy metals via synthesis of phytochelatin complexes. *Arch Microbiol.* 150:197-202.

Gupta, A, Kumar, M. and Goel, R. (2004). Bioaccumulation properties of Nickel, Cadmium and Chromium resistant mutant of *Pseudomonas aeruginosa* NBRI 4014 at alkaline pH. *Biol. Trace Ele. Res.* 98:1-9.

Gupta, A. (2006). Diversified arsenic resistant microbial population from industrial and ground water sources and their molecular characterization. Ph.D. Thesis submitted to G.B.P.U.A. and T., Pantnagar.

Gupta, A., Rai, V., Bagdwal, N. and Goel, R. (2005). *In situ* characterization of mercury-resistant growth promoting fluorescent pseudomonads. *Microbiol. Res.* 160:385-388.

Herawati, N., Susuki, S., Hayashi, K. and Koyama, H. (2000). Cadmium copper and zinc levels in rice and soil of Japan, Indonesia and China by soil type. *Bull. Environ. Contem. Toxical.* 64:34-39.

Higham, D.P., Sadler, P.J. and Scawen, M.D. (1984). Cadmium resistance in *Pseudomonas putida*: growth and uptake of cadmium. *J Gen Microbiol*. 131:2539–2544.

Hussein, H., Farag, S. and Moawad, H. (2003). Isolation and characterization of *Pseudomonas* resistance to heavy metal contaminants. *Arab. J. Biotechnol.* 7:13-22.

Inoube, M., Hiyama, M., Tohoyama, H., Joho, M. and Murayama, T. (1989). Cadmium binding protein in a cadmium resistant strains of *Saccharomyces cerevisiae. Biochim. Biophys. Acta.* 993:51-55.

Kachur, A.V., Koch, C. J. and Biaglow J. E. (1998). Mechanism of copper-catalyzed oxidation of glutathione. *Free Radical Res.* 28:259-269.

Kaewchai, S. and Prasertsan. (2002).Biosorption of heavy metal by thermtolerant polymer producing, bacterial cells and the bioflocculant. Songklanakarin. *J. Sci. Technol.* 24(3):421-430.

Khan, M. (2005).Cloning and Characterization of Cold Shock Gene/Protein of Fluorescent Pseudomonad Mutant $CRPF_1$. Ph.D. Thesis, Deptt. of Microbiology, G.B.P.U.A.T. Pantnagar, India.

Khan, M., Bajpai, V.K., Ansari, S.A., Kumar, A. and Goel, R. (2003).Characterization and localization of fluoresecent Pseudomonas cold shock protein (s) by Monospecific polyclonal antibodies. *Microbiol. Immunol.* 47(12):895-901.

Kumar, A. (2006). Isolation of diversified extremophilic cadmium resistant microbial population from "semra mines" palmau, Jharkhand and their functional characterization M.Sc. Thesis, Deptt. of Molecular biology and biotechnology, G.B.P.U.A.T. Pantnagar, India.

Kurek, E., Francis, A.J. and Bollag, M. (1991).Immobilization of cadmium by microbial extracellular products. *Arch. Environ. Contam. Toxicol.* 20:106-111.

Liu, W.T., Marsh, T.L., Cheng, H. and Forney, L. J. (1997). Characterization of microbial diversity by determining terminal restriction fragment length polymorphism of genes encoding 16S rRNA. *Appl. Environ. Microbiol.* 63:4516-4522.

Losi, M.E., and Frankenberger, W.T. (1997).Bioremediation of selenium in soil and water. *Soil Sci.* 162:692-702.

Loveley, D.R., Philips, E.J.P., Gorby, Y.A. and Landa, E.R. (1991). Microbial reduction of uranium. *Nature*, 350:413-416.

Lylod, J.R. (2003).Microbial reduction of metals and radionuclides. *FEMS Microbiol. Rev.* 27: 411- 425.

Macaskie, L.E., Empson, R.M., Cheetham, A.K., Grey, C.P. and Skarnulis, A. J. (1992).Uranium bioaccumulation by a Citrobacter sp, as a result of enzymatically-mediated growth of polycrystalline HUO_2PO_4. *Science* 257:782-784.

Macaskie, L.E., Jeong, B.C. and Tolley, M.R. (1994).Enzymatically- accelerated biomineralazation of heavy metals: application to the removal of americium and plutonium from aqueous flows. *FEMS Microbiol. Rev.* 14:351-368.

Mahesh, S., Ramesh, H.S., Sudhir, H.S. and Kumar, B.R. (2001). Microbial scavengers-fungi for Cu^{+2} removal from industrial waste water. *J. Indus. Poll. Conl.* 17 (1):135-140.

Meargeay, M., Nies, D., Schlegel, H.D., Gerits, J., Charles, P. and van Gijsegem, F. (1985). *Alcaligenes eutrophus* CH34 is a facultative chemolithtroph with plasmid-borne resistance to heavy metals. *J Bacteriol* 162:328-334.

Muyzer, G., De Wall, E.C. and Uitterlinden, A.G. (1993). Profiling of complex microbial population by denaturing gradient gel electrophoresis analysis of polymerase chain reaction- amplified genes coding for 16S rRNA . *Appl. Environ. Microbiol* . 59:695-700.

Naghma, N., Hilary, K.Y., Ahmed, N. and Gadd, G.M. (2005). Cadmium Accumulation and DNA homology with metal resistance genes in sulfate-reducing bacteria. *App. and Environ. Microbio.*: 71:4610-4618.

Nies, D.H. (1999). Microbial heavy- metal resistance. *Appl. Microbiol. Biotechnol.* 51:730-750.

Nies, D.H. (1992). Resistance to cadmium, codalt, zinc, nickel in microbes. *Plasmid*, 27:17-28.

Nies, D.H. (1995). The cobalt, zinc and cadmium efflux system czc ABC from *Alcaligens eutrophus* functions as a cation-proton antiportar in *Escherichia coli. J. Bacteriol.* 177:2707-2712.

Nies, D.H. (1995). The Cobalt, Zinc and Cadmium efflux system Czc ABC from *Alcaligenes eutrophus. J. Biol. Chem.* 265:5648-5683.

Nies, D.H. and Silver, S. (1995). Ion efflux systems involved in bacterial metal resistance. *J. Ind. Microbiol.* 14:186-199.

O'Connel, H.A., Kottkamp, G.S., Eppelbaum, J.L., Shibblefidd, B.A., Gilbert, S.E. and Gilbert, E.S. (2006). Influences of Biofilm Structure and Antibiotic Resistance Mechanisms on Indirect pathogenicity in a model polymicrobial biofilm. *Appl. Environ. Microbiol.* 72(7):5013-5019.

Ochiai, E.I. (1987). General principles of biochemistry of the elements. Plenum Press, New York.

Pandey, G., Paul, D. and Jain, R.K. (2005). Conceptualizing "Suicidal genetically engineered microorganism for bioremediation applications. *Biochem. Biophy. Res. Comm..* 327:637-639.

Paulsen I. T. and Saier M. H., Jr. (1997). A novel family of ubiquitous heavy metal ion transport proteins. *J. Membr. Biol.*, 156:99-103.

Paulsen, I. T., Sliwinski, M. K., Nelissen B., Goffeau, A. and Saier, M. H. (1998).Unified inventory of established and putative transporters encoded within the complete genome of *Saccharomyces cerevisiae*. *Febs Lett*, 430:116-125.

Postgate, J. R. (1979). *The sulfate reducing bacteria*. Cambridge University Press, Cambridge. UK.

Roane, T.M. and Pepper, I. L. (2000). Microorganism and metal pollution. *In*: *Environmental Microbiology*, edited by Maier, R. M., Pepper I.L. and Gebra, C.B., Academic Press, London

Saier, M. H. J. (1994). Computer-aided analyses of transport protein sequences: gleaning evidence concerning function, structure, biogenesis, and evolution. *Microbiol. Rev.* 58:71-93.

Saier M. H., Tam, R., Reizer, A. and Reizer J. (1994). Two novel families of bacterial membrane proteins concerned with nodulation, cell division and transport. *Mol Microbiol* 11: 841–847.

Sauge-Merle, S., Cuine, S., Carrier, P., Lecomte-Pradines, C., Luu, D. and Peltier, G. (2003). Enhanced toxic metal accumulation in engineered bacterial cells expressing *Arabidopsis thaliana* phytochelatin synthase. *App. Environ. Microbiol.*. 69: 490-494.

Scott, J.A. and Palmer, S.J. (1990). Sites of cadmium uptake in bacteria used for biosorption, *Appl. Microbiol. Biotechnol.* 33:221-225.

Silver, S. and Mishra, T.K. (1988). Plasmid mediated heavy metal resistance. *Annu. Rev. Microbiol.* 42:717-743.

Silver, S and Phung, L. T. (1996). Bacterial heavy metal resistance: New surprising. *Annu. Rev . Microbiol.* 50:753-789.

Small, J., Call, D.R., Brockman, F. J., Straub, T. M. and Chandler, D.P.(2001).Direct detection of 16S rRNA in soil extracts by using oligonucleotide microarrays. *Appl. Environ. Microbiol.* 67:4708- 4716.

Smith, C. J., Hopmons, P. and Cook, F.J. (1996). Accumulation of Cr, Pb, Cu, Ni, Zn and Cd in soil following irrigation with treated urban effluent in Australia. *Environ. Pollut.* 94:317-323.

Suzuki, M., Rappe, M.S. and Giovannoni, S.J. (1998). Kinetic bias in estimates of coastal picoplankton community structure obtained by measurements of small subunit rRNA gene PCR amplicon length heterogeneity. *Appl. Environ. Microbiol.* 64:4522 - 4529.

Tebo, B.N. (1995). Metal precipitation by marine bacteria: potential for biotechnological application. *Genetic Eng.* 17:231-263.

Tripathi, M. (2005). Isolation and characterization of cadmium and lead resistant plant growth promoting bacteria. Ph.D. Thesis submitted to G.B.P.U. A. & T., Pantnagar.

Tripathi, M., Munot, H.P., Shouche, Y., Meyer. J.M. and Goel, R. (2004). Isolation and functional characterization of siderophore producing lead and cadmium resistant *Pseudomonas putida* KNP9. *Cur. Microbiol.* 50:233-237.

Ueki, T., Sakamoto, Y., Yamaguchi, N. and Michibata, H. (2003). Bioaccumulation of copper ions by *Escherichia coli* expressing vanabib genes from the vanadium rich ascidran *Ascidia syndeiesis. Samea. Appl. Environ. Microbiol.* 69: 6442-6446

Van Roy, S., Peys, K. Dresselaers, T. and Diels, L. (1997). The use of an *Alcaligenes eutrophus* biofilms in a membrane bioreactor for heavy metal recovery. *Res. Microbiol.* 148:526-528.

Vasudevan, P., Padmavathy, V., Tewari, N. and Dhingra, S.C. (2001). Biosorption of heavy metal ions. *J. Scientific Indl. Res.* 60:12-120.

Volesky, B. and Prasetyo, I. (1994). Cadmium removal in a biosorption column. *Biotechnol. Bioeng.* 43:1010-1015.

Wang, C.L.., Michel, P.C., Dawson, S.C., Kitisakkul, S., Baross, J.A., Keasling, J.D. and Clark, D.S. (1997). Cadmium removal by a new strain of *Pseudomonas aeruginosa* in aerobic culture. *Appl. Environ. Microbiol.* 63(10): 4075-4078.

World Health Organization (1972). *Tech. Rep. Ser.* WHO No. 505.

Xian, X. (1989). Effect of chemical of lead, cadmium and zinc in polluted soil on their uptake by cabbage plants. *Pl. Soil.* 113:257-264.

□□□

CHAPTER **11**

Role of Bacterial Plasmid in Xenobiotic Degradation

Abdullah Altalhi, Bahig El Deeb and Youssuf Gherbawy*

Biology Department, Faculty of Science, Taif University, Taif, Saudi Arabia

ABSTRACT

Plasmids are circular, double-stranded DNA molecules which replicate autonomously from the chromosome of the host organism. Plasmids are found in many bacterial species and often confer upon the host cell phenotypic characteristics such as resistance to antibiotics, resistance to heavy metals, ability to produce particular protiens, conjugative (mating) ability and ability to degrade xenobiotc compounds. One of these degradative plasmids may encode a complete catabolic pathway (such as toluene to pyruvate, formate and acetaldehyde) or partial degradative steps, such as octane to octanol. In this later case, the remainder of the pathway may be encoded by chromosomal genes, by genes of other plasmid present in the host, or by other microorganisms. Many degradative plasmids possess a broad host range, especially among the gram-negative bacterial species.

There are different mechanisms which may be involved in the evolution of aromatic pathways in general and in the process of adaptation of microorganisms to xenobiotic substrates in particular. Gene transfer, the occurrence of plasmids in bacteria in the natural environment is an important pool of genetic information residing on plasmid vehicles may flow among indigenous microorganisms. Since many examples of self-transmissible plasmids that carry genes for degradation of aromatic or other xenobiotic compounds are known, and their role in spreading these genes to other microorganisms is well characterized. Early, studies of TOL, NAH, and SAL plasmids revealed strong homologies among these plasmids in regions both inside and outside the catabolic gene clusters. Plasmids, pJP4, pAC25, and PSS50, catabolic plasmids carrying genes for chloroaromatic degradation, have strongly homologous plasmid backbone that determine replication and transfer function. Furthermore, the chlorobenzoate plasmid p51 was also shown to be homologous to pJP4 outside regions of the catabolic genes. These observations suggest that a few common self-transmissible ancestor replicons (includes origin of replication and replication gene) may have been involved in acquisition and spread of different catabolic genes. DNA

*Email: youssufgherbawy@yahoo.com

rearrangements, the order of genes encoding the *ortho* cleavage pathways of *Acinetobacter calcoaceticus* and *Pseudomonas putida* differ from one another and from those of other organisms, suggesting that various DNA rearrangements have occurred. Further examples are found in a comparison of TOL and NAH encoded catabolic genes. The genes order of part of the *meta* cleavage operons are identical but are different in down stream genes. Gene duplication, has been considered an important mechanism for the adaptation of bacteria to different xenobiotics. For example, *Flavobacterium* sp. strain 172 harbors two isoenzyme of 6-aminohexanoate dimer hydrolase, one of the enzymes involved in the degradation of nylon oligomers. Both enzymes were encoded by plasmid pOAD2 in regions that were proposed to be duplicate. The two isoenzymes differ 100-fold in their activity toward 6-aminohexanoate dimer, despite their amino acid sequence identity of 88%. Studies with various TOL-type plasmids have shown that upper and lower pathway operons as well as *xylS* and *xyl R* regulatory genes, have sometimes inverted or increased their copy number. Furthermore, plasmid pJP4, which contains the genes for the degradation of 2,4-dichloroacetic acid appears to have undergone several genes duplication. Transposition, insertion elements have been shown to play an important role in rearrangement of DNA fragments, in gene transfer, and in activation or inactivation of silent genes. For catabolic pathways several examples of insertion elements are known. The TOL catabolic operons are part of large transposable element, Tn4651, which belongs to the family of Tn3-type transposon. This transposon was found to be of an even larger mobile element, Tn4653. A similar Tn3-type element, Tn4655 contains the catabolic genes on plasmid NAH7, although it is defective in its transposon function. *Pseudomonas cepacia* 249 was shown to carry at least nine different insertion elements, which were present in 1 to 13 copies in its genome. These IS elements are thought to be responsible for the extraordinary adaptability and catabolic potential of this strain. One particular element, IS931, was detected in multiple copies around genes locus for 2,4-dichorophenoxyacetic acid degradation. IS931 is presumed to have played a role in the development of the 2,4-dichorophenoxyacetic acid pathway, as it could mobilize adjacent DNA fragments

Genetic engineering techniques are used also for artificial construction of strains with broad degradative potentials. The use of genetic engineering techniques offers a considerable promise as an approach to accelerate the evolution of desired metabolic pathways. There are two basic techniques employed in the recent environmental biotechnology to enhance the degradative potential of the existing pathways, *in vivo* and *in vitro* genetic selection.

Key words: bacterial plasmid, xenobiotic degradation, xenobiotic compound, phenanthrene.

1. INTRODUCTION

Since the middle of the twentieth century, rapid developments in the chemical industry have led to the distribution of a wide variety of synthetic compounds into the environments. Hutzinger and Verkamp (1981) have developed a scheme for classification of all possible chemicals into two classes, either natural or xenobiotics. Natural compounds are those, which are long-time components of the biosphere, either present as products of biological organisms or present through other natural forces, such as volcanic action (Baross *et al.* 1982). These compounds have been present in the biosphere for a considerable length of time, and are part of the natural cycles. Xenobiotics are those that anthropogenic (man -made chemicals) and they do not (or seldom) occur in nature. Some of these xenobiotics e.g. halosubstituted and nitrosubstituted organics such as haloalkanes, haloaromatic compounds, pentachlorophenol, plastics, detergents, and pesticides, are resistant (recalcitrant) to degradation. The structural reasons for recalcitrance often involve carbon-halogen bonds, nitro-substitutents, tertiary and quaternary carbon atoms, and excessive molecular size (Alexander,1981).These recalcitrant or persistent compounds exert toxic, mutagenic or carcinogenic effects on biological system, and in certain instance of acute exposure, have been lethal (Salagovic *et al.*, 2000).

Most synthetic compounds have natural counterparts or are sufficiently similar to naturally occurring compounds to be subjected to microbial metabolism (Reineke, 1998). The other xenobiotics, having molecular structures and chemical bond sequences not recognized by existing degradative enzymes. However, degradation of these compounds in natural environments is mediated primarily by, bacteria and fungi. In particular, degradation of organic pollutants has been investigated mostly in bacteria. Although many bacteria are able to metabolize organic pollutants, a single bacterium does not possess the enzymatic capability to degrade all or even most of the organic compounds in a polluted soil (Reineke, 1998).Thus, mixed microbial communities have the most powerful biodegradative potential because the genetic information of more than one organism are necessary to degrade the complex mixtures of organic compounds present in contaminated areas (Harayama, 1997). The genetic potential and certain environmental factors such as temperature, pH, and available nitrogen and phosphorus sources, therefore, seem to determine the rate and the extent of degradation. The predominant bacteria of polluted soils belonging to the spectrum of genera and species have been described. Pseudomonads, aerobic Gram-negative rods that never show fermentative activities, seem to have the highest degradative potential, e.g., *Pseudomonas putida* and *P. fluorescens* (Reineke, 1998). Other important degraders of organic pollutants can be found within the genera *Comamonas* (Tralau *et al.*, 2001), *Burkholderia* (Rapp and Timmis, 1999), and *Xanthomonas* (Laurie and Lioyd, 1999). A second important group of degrading bacteria is the gram-positive *rhodococci* and coryneform bacteria (Saeki *et al.*, 1999). Many species, now classified as *Rhodococcus* spp. had originally been described as *Nocardia* spp., *Mycobacterium* spp., and *Corynebacterium* spp. *Rhodococci* are aerobic actinomycetes showing considerable morphological diversity (Larkin *et al.*, 1999). In addition, the immense potential of these microorganisms does not only depend on the catabolic enzymes, but also on their capability of metabolic regulation (De Schrijver and De Mot, 1999).

The isolation of adapted microorganisms with the required degradative capacities is usually preceded by an acclimation period; acclimation is generally accomplished by using classic batch and continuous enrichment techniques (Cook *et al.*, 1983; Grady, 1985).

Acclimation of microorganisms on xenobiotics compounds, which can be defined either as a decrease in the lag period (Lewis *et al.*, 1986; El-Deeb and Soltan, .2000) or functionally as an increase in the degradation rate (Aelio *et al.*, 1990), is the results of several mechanisms, including induction or derepression of enzyme, mutation or genetic exchange (gene transfer), adaptation to the toxins or inhibitors (heavy metals, toxic chemicals, antibiotics, bacteriocin) and competition with residual microflora for nutrients or ecological niche (Lewis *et al.*, 1986; Wiggins *et al.*, 1987; Swinddoll *et al.*, 1988; Aelio *et al.*, 1990).

Natural horizontal transfer of genes responsible for biodegradation has been described for several bacterial systems (Davison, 1998) and seems to play a major role in the acclimation of bacterial communities to the environmental pollutants. Furthermore, the versatility of adaptability of bacteria is based on the existence of catabolic plasmid (Janssen *et al.*, 1995).

Many catabolic pathways involved in the degradation of xenobiotic compounds have been found to be coded on catabolic Plasmids and biochemical and genetic studies of these catabolic plasmids extensively investigated (Johri *et al.*, 1999). In the well-studied catabolic TOL plasmid, the genes encoding the enzymes for oxidation of toluene via the *meta*-cleavage of catechol are organized into two operons (Worsey and Williams, 1975; Harayama *et al.*, 1989; Harayama and Rekik, 1990), the upper pathway operon is responsible for the conversion of toluene to benzoate, and the lower pathway operon is responsible for further conversion of benzoate to pyurvate and acetaldehyde. NAH7 another well-known catabolic plasmid, also carries two operons encoding catabolic enzymes for the oxidation of naphthalene via salicylate into central metabolic intermediates (Yen and Gunsalus, 1982; Schell, 1983; Eaton and Chapman, 1992). Furthermore, demonstrated that the catabolic genes coding for the Molluscicide (bayluscide) and nitrobenzene-degrading enzymes are located on two plasmids which are harbored by one bacterial strain (El-Deeb, 1999; Park and Kim, 2000). However, there is evidence for some catabolic pathways that host chromosome and plasmid both contribute genetic information specifying enzymes in degradative sequences of xenobiotic compounds (Wong and Dunn, 1974; El-Deeb *et al.* 2000).

Although soil and water microorganisms have a considerable evolution potential and, under appropriate selective conditions evolve pathways able to degrade many of synthetic compounds, some recalcitrant compounds, such as polychlorinated biphenyl and c-Hexa-chlorocyclohexane (c-HCH), are fully or considerably resistant to degradative activities in natural environment. This challenge has encouraged the use of gene manipulation techniques to construct the new metabolic pathway (namely, metabolic pathway engineering) able to attack recalcitrant compounds (Pieper and Reineke, 2000). Metabolic pathway engineering, by definition, is a new branch of biotechnology dealing with alteration and modification of existing metabolic pathways in the microorganisms in order to optimize the desired properties of the pathways. When applied to biodegradation, metabolic pathway engineering offers a considerable promise as an approach to the construction of a new microorganism with great degradative potentials. Two general strategies can be visualized for this new branch of environmental biotechnology: the constructing of an existing pathway and the assembly of a new route through the functional combination of appropriate sections of different pathways (Reineke, 1998). These strategies have been successfully used to construct the new pathways able to degrade polyhalogenated compounds and complex nitroaromatics.

2. XENOBIOTIC COMPOUNDS

Many xenobiotics are anthropogenic, or man-made, and they are recalcitrant, i.e. not degraded or degraded by processes not well understood, and very slowly (Reineke, 1998). Moreover, unusual substitutions (as with chlorine and other halogenes), unusual bond or bond sequences (such as in tertiary and quaternary carbon atoms), highly condensed aromatic rings and excessive molecular size (in case of polyethylene and other plastics) are some of the common reasons for recalcitrance (Alexander, 1981). Other more keen reasons, include the failure of compounds to induce degrading enzymes, even though it may be susceptible to their action, failure of the compound to enter the microbial cell for lacking suitable permease, unavailability of compounds due to insolubility or adsorption phenomena, and excessive toxicity of parent compound or its metabolic products: all could contribute the lack of biodegradation.

2.1 Types of Xenobiotic Compounds (the most recalcitrant compounds)

Many xenobiotic pollutants that have proven recalcitrant to microbial attack are halocarbons (Reineke, 1988).The carbon-halogen bond is highly stable. As a consequence halocarbons are chemically and biologically very stable. Stability is desirable in some of these products, but the same property causes persistence and environmental problems. The important groups of halocarbons include the following compounds:

2.1.1 Haloalkanes

For example, haloalkyl products and solvents are C1-C2 in which or nearly all hydrogen atoms have been replaced by fluorine-chlorine combinations (CCl_3F and CCl_2F_2). They serve as solvent and aerosol products in spray cans for such things as cosmetics, paints, and pesticides. In addition, dichloromethan, chloroform, carbon tetrachloride, dichoro-, trichloro- and tetrchloroethenes are important industrial and cleaning solvents. Spillage of these compounds seriously contaminated ground water (Fishbein, 1979; Esposito *et al.*, 1980).

2.1.2 Halobenzenes, halophenols, and halobenzoates

Chlorobenzenes are extensively used as industrial solvents. Chlorophenol, particularly the highly chlorinated pentachorophenol (PCPs) are used as preservatives for woods. Other chlorophenols are used in the synthesis of pesticide, resins, dyes, and pharmaceuticals (Fishbein, 1979).

2.1.3 Recalcitrant nitroaromatic compounds

Nitroaromatics are common military explosives, such as trinitrotoluene (TNT), solvents such as nitro benzene and pesticides, such as nitrophenol. Similar to halogen substitutents, nitro-substitutions of aromatic rings have a strong negative effect on biodegradability (Hallas and Alexander, 1983).

2.1.4 Polychlorinated biphenyl and dioxins

Polychorinated biphenyl (PCPs) is mixtures of biphenyl with one to ten chorine atoms per molecule. They are oily fluids with high boiling point, great chemical resistance. Because of these properties, they have been used as plasticizers in polyvinyl polymers, heat exchange fluids. The chemical inertness is paralleled by their great biologically stability (D'Itri and Kamrin, 1983).

2.1.5 Synthetic polymers

Synthetic polymers are used for plastic industries. More than 90% of plastic material consists of polyethylene, polyvinyl chloride, and polystyrene. Plastics in the environment have been regard as hazardous compounds because the materials are biologically inert (Mathur and Verma, 1992).

2.1.6 Petroleum hydrocarbon

Petroleum is one sense, a natural product, resulting from anaerobic conversion of biomass under high pressure and temperature. It has always entered the biosphere through production, transportation, and refining and causes serious problem to the all living organisms especially marine life. Petroleum is complex mixture of aliphatic, alicyclic, and aromatic hydrocarbons and the main source of the chemical production either in number or volume (Mehlman, 1992).

2.1.7 Recalcitrant halogenated pesticides

The chlorinated pesticides are relatively resistant to microbial attack. Biodegredation of halogenated compounds depends on the ability of bacterial dehalogenase enzymes to catalyze the cleavage of the carbon-chlorine bonds. There are seven known mechanisms for enzyme-catalyzed dehalogenation: reductive, oxygenolytic, hydrolytic, and thiolytic dehalogenation; intramolecular nucleophilic displacement; dehydrohalogenation; and hydration. (Fetzner, 1998 reviews the chemistry involved in these processes). Reductive dehalogenation in methanogenic conditions, as well as aerobic degradation with monooxidase of persistent compounds such as DDT and polychlorinated biphenyls has been observed (Ware, 1989; Choudhry and Chapalamadugu, 1991).

2.2 Degradation of Xenobiotic Compounds

The most important classes of xenobiotic compounds in the environment are mineral oil constituents and halogenated products of petrochemicals. Therefore, the capacities of aerobic microorganisms are of particular relevance for the biodegradation of such compounds and are exemplarily described with reference to the degradation of aliphatic and aromatic compounds as well as their chlorinated derivatives. Although, aerobic degradation processes are certainly not the only important biodegradation in the natural environments, the most rapid and complete degradation of the majority of xenobiotics is brought about under aerobic condition (Reineke, 1998).

2.2.1 The Major Pathway Involved in the Degradation of Several of the Classes of Xenobiotics, Which are Frequently of Environmental Concern

i. Aliphatic hydrocarbon

The initial attack on alkanes occurs by enzymes that have strict requirement for molecular oxygen, that monooxygenases or dioxygenases (Fig. 1, 2) (Britton, 1984). Long-chain *n*-alkanes (C10-C24) are degraded most rapidly by mechanisms demonstrated in Fig 2. Short-chain alkanes (less than

C9) are toxic to many microorganisms, but they evaporate rapidly from petroleum contaminated sites. Oxidation of alkanes is classified as being terminal or diterminal. The monoterminal oxidation is the main pathway. It proceeds via the formation of corresponding alcohol, aldehyde, and fatty acid. β-oxidation of fatty acids results in the formation of Acetyl-CoA (Fig. 1). The subterminal oxidation occurs with lower (C3-C6) and longer alkanes with formation of a secondary alcohol and subsequent ketone (Britton, 1984). Unsaturated 1- alkanes are oxidized at the saturated end of chains (Fox *et al.*, 1990; Saeki *et al.*, 1999). A minor pathway has been shown to proceed via an epoxide, which is converted to a fatty acid.

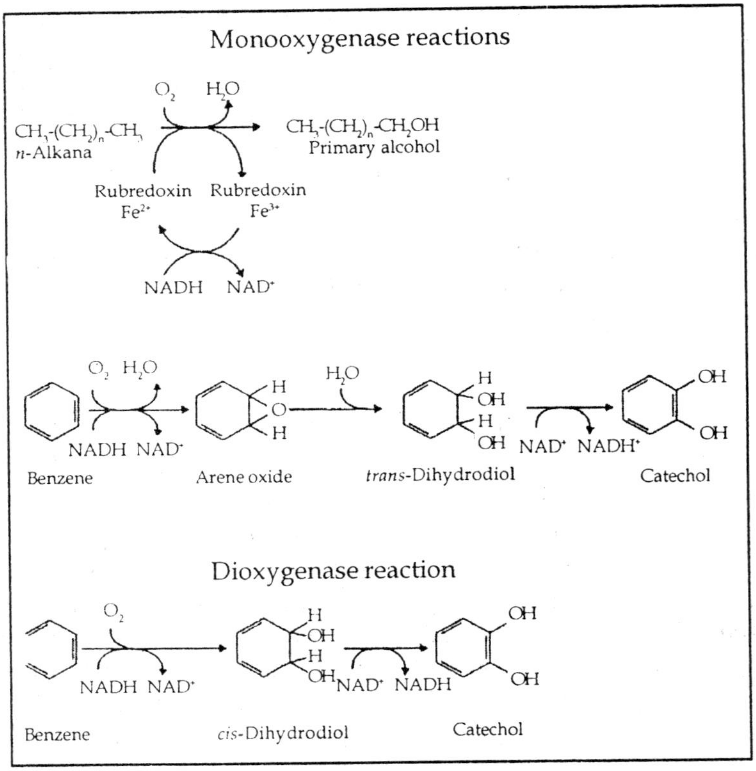

FIG. 1. Initial attack on xenobiotics by oxygenases. Monooxygenases incorporate one atom of oxygen of O_2 into the substrate, the second atom is reduced to H_2O. Dioxygenases incorporate both atoms into the substrate (Data from Gibson, 1984)

The presence of branching within the alkane molecule will prevent the β-cleavage reactions. Some highly branched alkanes such as pristan (2,6,10,14 tetramethylpentadecane) are extremely resistance to microbial degradation. Although the highly resistance of branching alkane to degradation, there are pathways by which the oxidation of branched alkane can occur (Britton, 1984).

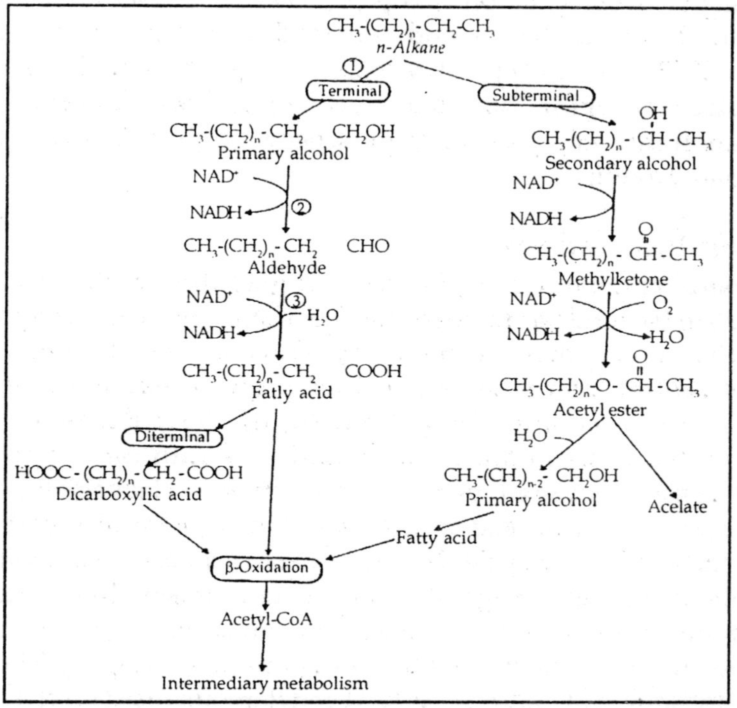

FIG. 2. Peripheral pathways of alkane degradation. The main pathway is the terminal oxidation to fatty acids catalyzed by (1) *n*-alkane monoxygenase, (2) alcohol dehydrogenase and (3) aldehyde dehydrogenase (Data from Gibson, 1984).

ii. Halogenated aliphatic hydrocarbons

Degradation of haloalkane by gram-negative and gram-positive (Scholtze, *et al.*, 1987; Sallis *et al.*, 1990, Kulakova *et al.*, 1995) bacteria has been demonstrated. Haloalkane dehalogenases are unique enzymes which catalyze hydrolytic dehalogenation of corresponding hydrocarbons without participation of oxygen or coenzyme (Curragh, 1994). Moreover, short-chain 1-haloalknes (C3-C8) are metabolized by the initial action of a hydrolytic dehalgenase to produce the corresponding alcohol (Curragh *et al.*, 1994; Kulakova *et al.*, 1995). However, growth on long-chain 1-haloalkane (C12-C18) results only in a low level of dehalogenase activity. The attack of long-chain 1-haloalkane is initiated by oxygenase action at the non-halogenated end to produce β-chlorofatty acids. These are then degraded by β-oxidation (Curragh *et al.*, 1994).

Halogenated hydrocarbons can be degraded under both anaerobic and aerobic conditions. Anaerobically, degradation proceeds through a process called reductive dechlorination. In this process, halogen atoms are sequentially removed from the molecule and replaced with hydrogen (Bouwer and McCarthy, 1983). In this type of reaction, halogenated hydrocarbon is used as an electron acceptor, not as a source of carbon. Therefore, reductive dechlorination can only proceed in presence of appropriate source of carbon for microbial growth.

iii. Aromatic hydrocarbons

A general comparison of the major degradation pathways of aromatic compounds in bacteria has revealed that the initial conversion steps are carried out by different enzymes but the compounds are transformed to a limited number of central intermediates, such as protoctechuate and (substituted) catechol (Choudhry and Chapalamadugu, 1991; Reineke, 1998). The dihydroxylated intermediates are channeled into one of two possible pathways, a *meta* cleavage-type pathway or an *ortho* cleavage type pathways. Both types of pathways lead to intermediates of central metabolic routes, such as tricarboxylic acid cycle. The generalized scheme of catabolic pathways for aromatic compounds suggests that microorganisms have extended their substrate range by developing peripheral enzymes, which are able to transform initial substrates into one of the central intermediates (Fig. 3). The peripheral pathways are shown in Figure 3. As has been demonstrated for the degradation of mononuclear aromatics such as aniline, benzene, benzoate, salicylate, and toluene and, moreover, for binuclear compounds such as biphenyl, dibenzofuran, dibenzo-*p*-dioxin, and naphthalene, analogous degradative sequences are present.

In these pathways, an aromatic compound without hydroxyl group is activated through a multicomponent dioxygenase such as benzene dioxygenase or biphenyl 2,3-dioxygenase using molecular oxygene (Fig. 3) (Cenci and Caldini, 1997). Subsequently, the resulting cis-dihydrodiol is converted by a dehydrogenase to give a 1,2-diphenolic structure, regaining the aromatic state. The resulting catechol is substrate for ring cleavage, which requires another mole of molecular oxygen. In the case of mononuclear aromatic, the ring-cleavage belongs to the central pathway.

During aniline degradation, the stage of the catechol is reached by the action of aniline dioxygenase which eliminates NH_3 at the same time. In this case no dehydrogenase is needed. Phenol will be converted by a monoxygenase reaction to produce catechol.

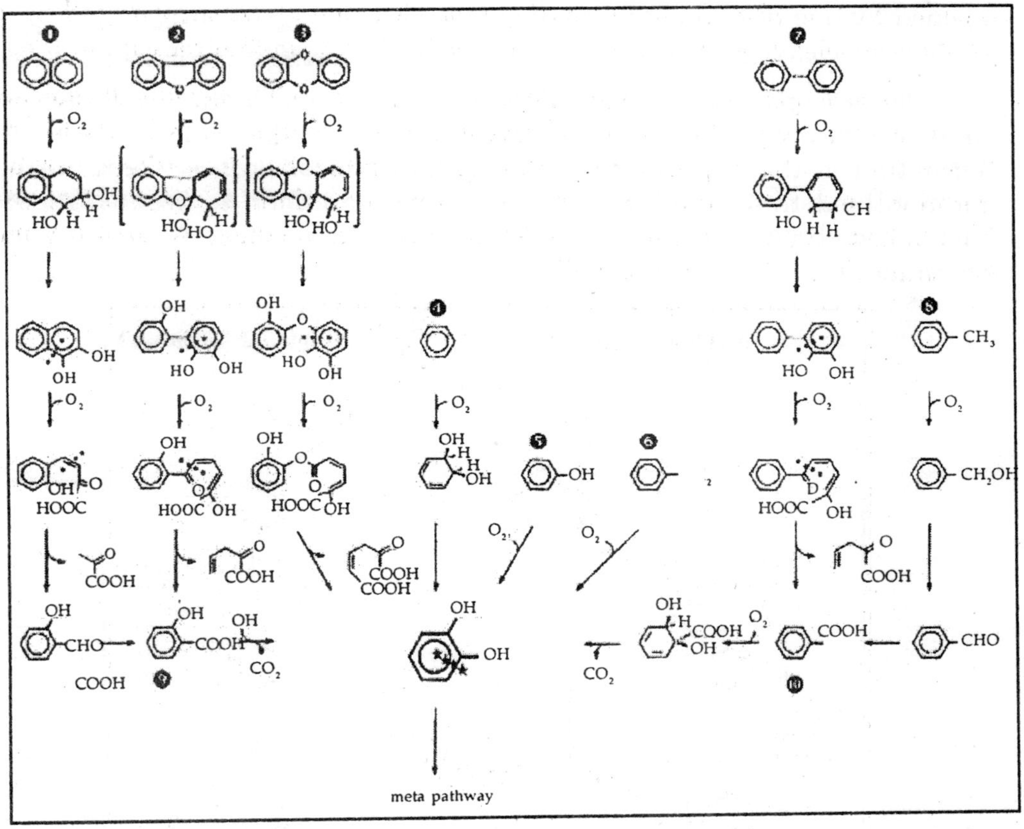

FIG. 3. Peripheral degradation sequences for aromatic with relevance for the development bacterial strains using chloroaromatics. 1, naphthalene; 2, dibenzofuran; 3, dibenzo-ñ-dioxin; 4, benzene; 5, phenol; 6, aniline; 7, biphenyl; 8, toluene; 9, salicylate; and 10, benzoate. Path 4 will also used for toluene and ethylbenzene while xylenes are degraded via path 8 so that methyl-or ethyl-substituted metabolites result, respectively. Catechol or methyl-and ethyl-substituted analogs, the central metabolities, will be degraded via the *meta* pathway. The cleavage of carbon-carbon is marked by *a line stars* (Reineke, 1998).

In the degradation of xylenes, three oxidative reactions-xylene monooxygenase, benzylalcohol, and benzylaldehyde dehydrogenase-initiate the pathway prior to the mentioned ring activation by toluate dioxygenase.

During the degradation of binuclear aromatics, the formation and cleavage of the 1,2-diphenol take place twice. The first cis-dihydrodiols, which result during degradation of dibenzo-p-dioxin and dibenzofuran, are unstable hemiacetals. Therefore, the either bond is cleaved spontaneously, and a dehydrogenase is not necessary to reach the catechol structure (Kanaly and Harayama, 2000). The aromatic nucleus in these compounds is subsequently opened by one of two pathways- the *ortho* cleavage (3-oxoadipate pathway) or the *meta*-cleavage pathway (Gibson, 1984; Gibson and Subramanian, 1984).

The *ortho* pathway involves cleavage of the aromatic nucleus of catechol or protocatechuate between the two-hydroxl groups. This leads to the formation of the respective muconates and mucolactones which further metabolized to 4-oxoadipate enol-lactone and then to 3-oxoadiapate. Metabolism finally proceeds to the Krebs cycle intermediate Acetyl-CoA and succinate (Fig. 4) (Cerniglia. 1992).

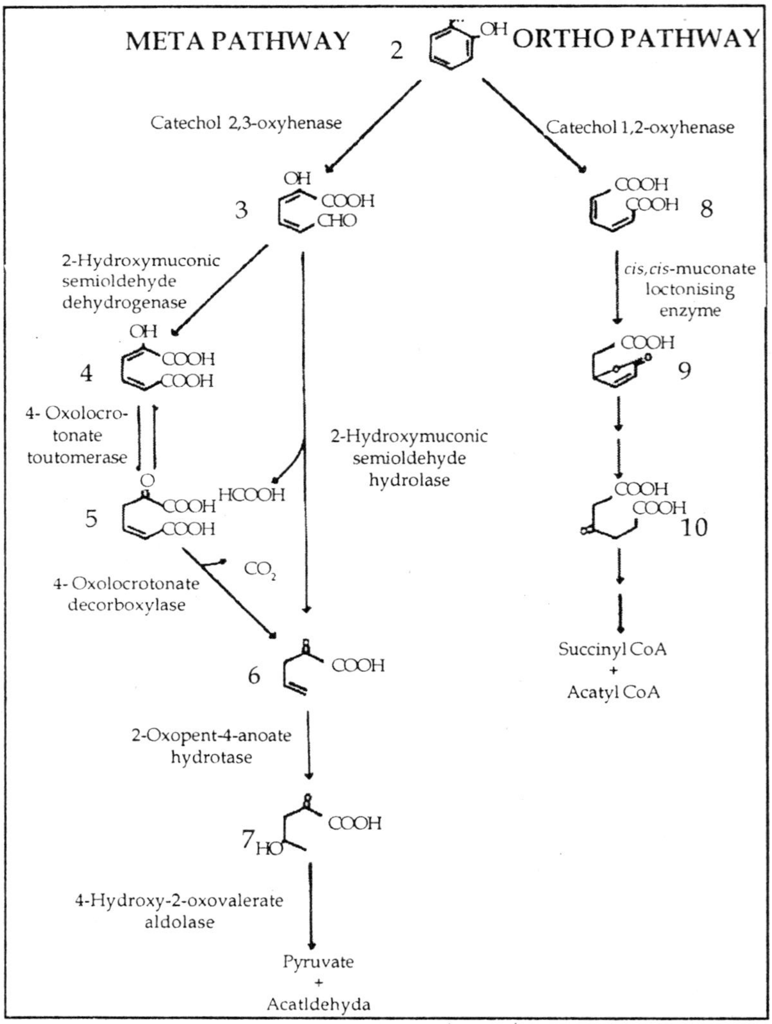

FIG. 4. Ortho (β-ketoadipate) and meta (α-keto acid) pathway utilized by microorganisms for catabolism of catechol (Data from Zuniga *et al.*, 1981).

In *meta* cleavage, initial ring cleavage occurs adjacent to the two-hydroxyl groups forming a 2-hydroxy-muconic semialdehyde as the initial ring cleavage product. Subsequent metabolism results in the formation of pyruvate, formate, and acetaldehyde, which feed into the Krebs cycle (Fig. 4) (Dagley, 1986).

iv. Halogenated aromatic compounds (modified *ortho* pathway)

Degradation of halogenated aromatic compounds proceeds through many of the pathways described for degradation of nonhalogenated aromatic compounds. Chlorobenzene, for example, is first converted to the analogous chlorocatechol (Chatterjee *et al.*, 1983; Gibson and Subramanian, 1984; Cenci and Caldini, 1997; Reineke, 1998). The reaction steps of central degradation pathway of chlorocatechols are shown in Figure 5. Ring cleavage by chlorocatechol 1,2-dioxygenase with consumption of molecular oxygen results in the formation of corresponding chloro-cis-cis-muconate. Further degradation by a chloromuconate cycloisomerase leads to a lactone, which eliminates chlorine as HCl from position 4 or 5, bringing about the formation of anexocyclic double bond. In this way, a (chloro) dienelactone is formed. It is transformed to the corresponding (chloro) maleylacetate via hydrolyic cleavage by (chloro) dienelactone hydrolase. A maleylacetate reductase form 3-oxoadipate, an intermediate shared with the degradation of various aromatic compounds.

Anaerobic degradation of halogenated aromatic compounds such as chlorobenzene, chlorophenol and PCPs has been found to occur in a wide variety of environments. As is the case with anaerobic degradation of halogenated aliphatic hydrocarbons, degradation generally proceeds through reductive dechorination reactions, and highly substituted compounds are more easily dehalogenated than monohalogenaled compounds (Coates *et al.*, 1997).

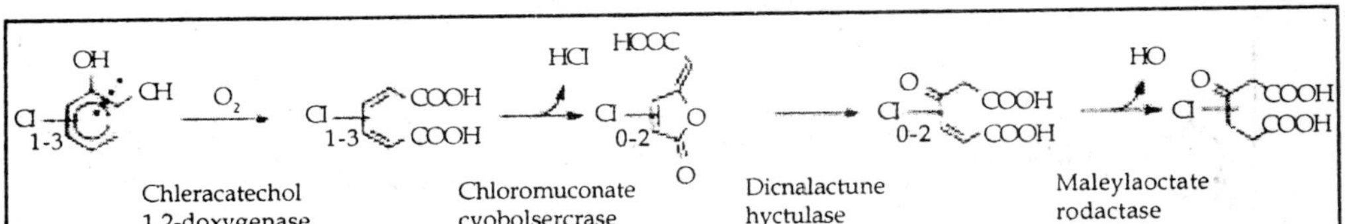

FIG. 5. Simplified presentation of the modified *ortho* pathway. The pathway for the degradation of di- and trichlorocatechols includes two dechlorinating steps. The *ortho* cleavage is marked by *a line of stars*. (Information was compiled from Reineke, 1998).

3. ORGANIZATION OF AROMATIC DEGRADATION PATHWAY GENES

Although aerobic degradation processes are certainly not the only important biodegradation processes in the natural environment, genetic studies have focused mainly on aerobic pathways, and many details of these metabolic routs have been described (Choudhary and Chapalamadugu, 1991; Reineke, 1998).

3.1 *Meta* Cleavage Pathway Genes

Tol plasmids are self-transmissible and contain two or more operons which encode enzymes required for the degradation of methyl benzene, such as toluene, xylenes, and 1,2,4-trimetylbenzene, via methylbenzoate (Franklin *et al.*, 1981; Frantze and Chakrabarty, 1986; Doten *et al.*, 1987; Burlage *et al.*, 1989; Assinder and williams, 1990; Harayama and Rekik, 1990). On the TOLplasmid pWWO, one operopn (*xylCABN*) codes for the "upper pathway" enzymes, which oxidize methylbenzenes to methylbenzoates (Harayama *et al.*, 1989), while the "lower pathway" or *meta* operon is composed of several genes that encode enzymes for conversion of methylbenzoates to pyruvate, acetaldehyde, and acetate via (methyl) catechols (Franklin *et al.*, 1981; Harayama *et al.*, 1984; Hayase *et al.*, 1990). The *meta* cleavage pathway enzymes are encoded by the genes *xylE, xylG, xylF, xylJ, xylQ, xylK, xylI,* and *xylH,* in order, which are part of the *meta* operon (Fig. 6).

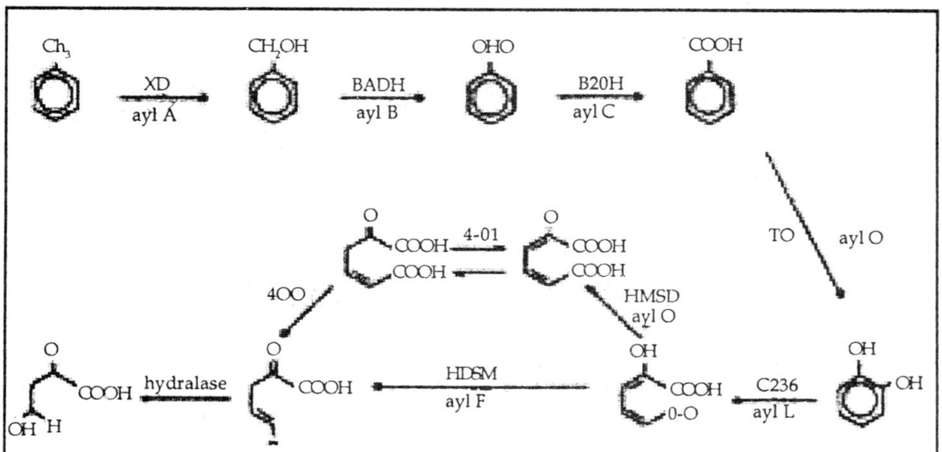

FIG. 6. Tol/xyl catabolic pathway encoded by plasmid pWWO of *P. putida* mt-2. XO, xylene oxygenase; BADH; benzyl alcohol dehydrogenase; BZDH, benzaldehyde dehydrogenase; TO, toluate/ benzoate oxygenase; C23O, catechol2,3-oxygenase; HMSH,hydroxymuconic semialdehyde hydrolase; HMSD, hydroxmuconic semialdehyde dehydrogenase; 4-OT, 4-oxalocrotonate tautomerase; 4-OD, 4-oxalocroonate decarboxylase; *xyl*A to *xyl*G, genes for the *Xyl/Tol* pathway enzymes (Data from Franklin *et al.*, 1981).

The catabolic genes of plasmid NAH7 which encode the enzymes required for the degradation of naphthalene via salicylate (Yen and Serdar, 1988) are clustered in two operon, the *nah* and *sal* operons (Yen and Serdar, 1988; Schell, 1990). The *nah* operon includes nah A-F genes coding for the conversion of naphthalene to salicylate (Fig. 7). The *sal* operon contains G-M coding for the metabolism of salicylate to tricarboxylic acid-cycle intermediate (Fig. 7).

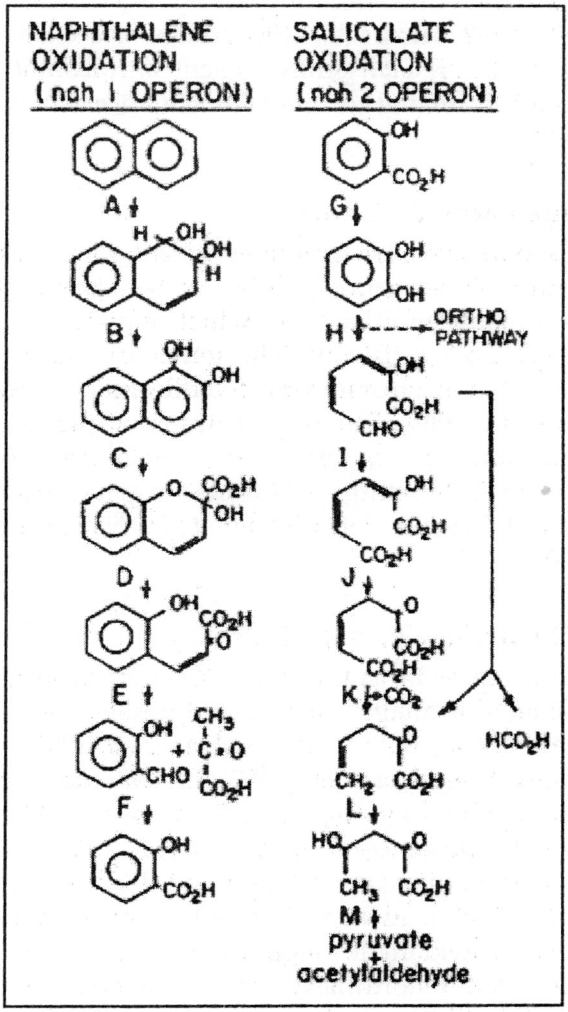

FIG. 7. Plasmid nah7 encoded naphthalene/salicylate oxidation. Capital letters indicate genes or products. A, naphthalene dioxygenase; B, cis-dihydrodiol naphthalene dehydrogenase; C, 1,2-dihydroxynaphthalene dioxygenase; D, 2-hydroxychromene-2-carboxylate isomerase; E, 2-hydroxybenzalpyruvate aldolase; F, salicyaldehyde dehydrogenase; G, salicylate hydroxylase; H, catechol dioxygenase; I, 2-hydroxymuconate semialdehyde dehydrogenase; J, 2-hydroxymuconate tautomerase; K-oxalcrotonate decarboxylase; L, 2-oxo-4-pentenoate; M, 2-oxo-4-hydroxypentanoate aldolase (Data from Yen and Gunsolus, 1982).

Moreover, The *sal* operons contains similar *meta* cleavage pathway genes to those present on pWWO, and DNA sequences of these parts of both operons (e.g *xylEGFJ* and *salGHINL*) are homologous. Another *meta* cleavage pathway was characterized in *Pseudomonas* strain CF600, which uses phenol, cresols, and 3, 4 dimethylphenol as the sole carbon source (Nordlund and Shingler, 1990; Shingler *et al.*, 1992; Kim *et al.*, 1994). This *meta* cleavage pathway is encoded by the *dmpQBCDEFGHI* genes present on plasmid pI150 and was shown to be very closely related to the *xyl*-encoded *meta* cleavage pathway (Shingler *et al.*, 1992).

3.2 Ortho Cleavage Pathway Genes

Ortho cleavage pathways are involved in the degradation of catechol and protocatechoute. These compounds are transformed to a common intermediate, 3-oxoadipate enol lactone, which is furthered converted into succinate and acetyl CoA pathway. The genes encoding *ortho* cleavage pathway enzymes located on chromosome (Neidle and Ornston, 1986; Hughes *et al.*, 1988; Zylstra *et al.*, 1989). The organizations of the *cat* genes encoding catechol degradation and the *pca* genes for protocatechoute degradation of different species show interesting variations in gene order and operonic clustering (Neidle and Ornston, 1986; Doten *et al.*, 1987; Hughes *et al.*, 1988; Ornston *et al.*, 1990).

3.3 Modified of the *ortho* Cleavage Pathways

The isolation and characterization of a number of bacterial strains capable of degrading chlorinated aromatics such as chlorinated benzenes (Spain and Nishino, 1987; Haigler *et al.*, 1988; Van der Meer *et al.*, 1991a; Reineke, 1998), chlorinated benzoates (Weisshaar *et al.*, 1987; Ghosal and You, 1989; Reineke, 1998), and 2,4-dchlorophenoxyacetic acid (Don *et al.*, 1985; Perkins *et al.*, 1990) demonstrated the presence of another set of *ortho* cleavage pathways enzymes. Most of these strains convert chlorinated compounds via a chloro-substituted catechol, which is always cleaved by a chatechol 1,2-dioxygenase (Fig.5). The enzymes involved in the degradation of chlorocatechol have wider substrate specificities than the ordinary *orth* cleavage pathway do. Therefore, the chlorocatechol degradative pathway is also called the modified *ortho* cleavage pathway. In *Pseudomonas* sp strain B13 (Schwien *et al.*, 1988; Reinek and Knackmus, 1988; Reinek, 1998), in *Alcaligenes eutrophus* JMP134 (Streber *et al.*, 1987; Kuhm *et al.*, 1990, Perkins *et al.*, 1990; Pieper *et al.*, 1991), and in other bacteria which metabolize chlorinated benzene (Reinek, 1998), this pathway has been well characterized. In contrast, to the ordinary *ortho* cleavage pathway genes, the genes for the modified *ortho* cleavage pathways are generally located on catabolic plasmid (Franzt and Chakrabarty, 1987; Van der Meer *et al.*, 1991b; Reinek, 1998), and their organization into operon structures differs substantially from that of the chromosomally encoded *cat* and *pca* genes. Modified *ortho* cleavage pathway genes from three bacteria have been extensively characterized (Fig. 8): (i) the *clcABDE* operon of *P. putida* (pAC27) (Chatterjee *et al.*, 1981; Chatterjee and Chakrabarty, 1983; Ghosal and You, 1989), (ii) the *tfdCDEF* operon of *Alcaligenes eutrophus* JMP134 (pJP4) (Don *et al.*, 1985; Mae *et al.*, 1993; Filer and Harker, 1997), and (iii) the *tcbCDEF* operon of *Pseudomonas* sp. strain P51 (pP51) (Van der Meer *et al.*, 1991a). In these three organisms, the chlorochatechol 1,2-dioxygenase genes and chlorocycloisomerase genes of these operons are significantly homologous to the same enzymes of the ordinary *ortho* cleavage pathway genes (Franzt and Chakrabarty, 1987; Perkins *et al.*, 1990; Van der Meer *et al.*, 1991b).

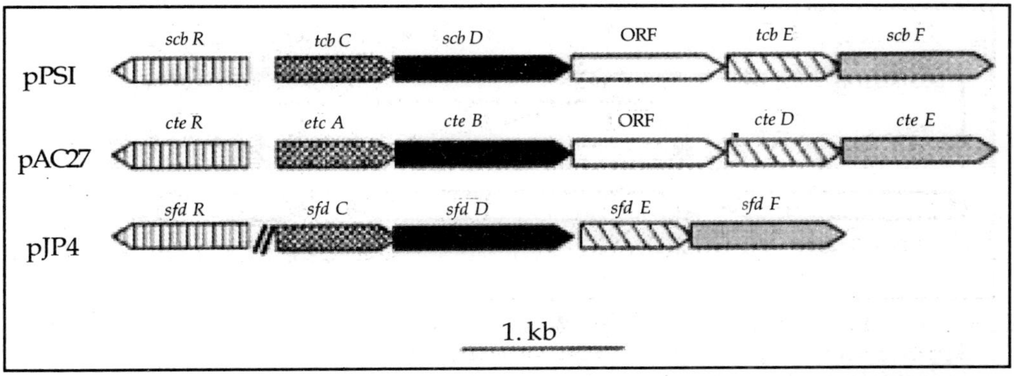

FIG 8. Operon structure of the genes of the modified *ortho* pathway with adjacent regulatory genes (tcb, trichlorobenzene; clc, chlorocatechol; tfd, two-four-D). Each arrow corresponds to a gene or an open reading frame (ORF). Homologous genes are marked the same way. The genes of the chlorocatechol 1,2-dioxygenases (tcbC, clcA, tfdC) are marked with crossed diagonals, those of the chloromuconate cycloisomerases (tcbD, clcB, tfdD) are colored black, those of the dienelactone hydrolases (tcdE, clcD, tfdE) are marked with diagonals, and those of the maleylacetate reductase (tcdF, clcE, tfdF) are colored gray. The regulator genes tcbR, clcR, and tfdR are marked with vertical stripes. tfdR is located several kilobases upstream of tfdCDEF. Arrows without tips denote an overlap of the reading frames (Data from Reineke, 1998).

3.4 Regulatory Genes

Catabolic pathways are controlled by specific regulatory proteins, which recognize inducer molecules and interact with promoter-operator regions of the catabolic operons. A number of regulatory genes of catabolic pathways and their products have been well characterized. In the TOL-encoded pathway, *xylR* and *xylS* (Fig. 9) coordinately regulate the full transcription of xyl operon (Inouye *et al.*, 1983; Inouye *et al.*, 1987; Nakazawa *et al.*, 1990; Filer and Harker, 1997). *XylR* responds to xylenes and methylbenzylalochols and activates transcription of the upper pathway genes (*xylCABN*) and of *xylS* gene (Holtel *et al.*, 1990). The transcription is mediated by an alternative RNA polymerase the RpoN sigma factor (Kohler *et al.*, 1989,). In addition, an integration host (IHF)-like protein is required for the expression of the upper operon (de Lorenzo *et al.*, 1991). *XylS* recognizes benzoate and its analogs and activates transcription of the *meta* cleavage genes.

The *nah* and *sal* catabolic operons on plasmid NAH7 are regualted by single regulator protein, *nah R* (Schell, 1985; Schell, 1991). *NahR*, having a size of approximately 1 kb, is transcripted in opposite direction from the target operon, and is separated from it by 200 bp. Moreover, the effector of *NahR* is salicylate (Schell, 1985; Schell, 1990).

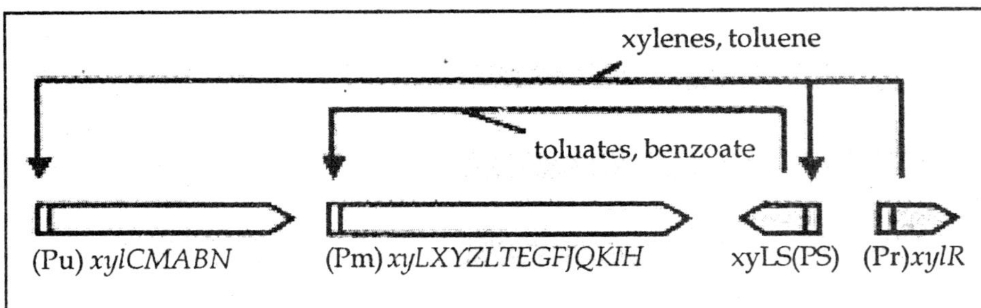

FIG. 9. Figure 14 Simplified model of the regulation of xyl gene operons (xyl, xylene). The upper operon (*xyl* CMABN) codes xylene monooxygenase, benzylalcohol, and benzaldehyde dehydrogenase, while 13 enzymes are encoded by the meta operon (*xylXYZLTEGFJQKIH*), including toluate1, 2- dioxygenase.

4. BACTERIAL PLASMIDS

Plasmids are circular, supercoiled DNA present in most species, but not all strains of bacteria. Most plasmids are small, from about 0.2 to 25% of the bacterial chromosome (Zillig *et al.*, 1994; Holmes *et al.*, 1995). Under most condition of growth, plasmids are dispensable to their host cells. Many plasmids, however, contain genes that have value in particular environments. Often these genes are the main indication that a plasmid is present. For example, R plasmid render their host cells resistant to certain antibiotics, so in nature a cell containing such a plasmid can survive better in the environments in which the antibiotic is present. Plasmid replicate autonomously from the chromosome of the host organism. However, plasmids depend on the metabolic functions of the host cell for their reproduction.

4.1 The Role Bacterial Plasmid in Degrading Xenobiotic Compounds

Bacteria have the ability to utilize a wide range of unusual and complexes organic compounds. Moreover, bacteria are able to rapidly evolve new metabolic activities in response to changes in environmental condition. Many degradative pathway are controlled by chromosomal genes, while some pathway, e.g. those for the catabolism of xenobiotic compounds, involve a large number of enzymatic steps many of which are specified by genes borne on degradative plasmids (Timmis *et al.*, 1985). Degradative plasmids specify biodegradation of diverse group of aliphatic compounds (Kulakova *et al.*, 1995; Saeki *et al.*, 1999), aromatic hydrocarbons (He and Spain, 1997; Reineke, 1998), Chlorinated hyydrocarbon (Kosono *et al.*, 1997) and pesticides (Mae *et al.*, 1994; Filler and Harker, 1997).

Degradative plasmids may encode either partial degradative sequences (naphthalene to salicylate (Yen and Serdar, 1988) or complete catabolic pathway (e.g, for toluene) (Frantz and Chakrabarty, 1986). Other plasmid or chromosomal genes may encode the enzymes for the completion of the pathway. For example, the plasmid OCT encodes enzymes catalyzing degradation of octane to its primary alcohol, then to the aldehyde. The aldehyde is converted to the aliphatic acid and, finally degraded by chromosomally-encoded enzymes (Frantz and Chakrabarty, 1986). The enzymes involved in the plasmid-specified degradative pathway are invariably inducible, and comprise coordinately regulated group of genes (operons). Furthermore, operons may be encoded by both chromosome and plasmid. However, the regulation of chromosomally-encoded degradative enzymes is different from that coded by the plasmids. Chromosomally-coded enzymes involved in the degradation of common intermediates, e.g. aliphatic alcohols in alkane degradation are induced by their substrates. In contrast, plasmid-determined enzymes are induced by the primary substrate of the plasmid coded pathway, i.e. the alkane (Frantze and Chakrabarty, 1986).

4.2 Plasmid-mediated Alkane Degradation

i. Octan

OCT plasmid is a catabolic large plasmid, which allows the host cell to grow with n-alkanes; hexane, octane, decane as carbon sources (Chakrabarty and Friello, 1974; Grund, 1975). The degradation starts with the conversion of alkane to its primary alcohol and then to the aldehyde (Fig. 1). The aldehyde is finally converted by chromosomally coded enzymes to the aliphatic acid which undergoes β-oxidation (Frantze and Chakrabarty, 1986). The octane-octanol conversion required a monooxygenase system composed of three proteins: ruberdoxin reductase, ruberdoxin, and β-hydroxylase. All of them are induced by octane whereas octanol and octanoate will not induce octane oxygenase (Chakrabarty *et al.*, 1973).

The OCT plasmid codes for the inducible alkane hydroxylating and primary alcohol-dehydrogenating activities, but not the entire alkane to fatty acid pathway. The existence of a plasmid-coded activity shows a genetic redundancy. In case of isobutyrate genes, the chromosomally coded genes are induced by isobutyrate, while OCT plasmid coded alcohol dehydrogenase is induced by alkane but not by the primary alcohol (Frantze and Chakrabarty, 1986). The genes *alkA, B,* and *C* are clustered and form an operon, which control by the gene *alkR*.

ii. Haloalkanes

Many haloalkanes persist in the environment to become major pollutants of the biosphere (Finerely, 1992). Degradation of haloalkane by gram-negative (Keuning *et al.*, 1985) and gram-positive (Scholtze *et al.*, 1987; Sallis *et al.*, 1990; Kulakova *et al.*, 1995) bacteria has been demonstrated. Haloalkane dehalogenases are unique enzymes which catalyze hydrolytic dehalogenation of corresponding hydrocarbons without participation of oxygen or coenzyme (Curragh, 1994).

Rhodococcus rhodococcus NCIMB 13064 can utilize a wide range of 1-haloalkne as a sole carbon and energy (Kulakova *et al.*, 1995). Moreover, short-chain 1-haloalknes (C3-C8) are metabolized by the initial action of a hydrolytic dehalgenase to produce the corresponding alcohol by strain NCIMB 13064 (Curragh *et al.*, 1994; Kulakova *et al.*, 1995). However, growth on long-chain 1-haloalkane (C12-C18) results only in a low level of dehalogenase activity. The attack of long-chain 1-haloalkane is initiated by oxygenase action at the non-halogenated end to produce *þ-chlorofatty acids.* These are then degraded by â-oxidation (Curragh *et al.*, 1994). Two plasmids, pRTL1 (100 Kb) and pRTL2 (80 Kb), have been found in strain NCIMB 13064 (Kulakova *et al.*, 1995).The genes for the dehalogenation of short-chain 1-haloalkanes are located on pRT1 (Kulakova *et al.*, 1995). However, no association was found between the utilization of 1-haloalkanes with chain length of C12-C18 (Kulakova *et al.*, 1997). This suggested that different genetic determinants are involved in the dehalogenation and utilization of short and long-chain 1-haloalkanes (Kulakova *et al.*, 1995; Kulakova *et al.*, 1997). In addition, three separate events led to the inability of NCIMB 13064 to dehalogenate and utilize short-chain 1-haloalkanes: either the complete loss of pRTL1; or the integration of pRTL1 into chromosome; or deletion of a 20 kb fragment from pRTL1 (Kulakova *et al.*, 1997).

Kulakova *et al.* (1997) also compared the nucleotide sequence of the haloalkane dehalogenase (*dhaA*) gene from *R. rhodococcus* NCIMB 13064 with the haloalkane dehalogenase (*dh1A*) gene from *Xanthobater autotrophicus* GJ10. There is no homology between the two genes. However, the amino acid sequences of the products of these genes showed approximately 30% identity and several of the catalytic amino acid residues were conserved in the NCIMB 13064. Also, numbers of repeated and inverted sequences which may cause genetic instability at the locus were found in the haloalkane dehalogenase genes region (Kulakova *et al.*, 1997).

iii. Haloalkene

Trichloroethene (TCE) is an extensively used industrial solvent and due to its broad application and high volatility it has become a major ground water contaminant. Since TCE is a potential health hazard, intensive efforts have been undertaken to analyze microbial TCE degradation and developing appropriate bioremediation processes (Ensley, 1991). Several organisms have been shown to degrade TCE via nonspecific catabolic oxygenase, such as methane monooxygenase of methanotrophs (Brusseau *et al.*, 1990; Fox *et al.*, 1990), toluene 2-monoxygenase of *Burkholderia* (Newman and Wakett, 1997), alkene monoxygenase of *Xanthobacter* Py2 (Ensign *et al.*, 1992), isopropylbenzene dioxygenase of *Pseudomonas* sp. JR1 and *Rhodococcus* Py2, *erythropolis* BD2 (Dabrock *et al.*, 1994; Kesseler *et al.*, 1996).

Rhodococcus corallinus (formally *Nocardia corallina*) B-276, isolated with propene as sole carbon source, is able to oxidize trichloroethene (TCE), and also initiates the oxidation of alkenes to the corresponding epoxides by an alkene monooxgenase (Furuhashi, 1992). Saeki *et al.* (1999) demonstrated that *R..corallinus* B-276 harbored four linear plasmid; pNC10 (70 kb), pNC20 (85 Kb), pNC30 (185 Kb), and pNC40 (235 Kb). The observation that pNC30-deficient strains had lost the ability to grow on propen suggested that the genes of the propene degradation pathway are encoded by the linear plasmid pNC30 (Saeki *et al.*, 1999). Furthermore, Southeren blot analysis with cloned alkene monooxygenase from strain B-276 revealed that the alkene monooxygenase is encoded by plasmid pNC30 (Saeki *et al.*, 1999).

In addition, eleven short-chain alkene oxidizing strains were screened for the presence of linear plasmids. Among these, two propene oxidizing *R.. ruber* and ethene oxidizing *Mycobacterium* strain were found to contain linear plasmids (Saeki *et al.*, 1999). Southern blot analysis of alkene monooxygenase gene from pNC30 revealed positive signals with linear plasmid of two propene-oxidizing *R. ruber*. These observations indicate that homologus alkene monooxygenases are encoded by linear plasmids in *R. corllinus* and *R. ruber*

(Saeki *et al.*, 1999). However, no hyberdization signal was detected with alkene monooxygenase gene located on the linear plasmid pBD2 harbored by propene oxidizing *R. erythropolis* (Dabrock *et al.*, 1994). This leads to the suggestion that the *R. erythropolis* gene responsible for the growth on propene differs from alkene monooxygenase gene in *R.corllinusi* B-276 and might even encode an enzyme in different family (Saeki *et al.*, 1999).

4.3 Plasmids-mediated Aromatic Compounds Degradation

i. Toluene/Xylene

Tol and XYL are different plasmids that specify the same degradative function. In other words, they appear to have identical xylene degradative genes, and the regulation of the two pathways appears to be the same.

The TOL plasmid originally described (Williams and Murray, 1974) for *P. putida* (arvilla) mt-2, represent a group of conjugative plasmids that specify degradation of hydrocarbons such as *m*-or *p*-xylene, 1,2,4-trimethylbenzene, 3-ethyltoluene, toluene, and their corresponding alcohol and acid derivatives through the *meta* pathway (Frantz and Chakrabarty, 1986). Moreover, the TOL plasmid specifies *meta* pathway enzymes and toluate oxygenase for the conversion of *m*- or *p*-toluate (Fig. 6).

The plasmid XYL has been characterized in *Pseudomonas* sp. Pxy, for which *p*-or *m*-xylene are metabolized through intermediate formation of the corresponding methylbenzene alcohol and acid and toluic acid also via *meta* pathway. Friello *et al.* (1976) suggested that the genes for the entire xylene-degradative pathway in *Pseudomonas* sp. pxy are borne on non-conjugative plasmid (Friello *et al.*, 1976) which is transmissible in presence of a helper plasmid (Chakrabarty and Friello, 1974).

A set of inducible enzymes specified by genes on the TOL plasmid enables *Pseudomonas* to grow on the aromatic hydrocarbon toluene and its alcohol, aldehyde and carboxylic acid derivatives (Spooner *et al.*, 1986). The genetic organization of the TOL plasmid has been analyzed by both restriction endonucleases (enzyme for DNA cutting in definite site), mapping, transposon mutagenesis, and gene cloning (Nakazawa *et al.*, 1985; Inouye *et al.*, 1983). Two operons (set of genes controlled by one promoter) separated by 10 kb organize the genes encoding the enzymes (Timmis, 1985). The *xylCABN* operon (op1) is responsible for the "upper catabolic pathway from toluene and xylene to benzoate through benzyl alcohol and benzaldehyde (Fig. 6) and the *xylDLEGF* operon (op2), which is required for the "lower" catabolic pathway, encodes enzymes for the oxidation of aromatic carboxylic acid to compounds that enter the tricarboxylic acid cycle, puruvate and acetaldehyde (Fig. 6) (Lebenze and Williams, 1985).

ii. Nitrobenzene

Nitrobenzene, one of the recalcitrant nitroaromatic compounds has been reported to be mineralized by microorganisms through oxidative (Nishino and Spain, 1993) and partial reductive pathways (He and Spain 1997). Park and Kim (2000) demonstrated that *P. putida* HS12, which is able to grow on nitrobenzene, was found to carry two plasmid, pNB1 and pNB2 (Fig. 13). The activity assay experiments of wild type HS12 (pBN1 and pBN2), a spontaneous mutant HS121 (pNB2), and a cured derivatives HS124 (pNB1) demonstrated that the catabolic genes coding for the nirobenzene-degrading enzymes, designated *nbz*, are located on two plasmids (Park and Kim, 2000). Furthemore, the genes *nbzA, nbzC, nbzd, and nbzE* encoding nitrobenzene nitroreductase, 2-aminophenol 1,6 dioxygenase, 2-aminomuconic 6-semialdehyde dehydrogenase, and 2-aminomuconate deaminase, respectively (Fig. 10), are located on pBN1 (59.1 Kb). Meanwhile, the *nbzB* gene encoding

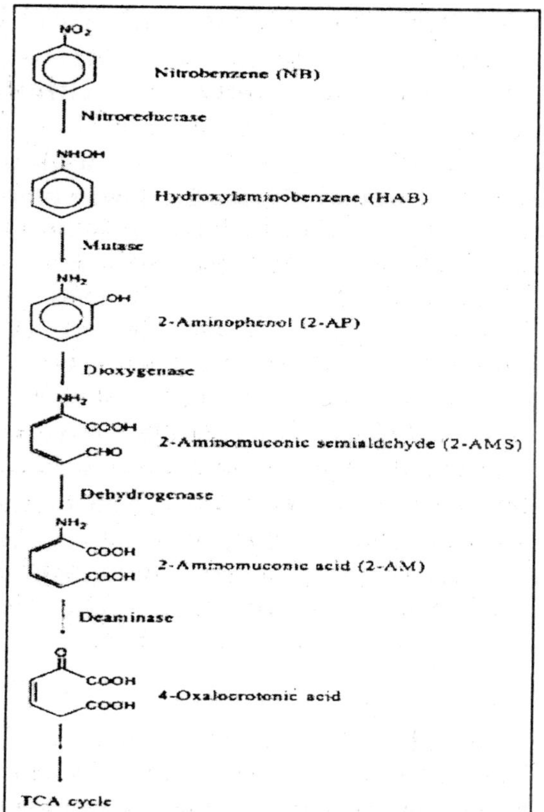

FIG. 10. A partial reductive pathway for the degradation of nitrobenzene (Data from Park and Kim, 2000)

hydroxyylaminobenzene mutase, a second step enzyme in nitrobenzene catabolic pathways, was found in pNB2 (43.8). Physical mapping, cloning, and functional analysis of the two plasmids are revealed in more detail the genetic organization of catabolic plasmids, pBN1 and pBN2. The genes *nbzA* and *nbzB* are located on the fragments 1.1 kb of pBN1 and 1.0 kb fragments of pBN2, respectively (Park and Kim, 2000). On the other hand, the genes *nbzC*, *nbzD*, and *nbzE*, involved in the ring cleavage pathways of 2-aminophenol are located on the 6.0 Kb fragment of pBN1 and clustered in the order *nbzC-nbzD-nbzE* as an operon.

iii. 1,2 dichloro-, 1,4 dichloro-, and 1,2,4 trichlorobenzene

Bacteria able to use chlorinated benzenes as sole carbon and energy substrates (Van der Meer *et al.*, 1991a) were found to oxidize the chlorinated benzene to a chlorocatechol by the action of a dioxygenase enzyme and a dehydrogenase. The chlorocatechol could then be degraded via a pathway similar to one described for 3-chlorobenzoate metabolism in *Pseudomonas* sp strain 13, i.e. ring cleavage by a catechol 1,2 dioxygenase enzyme, lactonization by a cycloisomerization II, and hydrolysis by hydrolyase II, yielding chloromaleylacetic acid (Fig. 5) (Reineke *et al.*, 1988). *Pseudomonase* sp. strain p51 is able to use 1,2 dichlorobenzene, 1,4 dichlorobenzene, 1,2,4 trichlorobenzene as a sole carbon and energy source (van Der Meer *et al.*, 1991a). Two gene clusters are involved in the degradation of these compounds were identified on a catabolic plasmid, p51, with size of 110 kb by using hyberdization (Fig. 11) (van der Meer *et al.*, 1991a). These genes were further characterized by cloning in *E. coli*, *P. pudita* kt244, and *Alcaligenes eutrophus* JMP222. Expression studies in these organisms showed that the upper pathways genes (*tcbA* and *tcbB*) code for the conversion of 1,2 dichlorobenzene and 1,2,4 trichlorobenzene to 3,4 dichlorocatechol and 3,4,6 trichlorocatechol respectively, by means of a dioxygenase and a dehydrogenase (van der Meer *et al.*, 1991a). Furthermore, the lower pathway genes have the order *tcbC-tcbD-tcbE* and encode a catechol 1, 2 dioxygenase II, a cycloisomeraseII and a hydrolyaseII, respectively (Fig. 8). The combined action of these enzymes degrade 3, 4 dichlorocatechol and 3, 4, 6 trichlorocatechol to a chloromaleylacetic acid (van der Meer *et al.*, 1991b).

iv. Phenol and its derivatives

Utilization and degradation of phenolic compounds by microorganisms were reported (Gibson, 1984; Shingler *et al.*, 1992; Dong *et al.* 1992, El-Deeb, 2005). The degradation pathways and regulatory mechanism in metabolism of phenol and its derivatives have been extensively studied (Inouye *et al.*, 1983, Ramos *et al.*, 1986; Dong *et al.*, 1992). The benzene ring of phenol or related compounds may be cleaved by either the *meta*-cleavage pathway or *othro*-pathway through the intermediate compound catechol or substituted catechol (Fig. 4) (Nakazawa and Yokota, 1973). However, hydroxylation of benzene ring is a rate-limiting step in the aerobic catabolism of aromatic compounds (Chakrabarty *et al.*, 1973; Dagley, 1986).

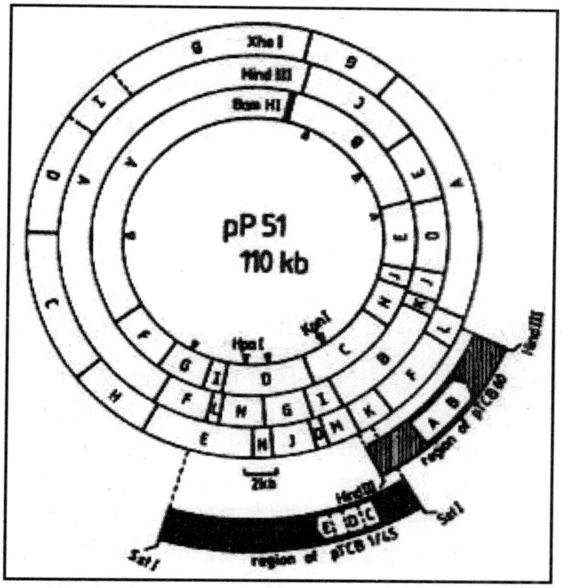

FIG. 11. Physical map of the catabolic plasmid pP51 from *Pseudomonas* sp. strain P51. Capital letters indicate the order of the generated restriction enzyme fragments according to size. The size of the total plasmid was estimated to be 110 kb. Shown are the two gene clusters that were cloned and characterized in this study: (i) the upper-pathway genes tcbA and tcbB, which are located on the HindB fragment of plasmid pP51, and (ii) the genes tcbC, tcbD, and tcbEof the lower or catechol pathway, which are located on the 13-kb Sstl fragment of pP51. Arrows in the gene clusters show the deduced orientations of the genes. Dotted lines (e.g., at XhoI-I) indicate possible ambiguity in locating the fragment (Data from van der Meer *et al.*, 1991b).

Existing genetic evidence suggests that a significant proportion of the pathway specifying catabolism of aromatic compounds by *Pseudomonas* is encoded on plasmid (Frantz and Chakrabarty, 1986). Shingler *et al.* (1992) isolated a phenol and 3,4-dimethyl phenol-degrading *Pseudomonas* sp. CF600 that harbors two plasmids, PVI45and pVI150 , with molecular mass 45 kb and 150 kb respectively. Biochemical and genetic analysis indicated that pVI150 encodes the phenol/3.4-dimethylphenol pathway.

Screening of a gene bank from pVI150 DNA, with DNA from the previously cloned catechol 2,3-dioxygenase gene of TOL plasmid pWWO

(Spooner *et al.*, 1986) indicate that catechol 2,3 dioxygenase, and phenol hydroxylase genes are located on pVI150 (Shingler *et al.*, 1992). Molecular cloning and polypeptide analysis demonstrated that a 1.2 kb region of DNA encoding 39.5 KD a polypeptide which mediated phenol hydroxylase. In addition, Plasmid pVI150 encodes a C23O gene and product that is homologous to that of TOL (Plasmids borne C2,3O genes) on both the DNA and functional level (Fig. 6) (Shingler *et al.*, 1992). Dong *et al.* (1992) cloned a catechol 2,3-dioxygenase gene from *Bacillus stearothermophilus* FDTP-3. The gene encoding catechol 2,3-dioxygenae, which is more thermostable than catechol 2,3 dioxygenase encoded by the other gene, shares rather limited homology with that from *P. putida*.

v. Aniline

Aniline and its derivatives are widely used in the chemical industry in a variety of processes such as the production of dyes, synthesis of chemical compounds, and development of aniline-based herbicides.

Bacterial degradation of aniline has been reported. Metabolization of anilines proceeds via primary oxygenation and ammonium elimination by an aniline oxygenase leading to the corresponding catechol (Fig. 3). The resultant catechol is degraded through either the *ortho*- (Zeyer *et al.*, 1985; Loidl *et al.*, 1990) or the *meta*-cleavage pathway (Fig. 6) (McClure and Venables, 1986; Konopka *et al.*, 1989; Fuchs *et al.*, 1991; El-deeb, 2001). Like other aromatic compound degradative plasmid, TOL, NAH, Sal and so on, plasmids encoding the ability to utilize aniline have been recognized (Fujii *et al.*, 1997; El-Deeb, 2001). Plasmid pCITI carried in *Pseudomonas* sp. strain CITI is about 100 Kb in size and contains the aniline oxidase and catechol 2,3-dioxgenase (C2C3) genes on a 42 kb *Sau*3A fragments (Anson and Mackinnon, 1984; Meyers, 1992). *P. putida* UCC2 bears plasmid (pTDN1) which contains the catabolic genes for aniline, toluidine and m-toluate within a 26 kb region bounded by direct-repeat sequences (Saint *et al.*, 1990). Furthermore, a 9.3 kb fragment of pTDN1 has been sequenced and found to encode several open reading frames (Fukumori and Saint, 1995). These open reading frames were predicted to encode large and small subunits of terminal dioxygenase, an electron-transfer protein, and a Lysr-regulatory protein.

Fujii *et al.* (1997) isolated an aniline-utilizing *Acinetobacter* sp. strain YAA that harbors large (plasmid pYAA). Cells that were cured with acridine orange lost the ability to metabolize aniline. The gene cluster involved in aniline oxidation was cloned in *E.coli* from plasmid DNA and the recombinant *E.coli* containing 18.5 kb insert fragment showed coloration on aniline-containing plates, indicating the formation of 2-hydroxymuconic semialdehyde from aniline (Fujii, 1997). In addition, subcloning of a 9.0 kb *sal*1 fragment from the insert in *E.coli* resulted in the accumulation of catechol. This indicates that C23O dioxygenase reside in large fragment (18.5 kb) not small fragment (9.0 kb).

vi. Biphenyl

The metabolic pathways and genetics of bacteria capable of degrading biphenyl or polychlorinated biphenyl has been described (Furukawa *et al.*, 1990; Chung, *et al.*, 1994; Seto, *et al.*, 1995). Kosono *et al.* (1997) isolated a *R. erythropolis* TA421 that utilizes biphenyl as a sole source of carbon and has seven *phbC* genes expressing 2,3 dihydroxybiphenyl dioxygenase activity. *R. erythropolis* TA421 harbored a large plasmid and probably linear plasmid on which three (*pbhC2, pbhC3 and pbhC4*) of seven *pbhC* genes were located (Kosono *et al.*, 1997). Furthermore, enzyme, i.e. multicomponent biphenyl dioxygenase, (BpbhA), dihydrodiol dehydrogenase (BphB), 2,3-dihydroxybiphenyldioxygenase (BphC), and Hydrolyase (BphD), catalyze biphenyl degradation (Fig. 12) (Asturis, 1991). Several studies have demonstrated that the genes for these enzymes are clustered and transcribed as an operon (Asturis *et al.*, 1994; Hayase *et al.*, 1990; Taira *et al.*, 1992; Masai *et al.*, 1995). In addition, Southeren blot (DNA-DNA hybridization) analysis

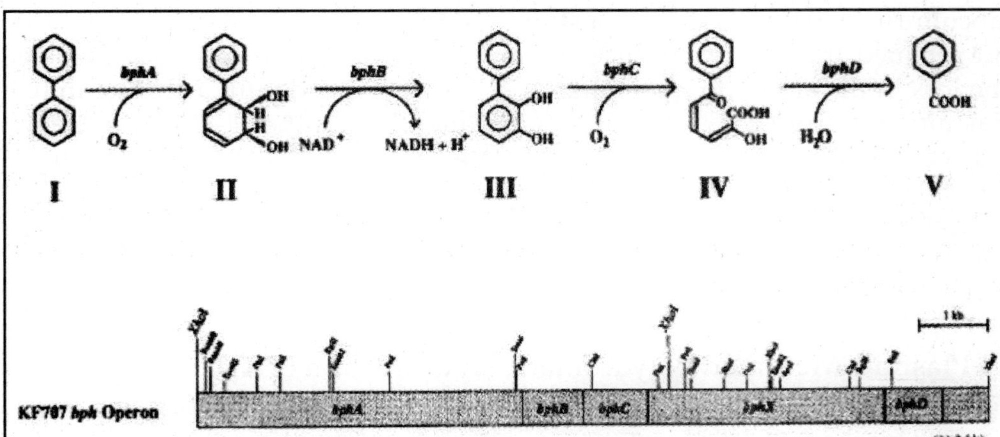

FIG. 12. Catabolic pathway for degradation of biphenyl and chlorinated biphenyls and organization of bph operon in *P. pseudoalcaligenes* KF707. The bphA region consists of at least four genes and their products convert biphenyl (I) to dihydrodiol (11). By the action of dihydrodiol dehydrogenase (product of bphB), I1 is converted to 2,3-dihydroxybiphenyl (111). I11 is then cleaved at 1, 2-position by the action of 2,3-dihydroxybiphenyl dioxygenase (product of bphC), yielding HPDA (IV) which is eventually hydrolyzed to benzoic acid (V) by HPDA hydrolase (product of bphD). bphX region is an extra DNA segment of unknown function (Data from Taira *et al.*, 1992).

showed that *R. erythropolis* TA421 and *R. .globerulus* P6 have similar set of *pbhC* genes and that the genes for biphenyl catabolism are located on plasmid of different sizes (Kosono *et al.*, 1997).

vii. 4-Chlorobiphenyl

Polychlorinated biphenyls (PCB) represent truly ubiquitous environmental pollutants and are considered to be not degradable. Despite this, there have been several reports of PCBs degradation in mixed and pure (Sayler *et al.*, 1990) bacterial culture. The genes for 4-chlorobiphenyl (4CB) degradation are mostly located on plasmid. Furukaw and Chakrabarty (1982) have reported the ability of *Acinetobacter* sp and *P. putida* mixed culture to completely degrade 4-CB. This was reportedly due to the three separated clusters of degradative genes. One clusters, on the pKF1 plasmid (78 Kb) of an *Acinetobacter* sp. encoded the oxidation of 4-chlorobiphenyl to 4-chlorobenzoate (as does pAC21). The other two genetic elements involved the TOL transposon and the pAC27 plasmid (*P. putida*) encoded enzymes for the subsequent mineralization of 4-chlorobiphenyl. Unlike pAC21, pKF1 and pAC27 plasmids, which encode only partial degradation of 4-chlorobiphenyl, another plasmid pSS50 (65 kb) was found to be responsible for the total mineralization of 4-chlorobiphenyl by *Acinetobacter* sp.A5. However, upon repeated subculturing, strain A5 (pSS50) lost the ability to mineralize 4-chlorobiphenyl but retained the ability produce the metabolize 4CB into 4-chlorobenzoate. The loss in the ability to mineralize 4CB was associated with a change in the plasmid size from 55 kb to 45kb. Subsequently, another 4-chlorobiphenyl degrading bacterium strain ICI demonstrated metabolism of 4CB to CO_2. Strain ICI had a plasmid (pSS60) that was very similar to pSS50 with exception of an additional 10 kb. Layton *et al.* (1994) suggested that the extra DNA fragment found in plasmid pSS60 is important in 4-chlorobenzoate, specifically in dehalogenation step.

4.4 Plasmid-Mediated Polyaromatic Hydrocarbon (PAH)

i. Naphthalene

The biochemistry and genetics of bacteria capable of utilizing naphthalene or other polycyclic aromatic hydrocarbons as sole growth substrates have been reviewed extensively (Yen and Serdar, 1988). The genes for naphthalene degradation are located on self-transmissible NAH plasmids in most bacteria (Yen and Gunsalus, 1982; Yen and serdar, 1988).

Yen and Gunsalus (1982) suggested that NAH plasmids can be transformed to other strain by conjugation and that the genes can be expressed in the recipient strains. The NAH7 plasmid has two operon involved in naphthalene metabolism: a *nah* operon that encodes an upper metabolic pathway, for metabolism of naphthalene to salicylate (Fig. 7), and a *sal* operon that encodes a lower pathway, for the metabolism of saliçylate via catechol to pyruvate and acetaldehyde (Fig. 7) (Yen and Gunsalus, 1982). Eleven genes encoded by plasmid NAH7 are responsible for the upper pathway and lower pathways of naphthalene metabolism (Yen and Gunsalus, 1982).

The genes for naphthalene catabolism located on several plasmid of *P. putida* have been analyzed for DNA sequence homology. A 15.0 kb EcoRI fragment of plasmid pDTG1 that included the *nah* operon was nick-translated and hybridized with *EcoR1* fragment of NAH7 and other plasmids (Serdar and Gibson, 1989; Eaton and Chapman, 1992). The extensive DNA sequence homology found between plasmids pTDG1 and NAH7 shows that enzymes encoded by each of the two plasmids are responsible for naphthalene degradation (Serdar and Gibson, 1989).

ii. Phenanthrene

Kim *et al.* (1986), studying an *Acinetobacter* sp. that utilizes phenanthrene as a sole source of carbon, found that the enzymes for phenanthrene metabolism are encoded by genes on two plasmids. These plasmids, of 4 kb and 40 kb, are lost when cells are cured with mitomycin C (Kim *et al.*, 1986).

Guerin and Jones (1988) isolated a *Mycobacterium* sp. that utilizes phenanthrene as a sole source of carbon and harbors three plasmids with molecular weights of 32, 88, and 117 kb. Kiyohara *et al.* (1990) isolated a phenanthrene-degrading *Alcaligenes faecalis* strain that harbors two plasmids, pHK1 and pHK2; the molecular sizes of the plasmids are 66.9 and 33.4 kb. Cells that were cured with mitomycin C lost the ability to metabolize phenanthrene. After the purified plasmid pHK2 had been used to transform several bacterial strains, the expression of phenanthrene dioxygenase in transformants was shown by oxygen consumption in the presence of phenanthrene and the production of indigo by cells incubated with indol (Kiyohara *et al.*, 1990).

4.5 Plasmids-Mediated Pesticide Degradation

4.5.1 Insecticide

i. Fenitrothion

Organophosphorus insecticides such as fenitrothion (O,O-dimethyl O-4-nitro-m-tolyl phosphorothioate) are used for controlling a wide range of insects. Microbial degradation is considered to be a major factor determining the fate of organophosphorus insecticide in the environment. Many bacteria that metabolize organophosphorus insecticide have been isolated (Rani and Lalithakumari, 1994; Hayatsu *et al.*, 2000). A *Burkhoderia* sp strain NF100 capable of utilizing fenitrothion as a sole carbon source was isolated from fenitrothion treated soil (Hayatsu *et al.*, 2000).

Metabolism of fenitrothion in NF100 was initiated by its hydrolysis to 3-methyl-4-nitrophenol, which was metabolized to nitrite and methylhydroquinone, the substrate for oxygenase-catalyzed ring fission (Spain and Gibson, 1991). NF100 was found to harbor two large-molecular size plasmids, designated pNF1 (105 kb) and pNF2 (33 kb). Hayatsu *et al.* (2000) concluded that pNF1 encodes for fenitrothion-hydrolyzing activity and pFN2 encodes for methylhydroquinone-degrading activity, whereas, the oxidative pathway for removal of the nitro group from nitrophenol is mediated by chromosome (Fig. 13). These conclusions were supported by several

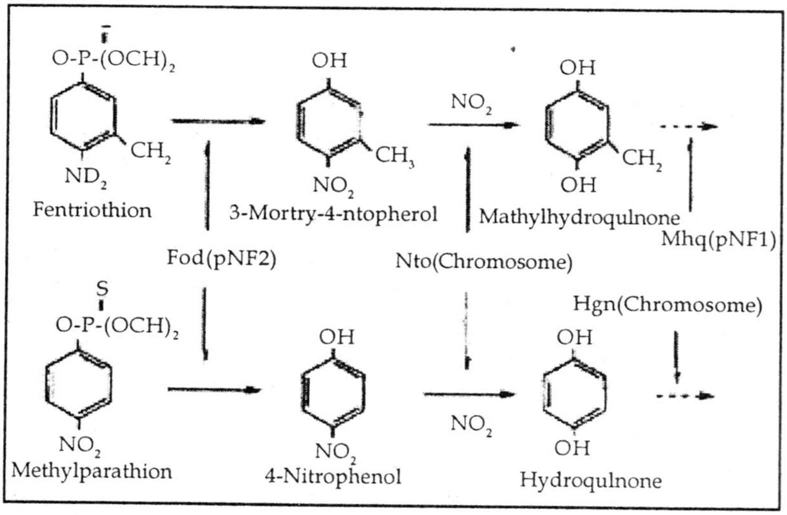

FIG. 13. Proposed pathways of fenitrothion and methyl parathion degradation by *Burkholderia* sp. strain NF100. Fed (pNF2), fenitrothion-hydrolyzing activity encoded on pNF2; Nto (chromosome), nitro group-removing oxidative pathway encoded on the chromosome; Mhq (pNF1), methylhydroquinone-degrading ability encoded on pNF1; Hqn (chromosome), hydroquinone-degrading ability encoded on the chromosome (Data from Hayatsu *et al.*, 2000).

observations; the loss of pNF1 and pNF2 from NF100 via mitomycin C treatment coincided with loss of methylhydroquinone-degrading and fenitrothion-hydrolyzing ability respectively, the subsequent introduction of pNF1 into NF103 (plasmid-free derivative of NF100) and pNF2 into NF1023 (plasmid-free derivative of NF100) resulted in reestablishment of the ability to degrade methylhydroquinone and the ability to hydrolyze fentitrothion respectively, and both cured strains (plasmid-free strains) retained the ability of oxidative removal of nitro group from nitrophenol. Furthermore, restriction analysis of pNF2 revealed that a portion of approximately 13 kb was deleted from pNF2, with the concomitant loss of fenitrothion property (Hayatsu *et al.*, 2000).

ii. Parathion

Microbial degradation of the organophosphate insecticide parathion (O,O-diethyl O-(4-nitrophenyl) phosphorothioate) has received considerable attention, due to in large part to the extensive use of this compound (Munnecke, 1981). Of the several microbial transformation of parathion known to occur, its hydrolysis to form nitrophenol and diethylthiophosphoric acid (Fig. 14) has received the most attention (Munnecke and Hsieh, 1976). Parathion hydrolase activity has been studied both in terms of its role in the fate of parathion in treated fields (Serdar *et al.*, 1985) and in its potential role in detoxifying various organophosphate wastes (Munnecke, 1981). *Flavobacterium* sp ATCC 27551 and *P. diminuta* were isolated for their ability to utilize parathion as a sole carbon source (Sethunathan and Yoshida, 1973; Serdar *et al.*, 1982).

In *P. diminuta,* parathion hydrolase is encoded on a plasmid (pCMS1) of approximately 66 kb (Serdar *et al.*, 1982; Serdar and Gibson, 1985). When the degradative host cells were cured of this plasmid by growth in presence of mitomycin C, parathion hydrolase activity was lost (Serdar *et al.*, 1982). Production of parathion hydrolase in the *Flavobacterium* sp ATCC 27551 was associated also with a plasmid (pPDL2) of approximately 43 kb confirmed by curing experiments (Mulbry *et al.*, 1987).

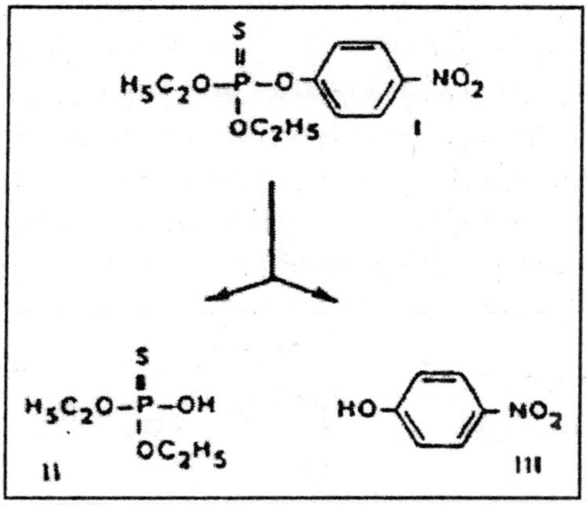

FIG. 14. Chemical structure of PAR (I) and its hydrolysis products diethylthiophoric (II) and PNP (III) (Data from Serdar *et al.*, 1982).

Construction of restriction endonuclease maps for the *opd* gene (gene encode for parathion hydrolase in both strains) -containing plasmids from *Flavobacterium* sp (pPDL2) and *P. diminuta* (pCMS1), coupled with more extensive hybridization experiments between their restriction fragments, revealed that homology between the two plasmids was limited to a region of 5.1 kb. This region extended approximately 2.6 kb upstream and 1.7 kb down stream of their *opd* genes. Restriction sites within the 5.1 kb conserved region were nearly identical in pCMS1 and pPDl2. In contrast, no shared restriction sites were evident in the two plasmids outside of this region (Mulbry *et al.*, 1987).

iii. Carbofuran

Carbofuran (2,3 dihydro-2,2-dimethyl-7-benzofuroyl N-methyl carbamate) is extensively used as a soil-incorporated N-methyl carbamate insecticide to control a variety of insect pests that infest crops such as corn and potatoes (Chapalamadugu and Chaudhry, 1992).

The biochemistry and genetics of carbofuran metabolism by bacteria have been characterized (Tomasek and Karns, 1989; Karns and Tomasek 1991; Head *et al.*, 1992; Feng *et al.*, 1997). First step in metabolism is generally thought to be hydrolysis of labile methylcarbamate linkage, yielding crabofurane-7-phenol and methylamine (Fig. 15) (Chapalamadugu and Chaudhry, 1992). Three different enzymes that catalyze carabamate hydrolysis have been isolated from different species (Karns and Tomasek, 1991; Mulbry and Eaton, 1991; Chapalamadugu and Chaudhry, 1993). Head *et al.* (1992), studying a *Flavobacterium* sp strain Mcd that utilizes carbofuran as a sole carbon source, found that the enzyme for cabofuran metabolism are encoded by genes on large plasmid pIH3 (199 kb). Furthermore, Feng *et al.* (1997) isolated a carbofuran-degrading *Sphingomonas* sp CF06 that mineralized both carbonyl group and aromatic ring of the carbofuran and used as a sole source of carbon and nitrogen. This strain has been shown to harbor five plasmids, at least some of which are required for the catabolism of carbofuran. However, utilization of carbofuran by CF06 as a nitrogen source may require only one enzyme (Karns and Tomasek, 1991; Mulbry and Eaton, 1991; Chapalamadugu

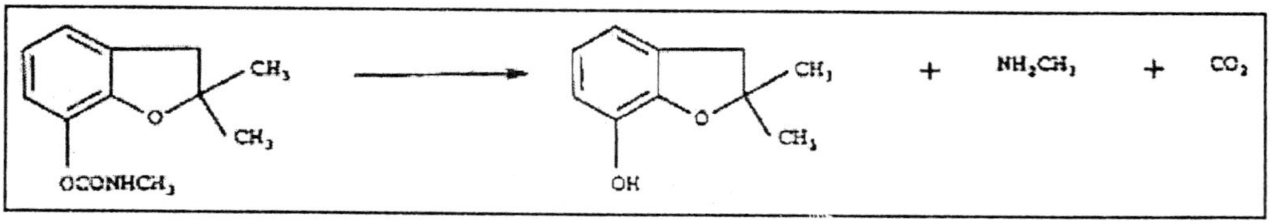

FIG. 15. Proposed conversion of carbofuran to carbofuran-7-phenol and methylamine (Data from Feng *et al.*, 1997)

and Chaudhry, 1993) that is likely to be encoded only by one or two plasmids (Feng *et al.* 1997). In addition, rearrangement, deletion, and loss of individual plasmids resulting in the loss of the carbofuran-degrading phenotype were observed.

The genetics of carbofuran metabolism by CF06 appears to be significantly more complex and probably less well organized than the most single-plasmid catabolic system. The apparent involvement of five plasmids with extensive regions of duplication between plasmids and the presence of several active IS elements (transposon elements) may be representative of a system at an early stage of its evolution. Such a system might be expected for metabolism of xenobiotics such as carbofuran that had been introduced during the past 30 years (Feng *et al.*, 1997).

iv. Carbamate (Carbaryl)

Carbamate pesticide such as carbaryl, comprise the major proportion of agriculture industry. The extensive use of carbamate pesticide in agriculture has induced the rapid evolution and dissemination of specific degradation pathways for the compound in soil bacteria (Sud *et al.*, 1972). A number of carbaryl degrading bacteria have been described, including several *Pseudomonas* (Larkin and Day, 1986; Chapalamadugu and Chaudhry, 1991; Chapalamadugu and Chaudhry, 1993), and *Achromobactor* (Sud *et al.*, 1972). *Arthrobacter* sp. Strain RC100 capable of utilizing carbaryl (1-naphthyl N-methylcarbamate) as the sole carbon source was isolated from carbaryl-treated soil (Hayatsu, 1999). A proposed pathway for the metabolism of carbaryl and 1-naphthol by *Arthrobacter* sp. strain RC100 is presented in Figure 16. Strain RC100 hydrolyzes the 1-naphthyl N-methylcarbamate linkage to 1-naphthol, which was further metabolized via salicylate and gentisate.

The biochemical and genetic studies indicated that the genes for carbaryl hydrolysis and 1-naphthol catabolism are located on two plasmid pRC1 (110 kb) and pRC2 (120 kb). Mutants (RC100 derivatives) unable to degrade carbaryl arose at a high frequency after treating the culture with mitomycin C. All carbaryl-hydrolysis-deficient mutants lacked pRC1 and all 1-naphthol-utilization-deficient mutants lacked pRC2, whereas, plasmid-free mutants grow on gentisate as a carbon source (Hayatsu *et al.*, 1999). These observations indicated that the pRC1 encodes the cararyl hydrolyzing abilities, and pRC2 encoded the metabolizing pathway of 1-naphthol to gentistic acid. By the way, the plasmid free strain lost the ability of 1-naphthol utilization but utilized gentisic as the sole carbon source, indicating that the gentisic acid metabolic pathway was encoded on the chromosome.

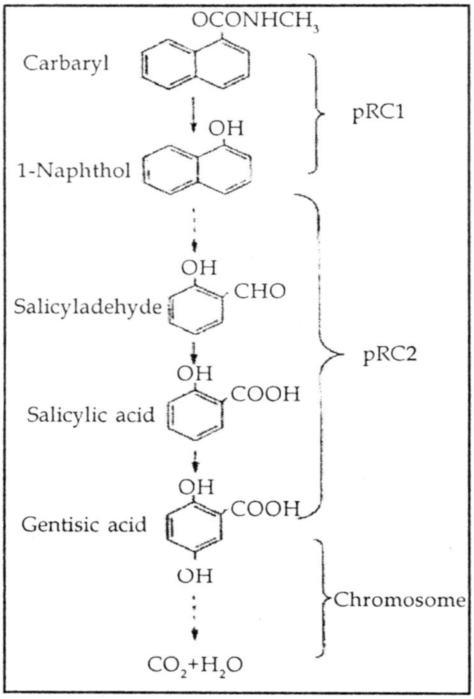

FIG. 16. Proposed pathway of carbaryl degradation by *Arthrobacter* sp. strain RC100 (Data from Hayatsu *et al.*, 1999)

4.5.2 b-Herbicide

i. 2, 4-Dichlorophenoxyacetic acid (2,4-D)

Chlorinated derivatives of phenoxyacetic acid have been released to the environment as as herbicide over the last 50 years. One widely used herbicide, 2, 4-dichlorophenoxyacetic acid (2,4-D) is rapidly degraded by soil bacteria. Several strains of multiple genera such as, *Alcaligenes, Arthrobacter, Corynebacterium,* and *Pseudomonas* are capable of degrading 2, 4-D (Tyler and Finn, 1974; Fisher *et al.*, 1978; Amy *et al.*, 1985).

Alcaligenes eutophus JMP 134 is one of the best-characterized organisms among this group. It harbors a conjugative plasmid, pJP4, of about 80 kb size, which carries genes essential for the degradation of 2, 4-D (Don and Pemberton, 1981). Recently, this plasmid has been isolated and well-characterized (Filler and Harker, 1997). Five genes involved in the degradation of 2,4-D have already been localized by transposon mutagenesis and cloned in *E. coli*. (Filler and Harker, 1997). Functions have been assigned by biochemical studies (Don *et al.*, 1985; filler and Harker, 1997): *tfdA, tfdB, tfdC, tfdD,* and *tfdE* encode 2, 4-monooxygenase, 2,4-dichlorophenol hydroxylase, dichlorocatechol 1,2-dihydroxygenase, chloromuconate cycloisomerase, and chlorodienelctone hydrolase, respectively (Fig.17). Furthermore, two regulatory

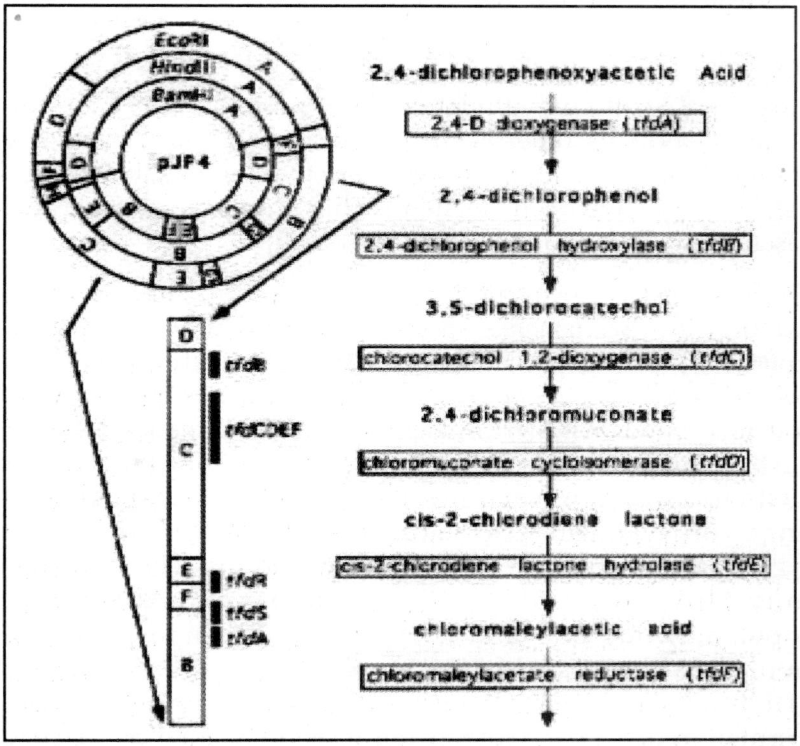

FIG. 17. Restriction digest map of the plasmid pJP4. The positions of the fragments and the relative positions of the genes are from data reported elsewhere (5, 9, 17, 21). The linear map represents the region encompassing the *tfd* genes. Intermediates produced and the respective genes and gene products responsible are shown on the right (Data from Filer and Harker *et al.*, 1997)

genes have been identified designated *tfdR* and *tfdS* are also localized on pJP4 plasmid. The product of these genes is thought to act as transcription activators as well as negative regulator, of the 2,4-degradative pathway (Filler and Harker, 1997). Mae *et al.* (1993) characterized a new plasmid pEST4011 isolated from *Pseudomonas* sp EST4002 which harbor genes for of 2,4-D catabolism and. The size (70 kb) and restriction pattern of plasmid pEST4011 differed considerably from those of pJP4. However, pEST4011has considerable homology with the regions of pJP4 containing the 2,4-D degradative genes (*tfd*) (Mae *et al.*, 1993). A further comparative study of the nucleotide sequence of catabolic genes in pJP4 (isolate originated from Australia) and pEST4011 (isolate originated from Estonia) and related plasmids would widen our understanding of the unique difference and evolutionary relationship between different 2,4-D degradative plasmids in strains isolated from different geographical regions of the world (Mae *et al.*, 1993).

ii. Atrazine

Atrazine (2-Chloro-4-ethylamino-6-isopropylamino-1,2,3-traizine) is one of the most heavily used herbicide for selective and non selective weed control (Seiler *et al.*, 1992). Several microorganisms that can metabolize atrazine during the last 15 years have been isolated (Behki and Khan, 1986; Yanze-Kontchou and Gschwind, 1994; Radosevich *et al.*, 1995).

De Souza *et al.* (1998) studied the genes and enzymes involved in atrazine degradation by *Pseudomonas* sp strain ADP, which harbored a 96 kb self-transmissible plasmid, pADP-1. Southern hybridization and plasmid transfer studies indicated that the *atzA*, *atzB*, and *atzC* genes are involved in atrazine catabolism, located on pADP-1 plasmid. These three genes encode the enzymes AtzA, -B, -and-C have been cloned and sequenced (De Souza *et al.*, 1998). The first enzyme AtzA, catalyzes the hydrolytic dechlorination of atrazine, yielding hydroxyatrazine (De Souza *et al.*, 1996)). The second enzyme, AtzB, catalyze hydroxyatrazine deamination, yielding N-isopropylammelide (Boundy-Mills *et al.*, 1997). The third enzyme, AtzC, transforms, N-isopropylammelide to syanuric acid (Fig. 18) (Sadowsky *et al.*, 1998). In addition, *Pseudomonas* sp ADP (Atr-), which lacked the plasmid ADP-1, able to utilize a cyanuric acid as a sole carbon source, indicating that transformation of atrazine into cyanuric acid encoded by plasmid, while the catabolic of cyanuric acid encoded by chromosome (De Souza *et al.*, 1998).

iii. Diuron

Diuron 3-(3, 4-dichlorophenyl)1,1-dimethylurea, is a selective pre- and post -emergence herbicide that is used for weed control in a wide range of crops and is an example of a substituted urea herbicide. The environmental fate of diuron is of great concern in many areas around the world, where it has been detected in ground water (Kathleen and Adency 1978). The biological activity of soil microorganisms is proposed to be the major factor for biodegradation of diuron in soil (Attaway *et al.*, 1982; Lella, 1988; Johnson and Camper, 1991; El-Deeb *et al.*, 2000). Previous work on the biodegradation pathway of urea-derivatives has shown that the basic of decomposition is N-demethylation, hydrolysis followed by oxidation of aromatic ring (Engesser *et al.*, 1990; Cernakova, 1995). *Pseudomonas* sp. strain Bk8 was isolated from field soil contaminated with different urea-herbicides (El-Deeb *et al.*, 2000a). This strain is a plasmid (pBk8)-harboring organism capable of complete

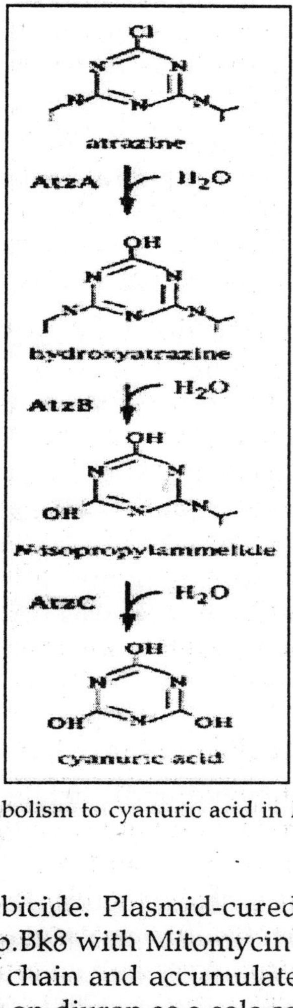

FIG 18. Pathway for atrazine catabolism to cyanuric acid in *Pseudomonas* sp. strain ADP (Data from De Souza *et al.*, 1998)

degradation of diuron herbicide. Plasmid-cured strain Bk8M obtained by treatment of *Pseudomonas* sp.Bk8 with Mitomycin C is capable of only partial degradation of diuron side chain and accumulated a phenolic compound in the medium during growth on diuron as a sole source of carbon and energy. Conjugation experiments were carried out using Bk8M as a recipient and Bk8 as a donor of pBk8 Plasmid. The transconjugant was able to degrade a diuron without accumulation of phenolic compound. It was proposed that plasmid pBk8 is self-transmissible and involved in the degradation of diuron aromatic ring but it is not connected with the transformation of diuron into diuron phenol compound (El-Deeb *et al.*, 2000 b).

iv. Plasmid-mediated crude oil degradation

Biodegradation of crude oil by natural population of microorganisms represents one of the primary mechanisms by which crude oil and other hydrocarbon pollutants are eliminated from the environment (Leahy, and

Colwell, 1990). The ability to degrade and/ or utilize hydrocarbon as growth substrates is exhibited by a wide variety of bacteria and fungi (Floodgate, 1984; Bossert and Bartha, 1984). Based on the number of published reports, the most important oil-degrading bacteria are *Arthrobacter, Bacillus, Flavobacterium, Rhodococcus* and *Pseudomonas* (Leahy, and Colwell, 1990). Indirect evidence for the role of plasmid in oil degradation has been provided by some studies in the greater frequency and/or multiplicity of plasmids observed among bacterial isolates from oil contaminated environments than among isolates from non contaminated sites (Hada and Sizemore, 1981).

Rusansky *et al.* (1987) isolated an *Acinetobacter calcoacetics* strain RA57 from oily sludge, which was able to grow on and disperse crude oil in liquid culture. Strain RA57 was found to contain four plasmids: pSR1 (5.1 kb), pSR2 (5.4 kb), pSR3 (10.5 kb), and pSR4 (20 kb). The cured and plasmid transfer studies indicated that derivative strain, which harbors pSR4 retained the ability to grow on and disperse the crude oil, whereas, cells cured of pSR4 lost the ability to grow on and disperse oil (Fig. 21). However, cured cells (plasmid-free, pSR4) were able to grow on hydrocarbon vapors, indicating that the defect associated with curing of pSR4 was related to the physical interaction of cells with hydrocarbon substrates rather than to its metabolism (Rusansky *et al.*, 1987).

v. Plasmid-mediated Quinoline and metylquinolines

Quinoline, methylquinolines, and other heterocyclic nitrogen compounds occur in shale oils. Moreover, quinoline and its derivatives have been shown to be hepatocarcinogen in mice and rats (LaVoie *et al.*, 1988). Several investigators have demonstrated microbial transformation or degradation of quinoline under both aerobic and anaerobic conditions. Most quinoline-degrading microorganisms have been identified as *Pseudomonas* species (Schwartz *et al.*, 1988; Brockman *et al.*, 1989), but quinoline-degrading strains of *Rhodococcus* (Schwartz *et al.*, 1988), *Moraxella*, (Al-Najjar and Al-Shakarchi, 1980) and *Desulfobacterium* spp. (Bak and Widdel, 1986) have also been reported.

Quinoline degradation has been shown to involve hydroxylation at the 2 position (Pereira *et al.*, 1987; Sutton *et al.*, 1996). Shukla, (1987) in a study of four bacteria capable of utilizing quinoline as sole source of carbon and nitrogen, found two alternative pathways for quinoline catabolism; one pathway proceeded via 2,6-dihydroxyquinoline, whereas after 2-hydroxyquinoline formation the other pathway proceeded via 2,8-dihydroxyquinoline and 8-hydroxycoumarin (Fig. 19).

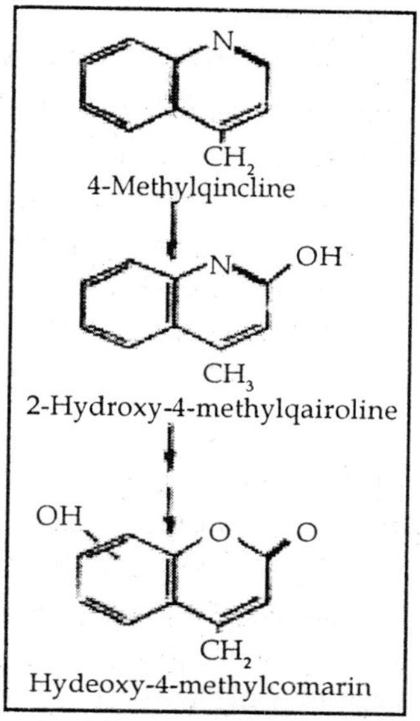

FIG. 19. Proposed pathway for 4-methylquinoline metabolism (Data from Sutton *et al.*, 1996).

Aislabie *et al.* (1990) isolated a bacterial strain, *P. aeruginosa,* from oil-contaminated soil, which is able to degrade quinoline and to transform or to degrade methylquinolines. The degradation of quinoline by strains of *P. aeruginosa* QP produced hydroxyquinoline, a transient pink compound, and other undetermined products. Moreover, this strain hydroxylated a limited number of methylquinolines but could not completely degrade them.

The genetic analysis indicted that *P. aeruginosa* QP contained four plasmids of 250, 320, 225, and 180 kb. Curing of *P. aeruginosa* QP of all four plasmids resulted in the complete loss of its ability to transform quinoline. On the other hand, curing *P. aeruginosa* QP of only 180 kb plasmid had no effect on the ability of the strain to degrade quinoline or on the pattern of product formation (Aislabie *et al.*, 1990), indicating that it is not involved in quinoline metabolism. The 225-, 250-, and 320-kb plasmid, on the other hand, all appear to be involved with quinoline metabolism. Moreover, strain containing both 225 and 250 kb plasmids completely degraded quinoline and did not accumulate significant quantity of hydroxylated quinoline metabolites. In contrast, strains with 320 kb plasmid alone or in combination with either only 225 or 250 plasmid transformed quinoline with the accumulation of significant amount of hydroxylated quinoline metabolites. Thus, the combined activities of genes contained on 225 and 250 kb plasmids appear to be involved in degradation of hydroxylated quinolines (Aislabie *et al.*, 1990).

vi. Plasmid-mediated dibenzothiophene

Burning sulfur-containing petroleum and coal contributes to environmental degradation in various ways (Fought *et al.*, 1990). Although removal of inorganic sulfur from these fuels may be accomplished by physical, chemical, or biological means, organically bound sulfur is difficult to remove. One possible strategy for reducing the organic sulfur content is to expose these substrates to microorganisms or enzymes that can specifically break carbon-sulfur bonds (Fought *et al.*, 1990; Kilbane, 1989; Kilbane and Jakowski, 1992; Izumi *et al.*, 1994), thereby releasing the sulfur in a water-soluble, inorganic form.

Rhodococcus sp. strain IGTS8 is able to extract sulfur from a variety of organosulfur compounds, including dibenzothiophene, sulfides, mercaptanes, sulfoxides, and sulfones (Kayser, *et. al.*, 1993). Furthermore, the organism is able to remove organic sulfur from petroleum and soluble coal-derived material without decreasing the caloric value of those substrates (Kilbane and Jakowski, 1992). IGTS8 possesses an enzymatic pathway that can remove sulfur from dibenzothiophene (DBT) to yield 2-hydroxybiphenyl (2HBP) as the final product (Fig. 20), meaning that there is no degradation of carbon-carbon bond in the DBT molecule (Denome *et al.*, 1993; Gallagher *et al.*, 1993; Kilbane and Jakowski, 1992). The ability of IGTS8 to desulfurize appears to be an exclusive property of a definite locus on a large plasmid of 150 kb. In this plasmid encoded pathway, three genes *SoxABC* arranged in an operonic manner and spanning a 4-kb region, are responsible for the transformation of

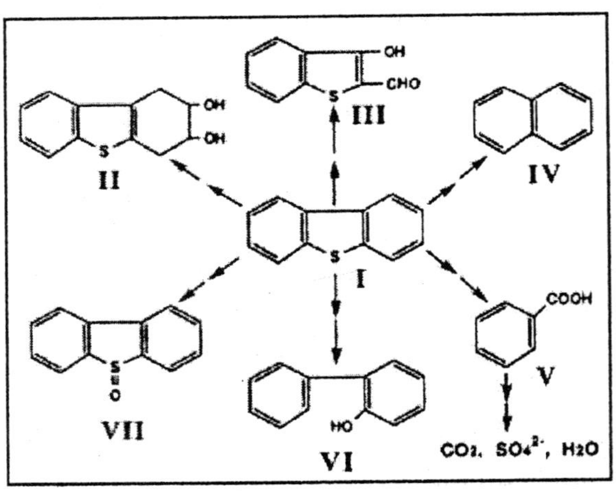

FIG. 20. Proposed pathways of DBT degradation. (I) DBT. (II) 1,2-Dihydroxy-1,2-dihydrodibenzothiophene. (III) 3-Hydroxy-2-formyl benzothiophene. (IV) Biphenyl. (V) Benzoic acid. (VI) 2-HBP, (VII) Dibenzothiophene 5-oxide (Data from Izumi *et al.*, 1994).

DBT to 2-hydroxybiphenyl and sulfate (Larose *et al.*, 1997). These genes encode for three enzymes, SoxC, (45-KDa protein), was found to mediate the initial oxidation of DBT to DBT sulfone (Lei and Tu, 1996). SoxA (40-KDa protein) and SoxB (40-KDa protein) are believed to act concertedly in transforming DBT sulfone to 2-hydroxybiphenyl and sulfate (Fig. 21) (Denome *et al.*, 1993;

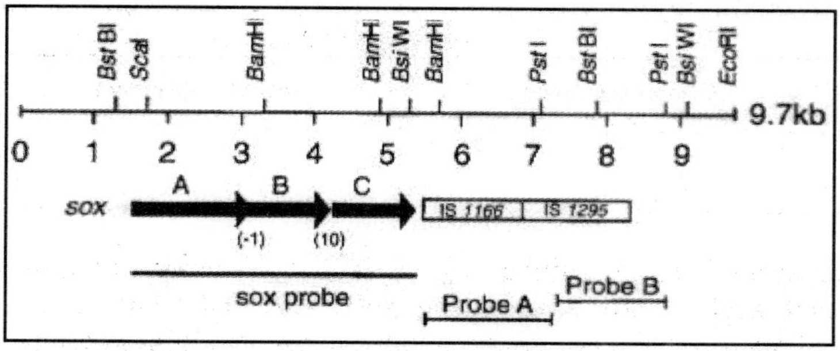

FIG. 21. Location of desulfurization (sox) genes and insertion sequence elements (IS1166 and IS1295) in *Rhodococcus* sp. strain IGTS8 (Data from Larose *et al.*, 1997).

Piddington *et al.*, 1995). On the other hand, the possibility of sulfur-specific removal from DBT by sulfate reducing bacteria has been examined (Finnery, 1992). Although, these bacteria metabolize DBT to hydrogen sulfide and biphenyl, no hybridization signal was found between *SoxABC* genes region of *Rhodococcus* and *Desulifovibrio dessulfuricans* DNA, indicating that a different pathway is operational in these anaerobic organisms (Larose *et al.*, 1997).

vii. Plasmid-mediated styrene degradation

Styrene is a chemical that is used extensively in the manufacturing of plastics and synthetic rubbers. This toxic compound is released into the environment mainly through factory wastewater, evaporation, and pyrolysis of polystyrene (Santos *et al.*, 2000). Different routes for styrene catabolism in different microorganisms have been described (Bestetti *et al.*, 1984; Hartmans *et al.*, 1989; Warhurst *et al.*, 1994; O'connor *et al.*, 1995). Two main routes of aerobic styrene break down have been described: (i) oxidation of the vinyl side chain with the formation of phenyl acetate (ii) initial oxidation of the aromatic ring by dioxygenase enzyme (Fig. 22) (Wierenga *et al.*, 1986).

Pseudomonas fluoresences, which is able to grow on styrene as a sole carbon source, has been characterized, and the upper pathway genes (*styABCD*), which code for the conversion styrene into phenylacetate have been determined (Velasco *et al.*, 1998; Santos *et al.*, 2000). The physiological and genetic studies indicted that the upper pathway genes (*styABCD*) were located on plasmid pEG (37 kb) (Bestetti *et al.*, 1984; Velasco *et al.*, 1998), while, the genes responsible for the catabolism of phenylacetate (lower pathway) might located on chromosome. Recently, Santos *et al.* (2000) concluded that the upper pathway genes (*styABCD*) encode for four enzymes; StyAB, styrene monooxygenase, which catalyze styrene into epoxystyrene, StyC, epoxyisomerase, catalyzes epoxystyrene into phenylaetaldehyde, and StyD, catalyzes, phenylaetaldehyde into phenylacetate (Fig. 22).

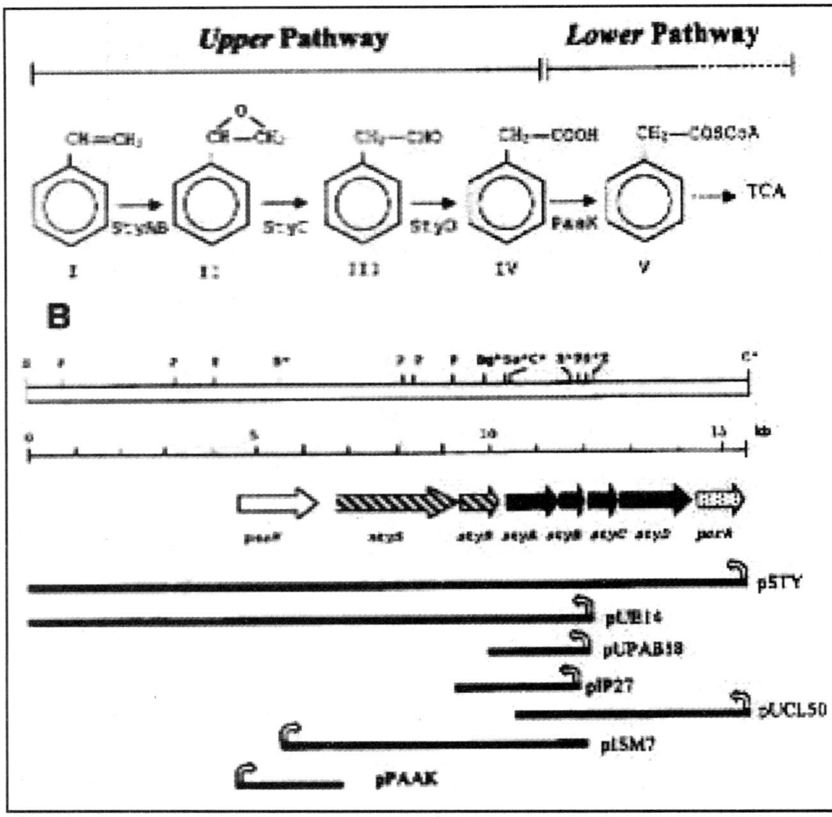

FIG. 22. Pathway for the catabolism of styrene in *Pseudomonas* sp. strain Y2 and genetic organization of the corresponding structural and regulatory genes. (A) Biochemistry of the pathway. Metabolites: styrene (compound I), epoxystyrene (compound II), phenylacetaldehyde (compound III), phenylacetate (compound IV), and phenylacetyl-coenzyme A (compound V). Enzymes: StyAB, styrene monooxygenase; StyC, epoxystyrene isomerase; StyD, phenylacetaldehyde dehydrogenase; PaaK, phenylacetyl-coenzyme A ligase. (B) Physical and genetic map of the chromosomal region encoding styrene catabolism. Locations of the genes are shown relative to those of some relevant restriction sites. Arrows indicate the direction of gene transcription. The cloned fragments (thick line) and their orientations with respect to the lacZ promoter (bent arrow) of plasmid pUC18 or pUC19 are indicated. Restriction sites: B, BamHI; Bg, BglII; C, ClaI; E, EcoRI; P, PstI; S, SmaI; Sa, SalI. Asterisks mean that other identical restriction sizes are present. Data from (Velasco *et al.*, 1998).

viii. Plasmid-mediated nylon degradation

Nylon-6 is produced from Ý-carpolactam by ring cleavage polymerization. It consists of more than 100 units of 6-aminohexanoic acid. During polymerization reaction, some molecules fail to polymerize and remain as oligemers, while other undergo head-to- tail condensation to form cyclic oligomers (Negoro, 2000). These nylon oligomer are by product rejected by nylon factories, thereby contributing to the increase in industrial waste material entering the environment (Kinoshita *et al.*, 1975; Kanagawa *et al.* (1989) isolated two bacterial strains, *Flvobacterium* sp. K127 and *Pseudomonas* sp. NK87, which grow on cyclic dimer of 6-aminohexanoate as a sole source of carbon and nitrogen, from soil and waste water from nylon factories, respectively.

ix. a-Biodegradation of nylon oligomers in *Flavobacterium*

Biochemical studies revealed that three enzymes produced by *Flvobacterium* sp. K127, 6-aminohexanoate cyclic-dimer hydrolase (E1), which is active only towards the cyclic dimer (kinoshita *et al.*, 1977), 6-aminohexanoate-dimer hydrolase (EII), which is active on aminohexanoate oligomers ranging from dimer to hexamer (Kinoshita *et al.*, 1981), and endotype 6- aminohexanoate-oligomer hydrolase (III), which active on cyclic tetramer and pentamer and on linear higher oligomers (Negoro *et al.*, 1992; Kakudo *et al.*, 1993,1995).

Curing studied indicated that *Flvobacterium* sp. K127 possessed three kinds of plasmids, pOAD1 (39 kb), pOAPD2 (44 kb), and pOAD3 (56 kb). Curing readily eliminated pOAD2 with loss of the enzyme EI and EII enzyme activities. However, pOAD1 was either conserved intact or partially deleted, even after curing. Furthermore, transformation of *Flvobacterium* sp. KI723 (free-plasmid derivative) with plasmid DNA of the wild type (KI27) showed that K723 readily accepted the pOAD2 and simultaneously recovered its EI and EII activities, indicating that The EI and EII enzymes encoded on plasmid pOAD2 (Negoro *et al.*, 1980). In addition, it was also shown that EIII enzyme activities were detected in the wild K127 and transformant KI723, but not in the cured strain KI723 (Negoro *et al.*, 1992). Moreover, genetic analysis indicated also that three nylon oligomerdegradation genes, nylA (EI gene), nylB (EII), and nylC (EII gene) were located on different parts of the plasmid (Fig. 23).

x. b-Biodegradation of nylon oligomers in *Pseudomonas*

Pseudomonas sp. NK87 harbored six kinds of plasmid, namely: pNAD1 (20 kb), pNAD2 (23 kb), pNAD3 (51 kb), pNAD4 (57 kb), pNAD5 (76 kb), and pNAD6 (80 kb). Cloning and hybridization experiments revealed that *nylA* and *nylB* genes of NK87 are located on pNAD2 and pNAD6 respectively (Kanagawa *et al.*, 1989). The EI enzyme from NK87 was composed of 493 amino acid residues, which was identical to the length of the EI from *Flvobacterium* sp. K127. However, the EII had more divergence sequences. The homology between the EII sequence of KI72 and that of NK87 was only 35% with respect to amino acid sequence (Kanagawa *et al.*, 1993).

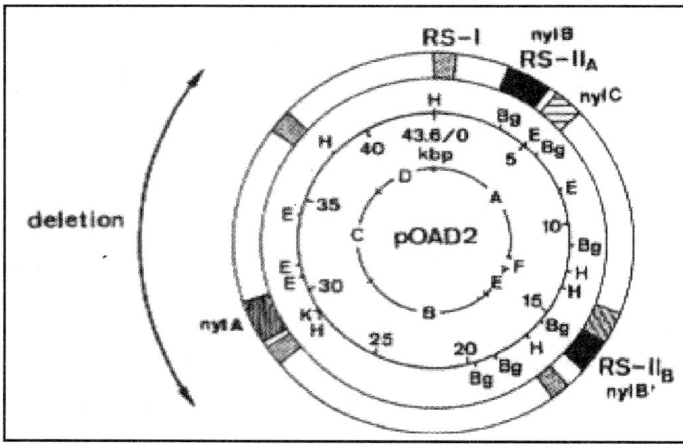

FIG. 23. Structural and functional map of plasmid pOAD2. The nylon oligomer degradation genes were at the following loci: nyUA, 29.1- to 30.6-kb region on the pOAD2 map; nylB, 2.7- to 4.1-kb region; nylB', 14.7- to 16.1-kb region; ny1C, 4.2- to 5.3-kb region. RS-I and RS-II are repeated sequences identified by Southern hybridization experiments. RS-I appeared five times on pOAD2 (RS-IA, 0 to 0.8 kb; RS-IB, 13.5 to 14.7 kb; RS-Ic, 16.8 to 17.4 kb; RS-ID, 28.2 to 28.8 kb; RS-IE, 37.1 to 37.9 kb). The nylB and nylB' genes are included in the RS-IIA and RS-IIB regions, respectively. Bg, E, H, and K represent BglII, EcoRI, HindIII, and KpnI restriction sites, respectively. pOAD21 was obtained from pOAD2 by homologous recombination between direct-repeat sequences RS-ID and RS-IE. The arrow indicates the region deleted in pOAD2.

4.6. The Role of Plasmids in the Adaptation of Microorganisms to Xenobiotics Degradation

Amino acid similarities of the different aromatic pathway enzymes and diverged organization of catabolic gene clusters discussed above suggest that several different gene clusters; modified *ortho* cleavage pathway genes (Fig. 8), *meta* cleavage pathway genes (Fig. 6, 7) and different dioxygenase genes may be combined as modules, to which other peripheral genes (genes responsible for transforming parent compound into catechols) may be added. There are different mechanisms which may be involved in the evolution of aromatic pathways in general and in the process of adaptation of microorganisms to xenobiotic substrates in particular:

4.6.1 Gene Transfer

Many examples of degradative plasmids illustrated above strongly suggest that common ancestral genes have spread among different microorganisms. The divergence of descendants in different microorganisms, however, does not necessarily correspond to the evolutionary distances determined for those organisms by 16S rRNA. This observation suggests that the occurrence of extensive horizontal gene transfer during evolution processes. Genetic interactions in microbial communities are affected by several mechanisms, such as conjugation, transduction and transformation (Davison, 1999).

The occurrence of plasmids in bacteria in the natural environment is an important pool of genetic information residing on plasmid vehicles which may flow among indigenous microorganisms (Neilson, *et al.*, 1994; Kruse and Sorum, 1994; DiGiovanni *et al.*, 1996; Madsen, 1997; Peter, *et al.*, 1997). Many examples of self-transmissible plasmids that carry genes for degradation of aromatic or other xenobiotic compounds are known (Sayler *et al.*, 1990; Sorensen, 1993; Peter, *et al.*, 1997), and their role in spreading these genes to other microorganisms is well characterized (DiGiovanni *et al.*, 1996; Madsen, 1997; Davison, 1998). Early studies of TOL, NAH, and SAL plasmids (Fig. 6, 7) revealed strong homologies among these plasmids in regions both inside and outside the catabolic gene clusters (Lehrbach *et al.*, 1983; Frantz and Chakrabarty, 1986; Assinder and williams, 1990). Recently, Burlage *et al.* (1989) have shown that pJP4, pAC25, and PSSS50 (Fig. 8), catabolic plasmids carrying genes for chloroaromatic degradation, have strongly homologous plasmid backbone that determine replication and transfer function. Furthermore, the chlorobenzoate plasmid p51 was also shown to be homologous to pJP4 outside regions of the catabolic genes (Reineke, 1998). These observations suggest that a few common self-transmissible ancestor replicons (includes origin of replication and replication gene) may have been involved in acquisition and spread of different catabolic modules.

The important of gene transfer for adaptation of host cells to new compounds has been illustrated in many studies on experimental evolution of novel metabolic activities (Reineke, 1998; Davison, 1999).

4.6.2 DNA Rearrangements.

The order of genes encoding the *ortho* cleavage pathways of *Acinetobacter calcoaceticus* and *P putida* differ from one another (Ornston *et al.*, 1990) and from those of other organisms, suggesting that various DNA rearrangement have occurred (Frantz and Chakrabarty, 1986; Perkins *et al.*, 1990; van der Meer *et al.*, 1991b). Further examples are found in a comparison of TOL and

NAH encoded catabolic genes. The genes order of part of the *meta* cleavage operons are identical (Fig. 6, 7) (Harayama and Rekik, 1990, Reineke, 1998) but are different in down stream genes.

4.6.3 Gene Duplication

Gene duplication has been considered an important mechanism for the adaptation of bacteria to different xenobiotics. For example, *Flavobacterium* sp. strain 172 harbors two isoenzyme of 6-aminohexanoate dimer hydrolase, one of the enzymes involved in the degradation of nylon oligomers (Negaro, 2000). Both enzymes were encoded by plasmid pOAD2 (Fig. 23) in regions that were proposed to be duplicate. The two isoenzymes differ 100-fold in their activity toward 6-aminohexanoate dimer, despite their amino acid sequence identity of 88% (Negoro, 2000). Studies with various TOL-type plasmids have shown that upper and lower pathway operons as well as *xylS* and *xyl R* regulatory genes, have sometimes inverted or increased their copy number (Osborne *et al.*, 1988). Furthermore, plasmid pJP4, which contains the genes for the degradation of 2,4-dichloroacetic acid (Fig. 17), appears to have undergone several gene duplication (Perkins *et al.*, 1990).

4.6.4 Transposition

Insertion elements have been shown to play an important role in rearrangement of DNA fragments, in gene transfer, and in activation or inactivation of silent genes. For catabolic pathways several examples of insertion elements are known. The TOL catabolic operons are part of large transposable element, Tn4651, which belongs to the family of Tn3-type transposon (Tsuda and Lino, 1988; Tsuda *et al.*, 1989). This transposon was found to be of an even larger mobile element, Tn4653 (Tsuda and Iino, 1988). A similar Tn3-type element, Tn4655 contains the catabolic genes on plasmid NAH7, although it is defective in its transposon function (Tsuda and Iino, 1990). *P. cepacia* 249 was shown to carry at least nine different insertion elements, which were present in 1 to 13 copies in its genome. These IS elements are thought to be responsible for the extraordinary adaptability and catabolic potential of this strain. One particular element, IS931, was detected in multiple copies around genes locus for 2,4-dichorophenoxyacetic acid degradation (Haugland *et al.*, 1990). IS931 is presumed to have played a role in the development of the 2,4-dichorophenoxyacetic acid pathway, as it could mobilize adjacent DNA fragments (Haugland *et al.*, 1990).

Other mobile and IS elements carrying catabolic genes include the DEH element, which carries the dehalogenase genes of *P. putida* PP3 (Nakatsu *et al.*, 1991), and Tn5271, which is present on plasmid pBRC60 and specifies the genes for 3-chlorobenzoate degradation by *Alcaligenes* sp. Strain BR60 (Nakatsu *et al.*, 1991). This transposable element is formed by two copies of a 3.2 kb repeated sequence. The chlorobenzene dioxygenase genes of *Pseudomonas* sp. strain P51.were shown to be flanked by two copies of iso-inseration elements, IS1066 and IS1066. The composite element called TN5280, consists of IS1066, the chlorobenzendioxygenase genes, and IS1067 and was functional transposon capable of inserting at random into chromosome (van der Meer *et al.*, 1991c).

5. ENGINEERING BACTERIA FOR BIODEGRADATION

As described above, the main agents for returning natural and, presumably, synthetic organic compounds to the carbon cycle are indigenous microorganisms, providing that catabolic enzymes with appropriate specificities, transport systems and regulatory mechanism can be activated (Genetic Eng.). One of several of these conditions for total degradation does not seem to be fulfilled for certain types of xenobiotic compounds, e.g. haloaromatics (Reinke, 1998). Although soil and water microorganisms have a considerable evolution potential and, under appropriate selective conditions, evolve pathways able to degrade many synthetic compounds. These conditions do not always arise in nature and the evolution of effective pathways for certain compounds may be thus relatively slow given the condition that large amount of novel organic compounds are released into the environment by rapidly growing chemical industry.

These challenges have encouraged the use of genetic engineering techniques for artificial construction of strains with broad degradative potentials. The use of genetic engineering techniques offers a considerable promise as an approach to accelerate the evolution of desired metabolic pathways: There are two basic techniques employed in the recent environmental biotechnology to enhance the degradative potential of the existing pathways:

5.1 *In vivo* Genetic Selection

The range of biodegradative capacity of *Pseudomonas* sp. 13 was horizontally expanded from 3-chlorobenzoate to 4-chlorobenzoate and 3,5 and dichlorobenzoate by transfer of the TOL plasmid pWWO from *P. putida* mt-2 (Jeenes *et al.*, 1982), which led to hybrid strain able to grow with 4-chlor- and 3,5-dichlorobenzoate. Reinke (1998) described that strain B13 was enriched with 3-chlorobenzoate (Rrineke, 1988) and converts this substrate to 3-chlorocatechol, which is mineralized via the modified *ortho* pathway. Strain B13 is not able to use 4- and 3, 5-dichlorobenzoate as growth substrates because its benzoate 1,2-dioxygenase has narrow specificity and does not convert these compounds. In addition, both compounds fail to induce the initial enzymes in strain B13 (Reinke, 1998). The toluate 1,2-dioxygenase of strain *P. putida* PaW1, which encode on the TOL plasmid, has a broader substrate specificity than the corresponding B13 enzyme and accepts 4-chloro- and 3,5 dichlorobenzoate as substrates. Transconjugants of the mating between *Pseudomonas* sp 13 as acceptor and *P putida* PaW1 as a donor showed the phenotype of strain B13 with additional ability to grow on 4-chloro and 3,5-dichloro benzoate (Reinke, 1998). This transfer provided strain B13 with the *xylXYZ* toluate dioxygenase of the TOL plasmid, an enzyme with broader substrate range than chlorobenzoate dioxygenase of strain B13. Moreover, in hybrid strain which carried, *meta* and *orhto* cleavage pathways simultaneously, catechol 2, 3-dioxygenase had to be inactivated for proper growth on chlorinated aromatic compound (Reineke, 1998).

The role of TOL plasmid and strain B13 in the degradation of various mono- and dichlorobenzoates is not only example of *in vivo* genetic manipulations for the degradation of novel compounds. Schewien *et al.* (1988) have also demonstrated that transfer of the chlorocatechol degradative potential from *Pseudomonas* B13 to an *Alcaligenes* A7 capable of growing on phenol allowed exconjugant *Alcaligenes* A7-2 to utilize all three isomers of chlorophenol, which were not attacked by any of the parents. Similar gene transfer to *Pseudomonas* sp. WR401 that is capable of growing with salicylate but unable to attack chlorosalicylate enables the exconjugant cells to utilize chlorosalicylate (Lehrbach *et al.*, 1984; Reineke, 1998).

5.2 *In vitro* Gene Cloning

It has been mention that the benzoate dioxygenase present in *Pseudomonas* sp. B13 can use 3-chlorobenzoate as a substrate but not 4-chlorobenzoate. Instead of introducing the entire TOL plasmid into *Pseudomonas* B13, Lehrbach *et al.* (1984) cloned xylL gene into *Pseudomonas* sp B13, which enabled B13 to grow on 4-chlorobenzoate. This method is called in vitro gene cloning.

The following description of this cloning will serve as meaningful information on the molecular cloning and analysis of degradative genes.

5.2.1Cloning of the 2,3 dihydroxybiphenyl dioxygenase gene (bphC) from Pseudomonas sp. KKS102

Kimbara *et al.* (1989), isolated an biphenyl-degrader *Pseudomonas* sp. KKS102, which metabolize biphenyl/PCP to benzoate /chlorobenzoate.

A genomic library of KKS102 was constructed in *E.coli* HB101, using the broad host range cosmid vector pKS13 (Fukuda and Yano, 1985). Transformants were screened for the presence of yellow color after spraying the colonies with 2, 3-dihydroxybiphenyl, because the product of *bphC*, 2, 3-dihydroxybiphenyl dioxygenease converts 2, 3-dihydroxybiphenyl to the *meta*-cleavage compounds, which is bright yellow. Furthermore, the genomic library mobilized into *P. putida* by triparental mating (conjugation), because of poor expression of *bphC* in *E.coli*, one of these transformants was turned yellow and found to be harbored one pasmid (pKH1), which had BamH1 insert of 18 kb. Subcloning and genetic analysis indicated that a 3.2 kb Pst1 fragment, containing the *bphC* gene (Fig 23) (Kimbara *et al.*, 1989).

The accumulation of information on degradative pathways and degradative genes for recalcitrant compounds enabled the researchers to create an improved pathway or a new forceful pathway through DNA cloning. Several examples have illustrated that single-site mutation (*in vitro* mutagenesis) can alter substrate specifities of enzymes or effector specificities: (1) the substrate range of catechol 2, 3-dioxygenase enzyme by TOL plasmid pWWO has an extended to 4-ethylcatechol by single substitutions of amino acids (Ramos *et al.*, 1986). (2) *P. putida* strain GJ31 containing a novel chlorocatechol 2,3-dioxygenase that can efficiently cleave 3-chlorocatechol at the position, 2,3 leading to simultaneous ring cleavage and dechlorination, thereby allowing the strain to degrade chlorobenzene via a meta-cleavage pathway, has recently been described (Mars *et al.*, 1999). (3) The substrate specificity of xylene monooxygenase, encoded by *xylAM*, was extended to *p*-ethyltoluene, which is not a natural substrate (Abril *et al.*, 1989). (4) the specificity of the XylS regulatory protein was modified by mutagenesis, resulting in recognition of other compounds, such as 4-ethylbenzoate, salicylate, 3,5-, 2,5-, and 2,6 dichlorobenzoate, as effector molecules.

There are several advantages and some disadvantages of the in vitro technique over the in vivo selection mechanism (Reineke, 1998). The major advantage is the fact that introduction of well defined cloned genetic elements allows detailed study of the conversion processes that occur in the cell. This is particularly worthwhile since it is possible to make useful mutations in the cloned derivative by site-directed or oligonucleotide-directed mutagenesis which also avoids complications due to operation of non-productive pathways or pathways leading to accumulation of toxic intermediates. Gene cloning also allows introduction of completely foreign genes into the cell, which can be expressed at high level due to its location under a strong promoter; such foreign genes cannot normally be introduced by natural plasmid transfer processes (Hrywna *et al.*, 1999). On the other hand, a major disadvantage of the *in vitro* technique is that it necessitates a through understanding of the location of genes and its regulation. Also, a single gene or genes that are clustered together only can be cloned. Moreover, recombinant plasmids (e.g. pHK1, see above) are unstable in the new cellular environment and tend to get lost or undergo deletion spontaneously at a high frequency (Fukuda, et al 1985; Reineke, 1998). However, the problem of stability has been overcome by the use of so-called mini-transposons for the stable integration of cloned genes into chromosome of recipient strains (de Larenzo *et al.*, 1998). Finally, the combination of in vitro and in vivo selective techniques is expected to provide a rational approach to construction of improved strains for enhanced degradation of toxic, persistent compounds (Hrywna *et al.*, 1999).

6. SURVIVAL AND EFFECTIVENESS OF LABORATORY DEVELOPED STRAINS IN AN OPEN ENVIRONMENT

A major reason for the construction of genetically-engineered strains capable of utilizing toxic, persistent pollutants is to examine their effectiveness for degrading such chemicals in toxic waste dump and other contaminated sites. Although, microorganism have been used for decades for sewage sludge digestion and have occasionally been used for cleaning up the oil polluted area (Pieper and Reineke, 2000), no definitive information exists regarding the ability of genetically-engineered microorganisms to decompose toxic chemicals in an open environment.

Various arguments have been put forward as to why genetically engineered microorganisms may not produce desired effects in terms of removing recalcitrant pollutants from contaminated environments (Pieper and Reineke, 2000). The major two arguments state that (1) most genetically-engineered microorganisms would be unable to compete with indigenous microorganisms and would not therefore survive for too long to utilize the toxic chemicals; (2) he genetically engineered microorganisms would have no encouragement to attack the toxic chemicals because of the presence of a large amounts of other easily assimilable carbon source.

7. CONCLUSION

The use of synthetic chemicals, while essential in moderen industrial society, has created problems of environmental pollution that must be recognized and remedied. Microorganisms have evolved the capacity to dissimilate a number of herbicides and pesticides by altering substrate specificity of enzymes, already encoded in the genome, as seen in the evolution of the different pathways. However, many synthetics, especially those that are highly halogenated, remain recalcitrant to biodegradation. It is possible, nonetheless, to use selective evolution (through use of a chemostat) to develop strains capable of degrading recalcitrant compounds, such as 2,4,5-T. This type of evolution occurs at a much accelerated rate compared to that in nature and is facilitated by the presence of a large gene pool, so organisms can readily take up foreign DNA. It is possible that bacteria in nature are able to evolve rapidly through the uptake of foreign DNA, which is the likely mode for evolution in some kindes of bacteria. This ability can be utilized in a laboratory setting for further development of strains capable of enhanced degradation of toxic, synthetic pollutants in the environment.

REFERENCES

Abril, M. A., Michan, Timmis, K. N. and Ramos, J. L. (1989). Regulator and enzyme specificities of the Tol plasmid-encoded upper pathway for degradation of aromatic hydrocarbons and expansion of the substrate range of the pathway. *J. Bacteriol.* 171:6782-6790.

Aelio, C. M., Swindoll, C. M. and Pfaender, K. F. (1990). Adaptation to and biodegradation of xenobiotic compounds by microbial communities from pristine aquifer. *Appl. Environ. Microbial.* 53:2212-2217.

Aislabie, J., Bej, A. K., Hurst, B. H., Rothenburger, S. and Atlas, R. M. (1990) Microbial degradation of quinoline and methylquinolines. *App. Environ. Microbiol.* 56:345-352.

Alexander, R. M. (1981). Biodegradation of chemicals of environmental concern. *Science* 211:132-138.

Al-Najjar, T. R. and Al-Shakarchi, A. M. (1980). Effect of quinoline and carbostyril concentrations on their oxidation by a *Moraxella. Bull. Biol. Res. Cent.* (Baghdad) 12:99-108.

Amy, P. S., Schulke, J.W., Frazier, L. M, and Seidler, R. J. (1985). Characterization of an aquatic bacteria and cloning of genes specifying partial degradation of 2,2-dichlorophenoxyacetic acid. *Appl. Environm. Microbiol.* 49:1237-1245.

Assinder, S. J. and Williams, P. A. (1990). The TOL plasmids: determination of catabolism of toluene and the xylenes. *Adv. Microb. Physiol.* 31:169

Anson, J. and Mackinnon, G. (1984).Novel *Pseudomonas* plasmid involved in aniline degradation. *Appl. Environ. Microbiol.* 48: 868-869.

Asturis, J. A., and Timmis, K.N. (1991). Three different 2,3-dihydroxybiphenyl-1,2-dioxygenase genes in the gram-positive polychorobiphenyl-degrading bacterium *Rhodococcus globerulus* P6. *J. Bacteriol.*175:4631-4640.

Asturis, J. A., Eltis, L. D., Prucha, M. and Timmis, K. N. (1994).Analysis of three 2,3-duhydroxybiphenyl 1,2-dioxygenase found in *Rhodococcus globerulus* p6. *J. Biol. Chem.* 269:7807-7815.

Attaway N. D., Camper, N. D. and Painter, M. J. B. (1982). Anaerobic microbial degradation of diuron by pond sediment, *Pest. Biochem. Physiol.*17:96 – 101.

Bak, F. and Widdel, F. (1986). Anaerobic degradation of indolic compounds by sulfate-reducing enrichment cultures, and description of desulfobacterium indolicum gen. Nov., Sp. Nov. *Arch. Microbiol.*146:170-176.

Baross, J. A., Dahm, C. N., Ward, A. K. Lilley, M. D. and Sedell, J. R. (1982). Initial microbial response in lakes to the Mt. St. helens eruption. *Science* 296: 49-52.

Behki, R. M. and Khan, S. U. (1986). Degradation of atrazine by *Pseudomonas*; N-dealkylation and dehalogenation of atrazine and its metabolites. *J. Agric. Food Chem.* 34:746-749.

Bestetti, G., Galli, Ruzzi, M., Baldacci, G., Zennaro, E. and Frontali, L. (1984). Molecular characterization of a plasmid from *Pseudomonas fluorescens* involved in styrene degradation. *Plasmid.* 12:181-188.

Boundy-Mills, K. L., De Souza, M. L., Mandelbaum, R. T., Wackett, L. P. and Sadowsky, M. J. (1997). The *atzB* gene of *Pseudomonas* sp strain ADP encode the second enzyme of a novel atrazine degradation pathway. *Appl. Environ. Microbiol.* 63:916:916-923.

Bouwer, E. J. and McCarthy, P. L. (1983) Transformation of halogenated organic compounds under denitrifying conditions. *Appl. Environ. Microbial.* 45:1286-1294.

Bossert, I., and Bartha, R. (1984). The fate of petroleum in soil ecosystem, p. 434-476. In R. M. Atlas (ed.), Petroleum microbiology. Macmillan Publishing Co., New York.

Britton, L. N. (1984). Microbial degradation of aliphatic hydrocarbons in Microbial degradation of organic compounds, d. T. Gibson (ed.), Marcel Dekker, Inc., New York, Pp.89-129.

Brockman, F. J., Denovan, B. A., Hicks, R. J., and Fredrickson, J. K. (1989). Isolation and characterization of quiloline-degrading bacteria from subsurface sediments. *Appl. Environ. Microbiol.* 55:1029-1032.

Brusseau, G. A., Tsien, H. C., Hanson, R. S. and Wackett, L. P. (1990). Optimization of trichloroethylene oxidation by methanotrophs and the use of a calorimetric assay to detect soluble methane monooxygenase activity. *Biodegradation.* 1:19-29.

Burlage, R. S. S., Hooper, S. W. and Sayler, G. S. (1989). The Tol (pWWO) catabolic plasmid. *Appl. Environ. Microbial.* 55:1323-1328.

Cenci, G. and Caldini, G. (1997).Catechol dioxygenase expression in a *Pseudomonas fluorescens* strain exposed to different aromatic compounds. *Appl. Environ. Biotechnol.*47:306-368.

Cernakova, M. (1995). Biological degradation of isoproturon, chlortoluron and fenitrothion. *Folia Microbiol.* 40:201-206.

Cerniglia, C. E. (1992). Biodegradation of polycyclic aromatic hydrocarbons. *Biodegradation* 3:351-368.

Chakrabarty, A. M. and Friello, D. A. (1974). Dissociation and interaction of individual components of a degradative plasmid aggregate in *Pseudomonas*. *Proc. Natl. Acad. Sci.* USA, 71:3410-3414.

Chakrabarty, A. M., Chou, G. and Gunsalus (1973).Genetic regulation of octan dissimilation plasmid in *Pseudomonas*. *Proc. Natl. Acad. Sci.* USA, 70:1137-1140.

Chapalamadugu, S. and Choudhary, G. R. (1991). Hydrolysis of carbaryl by a *Pseudomonas* sp. and construction of a microbial consortium that completely metabolizes carbaryl. *Appl. Environm. Microbiol.* 57:744-750.

Chapalamadugu, S. and Choudhary, G. R. (1992). Microbiological and biotechnological aspects of metabolism of carbamates and organophosphates. *Crtit. Rev. Biotech.* 12:357-389.

Chapalamadugu, S. and Choudhary, G. R. (1993). Isolation of a constitutively expressed enzyme for hydrolysis of cararyl in *Pseudomonas aeruginosa*. *J. Bacteriol.* 175: 6711-6716.

Chatterjee, D. K. and Chakrabarty, A. M. (1983). Genetic homology between independently isolated chlorobenzoate degradative plasmids *J. Bacteriol.* 153:532-534.

Chatterjee, D. K., kellogg, S. T., Hamada, S. and Chakrabarty, A. M. (1981). Plasmid specifying total degradation of 3-Chlorobenezoate by a modified *ortho* pathway. *J. Bacteriol.* 146:639-646.

Choudhary, G. R. and Chapalamadugu, S. (1991).Biodegradation of halogenated organic compounds. *Microbiol. Rev.* 55:59-79.

Chung, S. Y., Maeda, M., Song, E., Horikoshi, K., and Kudo, T. (1994). Isolation and characterization of a grame-positive polychlorinated biphenyl-degrading bacterium, *Rhodococcus erythropolis* TA421, from a termite ecosystem. *Biosci. Biochem.* 85:2111-2113.

Coates, J. D., Woodward, J, Allen, J., Philp, P. and Lovley, D. R. (1997). Anaerobic degradation of polycyclic aromatic hydrocarbons and alkanes in petroleum-contaminated marine sediments. *Appl. Environ. Microbiol.* 63:3589-3593.

Cook, A. M., Grossenbacher, H. and Huter, R. (1983).Isolation and cultivation of microbes with biodegradative potential. *Experientia.* 39:1191-1198.

Curragh, H., Flynn, O., Larkin, M. J. Stafford, T. M., Hamilton, T.G. and Harper, D.B. (1994). Haloalkane degradation and assimilation by *Rhodococcus rhodochrous* NCIMB 13064. *Microbiology,* 140:1433-1442.

Dabrock, B., Keeler, M., Averhoff, B. and Gottschalk, G. (1994). Identification and characterization of a transmissible linear plasmid from *Rhodococcus erythropolis* BD2 that encodes isopropylbenzene catabolism. *Appl. Environ. Microbiol.* 60:853-860.

Dagley, S. (1986). Biochemsitry of aromatic hydrocarbon degradation in *Pseudomonas. In The bacteria* Vol. X, *The biology of Pseudomonas*, Pp. 527-556 Edit by J. R. Sokatch. London: Academic Press.

Davison, J. (1998). Genetic exchange between bacteria in the environment. *Plasmid* 42:73-91.

De Lorenzo, V., Herrero, M., Sanchez, J. M. and Timmis, K. N. (1998). Mini-transposons in microbial ecology and environmental biotechnology. *FEMS. Microbiol. Ecol.* 27:211-224.

Denome, S. A., Olson, E. S. and Young, K. D. (1993). Identification of cloning of genes involved in specific desulfization of dibenzothiophene by *Rhodoccous* sp. Strain ITGS8. *Appl. Environ. Microbiol.* 59:2837-2843.

De Schrijver, A. and De Mot, R. (1999). Degradation of pesticides by actinomycetes. *Crit. Rev. Microbiol.*25:85-119.

De Souza, M. L., Sadowsky, M. J. and Wackett, L. P. (1996). Atrazine chlorhydrolase from *Pseudomonas* sp Stain ADP: gene sequence, enzyme purification, and protein characterization. *J. Bacteriol.* 178:4894-4900.

De Souza, M. L., Wackett, L. P., and Sadowsky, M. J. (1998). The *atzABC* genes encoding atrazine catabolism are located on a self-transmissible plasmid in *Pseudomonas* sp strain ADP. *Appl. Environ. Microbiol.* 64:2323-2326.

DiGiovanni, G. D., Neilson, J. W., Pepper, I. and Sinclair, N. A. (1996).Gene transfer of *Alcaligenes eutrophus* JMP134 plasmid pJP4 to indigenous soil recipients. *Appl. Environ. Microbiol.* 62:2521-2526.

D'Itri, F. M. and Kamrin, M. A. (1983). PCPs: Human and Environmental hazords. Boston, MA:Butterworth Publishers, Boston.USA

Don, R. H., and Pemberton, J. M. (1981). Properties of six pesticide degrading plasmids isolated from *Alcaligenes eutrophus* and *Alcaligenes paradoxus. J. Bacteriol.* 145:681-686.

Don, R. H. and Pemberton, J. M. (1985). Genetic and physical map of the 2,4-dichlorophenoxyacetic acid-degradative plasmid pJP4. *J. Bacteriol.* 161:466-468.

Don, R. H., Weightman, A. J., Knackmuss, H. J. and Timmis, K. N. (1985). Transposon mutagenesis and cloning analysis of the pathways for degradation of 2,4-dichlorophenoxyacetic acid and 3-chlorobenzoate in *Alcaligenes eutrophus* JMP134 (pJP4). *J. Bacteriol.* 161:85-90.

Dong, F., Wang, L., Wang, C., Cheng, J., He, Z., Sheng, Z. and Shen, R. (1992). Molecular cloning and mapping of phenol degradation genes from *Bacillus stearothermophilus* FDTP-3 and their Expression In *Escherichia coli. Appl. Environ. Microbiol.* 58:2531-2535.

Doten, R. C., Ngai, K. L., Mitchell, D. J. and Ornston, L. N. (1987). Cloning and genetic organization of the *pca* gene cluster from *Acinetobacter calcoaceticus*. *J. Bacteriol.* 169:3168-3174.

Eaton, R. W. and Chapman, P.J. (1992). Bacterial metabolism of naphthalene: construction and used recombinant bacteria to study ring cleavage of 1,2 -dihydroxynaphthalene and subsequent reactions. *J. Bacteriol.* 174:7452-7554.

El-Deeb, B.A. (1999). Isolation and characterization of bacterial strain, able to utilize a molluscicide bayluscide as a carbon and nitrogen sole source. The role of plasmids in bayluscide degradation Bull. *Fac. Sci. Assiut Unvi.*28:133-146.

El-Deeb, B. A. (2001). Plasmid-mediated aniline assimilation by *Pseudomonas* sp.B10. *J. Microbiol. Biotechnol.* 11:79-84.

El-Deeb, B. A. (2006) El-Deeb, A. B.(2006) Plasmid and Chromsome mediated phenol and cyanide assimilating by *Pseudomonas* species. *J. Microbiol and Biotech.* 16:1068-1077

El-Deeb, B. A., Ali, A. M. and Ali, K, A. (2000). Some evidences for the involvement of plasmid in diuron herbicide degradation. *Acta microb. Et Immun. Hungarica.* 47:63-73.

El-Deeb, B. A., Ali, A. M., and Ali, K, A. (2000).Detoxication of diuron by *Pseudomonas* sp. *Folia Microbiol.* 45:211-216.

El-Deeb, B. A. and Soltan, E. S. (2000). Isolation and characterization of a soil bacterium capable of utilizing aniline as a sole source of carbon and nitrogen. *Bull. Fac. Sci. Assiut Unvi.* 29:47-56.

Ensign, S. A., Hyman, M. R. and Arp, D. J. (1992).Cometabolic degradation of chlorinated alkene monooxygenase in a propylene-grown *Xanthobacter* strain. *Appl. Environ. Microbiol.* 58:3038-3046.

Ensley, B. D. (1991). Biochemical diversity of trichloroethylene metabolism. *Ann. Rev. Microbiol.* 45:283-299.

Esposito, M. P., Drake, H. M., Smith, J. A. and Owens, T.W. (1980). Dooxins. I. Sources, exposure, transport and control. (EPA 600/2-80-156: June 1980) Cincinnati, U. S. Environmental Protection Agency.

Feng, X., OU, L-T. and Ogram, A. (1997). Plasmid-mediated mineralization of Carbofuran by *Sphingomonas sp*. Stain CF06. *Appl.Environ.Microbiol.* 63:1332-1337.

Fetzner, S. and F. Lingens. 1994. Bacterial dehalogenases: biochemistry, genetics and biotechnology applications. *Microbiol. Rev.* 58::641–685

Filler, K. and Harker, A. (1997).Identification of the inducing agent of the 2,4-dichlorophenoxyacetic acid pathway encoded by plasmid pJP4. *Appl. Environ. Microbiol.* 63:317-320.

Finnery, W. R. (1992). The biology and genetics of the geneus *Rhodococcus*. *Ann. Rev. Microbiol.* 46:193-218.

Fishbein, L. (1979). Potential halogenated industrial carcinogenic and mutagenic chemicals. *Sci.Total.Environ.* 11:111-161.

Fisher, P. R., Appleton, J. and Pemberton, J. M. (1978). Isolation and characterization of pesticide-degrading plasmid pJP1 from *Alcaligenes paradoxus*. *J. Bacteriol.* 135:798-804.

Floodgate, G. (1984). The fate of petroleum in marine ecosystem, Pp.355-398. In: R. M. Atlas (ed.), Petroleum microbiology. Macmillan Publishing Co., New York.

Fought, J. M., Fedorak, P. M., Gray, M. R. and Westlake, D. W. (1990).Microbial desulfurization of petroleum, Pp.379-407. *In* H. L. Ehrlich and C. L. Brierly (ed.), microbial mineral recovery. McGraw-Hill, New York.

Fox, B. G., Borneman, J. G., Wackett, L. P. and Lipscomb, J. D. (1990). Haloakene oxidation by the soluble methane mono oxygenase from *Methylosinus trichosporium* OB3b: mechanistic and environmental implications. *Biochemistr*,29:6419-6427.

Fuchs, K., Schreiner, A., and Lingens, F. (1991) Degradation of 2-methylaniline and chlorinated isomers of 2-methylaniline by *Rhodococcus rhodochrous* strain CTM. *J. Gen. Microbiol.* 137:2033-2039.

Fujii, T., Takeo, M. and Maeda, Y. (1997). Plasmid-encoded genes specifying aniline oxidation from *Acinetobacter* sp. strain YAA. *Microbiology* 143:93-99.

Fukuda, M. and Yano, K. (1985). Construction of Broad host range cloning vectors for gram-negative bacteria *Agric Biol. Chem.* 49:2719.

Fukumori, F. and Saint, C. P. (1995). Analysis of aniline degradation pathway of *Pseudomonas putida* UCC2 (pTDN1). Fifth international symposium on *Pseudomonas*: *Biotech Mol. Biol.*, abstract, P.70.

Furuhashi, K. (1992). Biological routes to optically active epoxides . In chirality in industry, Pp. 167-186. Edited by A.N. Collins, G.N. sheldrake and Crosby, J. Chichester and New Yourk: Wiley.

Furukawa, K. and A. M. Chakrabarty. 1982. Involvement of plasmids in total degradation of chlorinated biphenyls. *Appl. Environ. Microbiol.* 44:619-626.

Furukawa, K., Matsumura, F. and Tonomura, K. (1990). *Alcaligenes* and *Acinetobacter* strains capable of degrading polychlorinated biphenyls. *Agric. Biol. Chem.* 42:543-548.

Franklin, F. C. H., Bagdasarian, M. M. and Timmis, K. N. (1981). Molecular and functional analysis of the TOL plasmid pWWO from *Pseudomonas putida* and cloning of genes for the entire regulated aromatic-ring cleavage pathway. *Proc. Natl. Acad. Sci.* USA 78:7458-7462.

Frantz, B. and Chakrabarty, A. M. (1986). Degradative plasmid in *Pseudomonas. In The bacteria,* Vol. X, The biology of *Pseudomonas,* Pp.295-317. Edited by J.R. Sokatch. London: Academic press.

Frantz, B. and Chakrabarty, A. M. (1987). Organization and nucleotide sequence determination of a gene cluster involved in 3-chlorochatecol degradation. *Proc. Natl. Acad. Sci.* USA 84:4460-4464.

Freillo, D. A., Mylroe, J. R., Gibson, D. T., Rogers, J. E. and Chakrabarty, A. M. (1976). *Xyl,* a nonconjugative plasmid in *Pseudomonas Pxy. J. Bacterio.* 127:1217-1224.

Gallagher, J. R., Olson, E. S. and Stanley, D. C. (1993). Microbial desulfurization of dibenzothiophene; a sulfur specific pathway. *FEMS Microbiol. Lett.* 107:31-36.

Ghosal, D., and You, I. S. (1989). Operon structure and nucleotide homology of the chlorocatechol oxidation genes of plasmids pAC27 and pJP4. *Gene* 83:225-232.

Gibson, D.T. (1984). Microbial degradation of organic compounds Vol. 13. New York: Marcel Dekker.

Gibson, D.T. and Subramanian, V. (1984). Microbial degradation of aromatic hydrocarbons, Pp.181-252. In D. T. Gibson (ed.), Microbial degradation of organic compounds Vol. 13. New York: Marcel Dekker.

Guerin, W. F. and Jones, G. E. (1988). Mineralization of phenanthrcene by *Mycobacterium* sp. *Appl. Environ. Microbiol.*54:937-944.

Grady, C. P. L. (1985). Biodegradation: its measurement and microbiological basis. *Biotechnol. Bioeng.* 27:660-674.

Grund, A. D. (1975). Regulation of alkane oxidation in *Pseudomonas putida. J. Bacteriol.* 123:546-556.

Hada, H. S. and Sizemore, R. K. (1981). Incidence of plasmids in marine *Vibrio spp.* Isolated from an oil field in the northwestern Gulf of Mixico. *Appl. Environ. Microbiol.* 41:199-202.

Haigler, B. E., Nishino, S. F. and Spain, J. C. (1988). Degradation of 1,2-dichlorobenzene by a *Pseudomonas* sp. *Appl.Environ. Microbiol.* 54:294-301.

Haugland, R. A., Sangodkar, U. M. X. and Chakrabarty, A. M. (1990). Repeated sequence including RS1100 from *Pseudomonas cepacia* AC1100 function as IS element. *Mol Gen. Genet.* 220:222-228.

Harayama, S. (1997). Polycyclic aromatic hydrocarbon bioremediation design. *Curr. Opin. Biotechnol.* 8:268-273.

Harayama, S., and Rekik, M. (1989) Bacterial aromatic ring cleavage enzymes are classified into two different gene families. *J. Biol. Chem.* 264:15328-15333

Harayama, S. and Rekik, M. (1990). The *meta*-cleavage operon of TOL degradative plasmid pWWO comprises 13 genes. *Mol.Gen.Genet.* 221:113-120.

Harayama, S., Rekik, M., Wubbolts, M., Rose, K., Leppik, A. R. and Timmis K. N. (1989). Characterization of five genes in the upper-pathway operon of TOL plasmid pWWO from *Pseudomonas putida* and identification of the gene products. *J. Bacteriol.*, 171:5048-5055

Hartmans, S., Smits, J. P., Van der Werf, M. J., Volkering, F. and de Bont, J. A. M. (1989). Metabolism of styrene oxide and 2-phenylethanol in the styrene-degrading *Xanthobacter* strain 124X. *Appl. Environ. Microbiol.* 55:2850-2855.

Hayase, N., Taira, K. and Furukawa, K. (1990). *Pseudomonas Putida* KF715 *hbhABCD* operon encoding biphenyl and polychlorinated biphenyl degradation: cloning, analysis, and expression in soil bacteria. *J. Bacteriol.* 172:1160-1164.

Hayatsu, M., Hirano, M. and Nagata, T. (1999). Involvement of two plasmids in the degradation of carbaryl by *Arthrobacter* sp strain NC100. *Appl. Environ. Microbiol.* 65:1015-1019

Hayatsu, M., Hirano, M., and Tokuda, S. (2000).Involvement of two plasmids in Fenitrothion degradation by *Burkholderia* sp strain NF100. *Appl. Environ. Microbiol.* 66:1737-1740

He, Z. and Spain, J.C. (1997). Studies of the catabolic pathway of degradation of nitrobenzene by *Pseudomonas sp.* JS45: removal of the amino group from 2-aminonmuconic semialdehyde. *Appl. Environ. Microbiol.* 63:4839-4843.

Head, I. M., Cain, R. B. and Suett, D. L. (1992). Characterization of a carbofuran-degrading bacterium and investigation of the role of plasmids in catabolism of the insecticide carbofuran. *Arch. Microbiol.* 158: 302-308.

Holmes, M. L, Pfeifer, F. and Dyall-Smith, M. L. (1995). Analysis of the halobacterial plasmid pHK2 minimal replicon. *Gene*, 153:117-121.

Holtel, A., Abril, M. A., Marques, S., Timmis, K. N., and Ramos, J. L. (1990). Promoter-upstream activator sequences are required for expression of the *xylS* gene and upper-pathway operon on the *Pseudomonas* TOL plasmid. *Mol. Microbiol.* 4:1551-1556.

Hughes, E. J., Shapiro, M. K., Houghton, J. E. and Ornston, L. N. (1988). Cloning and expression of *pca* genes from *Pseudomonas putida* in *E. coli. J. Gen. Microbiol.* 134:2877-2887

Hutzinger, O. and Veerkamp, W. (1981). Xenobiotic chemicals with pollution potential. In microbial degradation of Xenobiotic and recalcitrant compounds, eds. T. Leisinger, A. M. Cook, R. Hutter, and J. nuesh, Pp. 3-45. New York: Academic Press.

Hrywna, Y., Tsoi, T. V., Maltseva, O.V., Quensen III, J. F. and Tiedje, J. M. (1999). Constructure and characterization of two recombinant bacteria that grow on ortho- and -para-substituted chlorobiphenyls. *Appl. Environ. Microbiol.* 65:2163-2169.

Inouye, S., Nakazawa, A. and Nakazawa, T. (1983). Molecular cloning of regulatory gene *xylR* and operator-promoter regions of the *xylABC* and *xylDEGF* operons of the TOL plasmid. *J. Bacteriol.* 155:1192-1199.

Inouye, S., Nakazawa, A. and Nakazawa, T. (1987). Expression of the regulatory gene *xylS* on the TOL plasmid is positively controlled by the *xylR* gene product. *Proc. Natl. Acad. Sci.* USA 84:5182-5186.

Izumi, Y., Ohshiro, T., Ogino, H., Hine, Y., and shimao, M. (1994). Selection desulfurization of dibenzothiophene by *Rhodococcus erythropolis* D-1. *Appl. Environ. Microbiol.* 60:223-226.

Janssen, D. B., Van der Ploeg, J. R. and Pries, F. (1995).Genetic adaptation of bacteria to halogenated aliphatic compounds. *Environ. Health. Perspect.* 5:29-32.

Jeens, D.J., Reineke, W., Knackmuss, H. J. and Williams, P. A. (1982). Tol plasmid pWWO in constructed halobenzoate-degrading *Pseudomonas* strains: Enzyme regulation and DNA structure. *J.Bacteriol.*150:180-187.

Johnson W H. and Camper, N. D. (1991). Microbial degradation activity pretreated soil. *J.Environ. Sci. Health*, 26B:1-4.

Johri, A. K., Dua, M., Sethunathan, N. and Legge, R. L. (1999). Characterization and regulation of catabolic genes. *Crit. Rev. Microbiol.* 25:245-273.

Jiang, S. C. and Paul, J. H. (1998). Gene transfer by transduction in marine environment. *Appl. Environ. Microbiol.* 64:2780-2787.

Kakudo, S., Negoro, S., Urabe, I. and Okada, H. (1993). Nylon oligomer degradation gene, *nylC* on plasmid pOAD2 from a flavobacterium strain encodes endo-type 6-aminohexanoate oligomer hydrolase: purification and characterization of the *nylC* gene product. *Appl. Environ. Microbiol.* 59:3978-3980.

Kanagawa, K, Negoro, S., Takada, N. and Okada, H. (1989). Plasmid dependence of *Pseudomonas* sp. Strain NK87 enzymes that degrade 6-aminohexanoate-cyclic dimer. *J. Bacteriol.* 171:3181-3186.

Kanagawa, K., Oishi, M., Negoro, S., Urabe, I. and Okada, H. (1993). Characterization of the 6-aminohexanoate- dimer hydrolase from *Pseudomonas* sp. NK87. *J. Gen. Microbiol.* 139:787-795.

Kanaly, R. A., and Harayama, S. (2000). Biodegradation of high-molecular weight polycyclic aromatic hydrocarbon by bacteria. *Appl. Environ. Microbiol.* 182:2059-2067.

Kaufman, D. D. and Kearney, (1976). Microbial transformation in the soil, Pp. 26-46. In L.J. Audus (ed.), herbicides: physiology, biochemistry, and ecology. Academic Press, Ltd., London, England.

Karns, J. S. and Tomasek, P. H. (1991). Carbofuran hydrolase: purification and properties. *J. Agric. Food Chem.* 39:1004-1008.

Kathleen, H. B. and Adency, A. J. (1978). Residues of diuron and phytotoxic degradation products in aquatic situations. I. Analytical methods for soil and water. *Pestic. Sci.*9:342-353.

Kayser, K. J., Bielaga-Jones, B. A., Jackowski, K., Odusan, O. and Kilbane, J.J. (1993). Utilization of organosulfur compounds by axenic and mixed cultures of *Rhodoccous Rhodochrous* IGTS8. *J. Gen. Microbiol.* 139:3123-3129.

Kesseler, M., Dabbs, E.R., Averhoff, B. and Gottschalk, G. (1996). Studies on the isopropylcatechol 2,3-dioxygenase genes encoded by the linear plasmid of *Rhodococcus erythropolis* BD2. *Microbiology.* 142: 3241-3251.

Kidambi, S. P., Ripp, S. and Miller, R. V. (1994).Evidence for phage-mediated gene transfer among *Pseudomonas aeruginosa* strains on phylloplane. *Appl. Environ. Microbiol.*60: 496-500.

Kilbane, J.J. (1989) Desulfurization of coal: the microbial solution. *Trends Biotechnol.* 7:97-101.

Kilbane, J.J. and Jackowski, K. (1992). Biodesulfurization of water-soluble coal-derived material by *Rhodoccous Rhodochrous* IGTS8. *Biotechnol. Bioeng.* 40:1107-1114.

Kim, C. K., Kim, J. W. and Mheen, T. I. (1986). Isolation of aromatic hydrocarbon-degrading bacteria and genetic characterization of their plasmid genes [in Korean]. *Kor. J. Microbiol.* 24:67-72.

Kim, Y., Ayoubi, P. and Harker, A.R. (1996) Constitutive expression of the cloned phenol hydroxylasegenes from *Alcaligenes eutrophus* JM134 and concomitant trichloroethylene oxodation. *Appl. Environ. Microbiol.* 62:3227-3233.

Kimbara, K., Hasimoto, T., Fukuda, T., Koana, M., Takagi, M., Oishi, M. and Yano, K. (1989). Cloning and sequencing of two tandem genes involved in degradation of 2,3-dihydroxybiphenyl to benzoic acid in the polychlorinated biphenyl-degrading soil bacterium *Pseudomonas* sp. strain KKS102, *J. Bacteriol.* 171:2740-2746.

Kinoshita, S., Negoro, S., Muramatsu, M., Bisaria, V. S., Sawada, S. and Okada, H. (1977). 6-aminohexanoic acid-cyclic dimer hydrolase: a new cyclic amid hydrolase produced by *Achromobacter guttatus* KI72. *Eur. J. Biochem.* 80:489-495.

Kiyohara, H., Takizawa, N., Date, h., Torigoe, S. and Yano, K. (1990). Characterization of a phenanthrcene degradation plasmid from *Alcaligenes faecalis* AFK2. *J. Ferment. Bioeng.* 69:54-56.

Kohler, T., Haramaya, S., Ramos, J. L. and Timmis, K. N. (1989). Involvement of *Pseudomonas putida* RponN ä factors in regulation of various metabolic functions. *J. Bacteriol.* 171:4326-4333

Konopka, A., Knight D. and Turco, R. F. (1989). Characterization of a *Pseudomonas* sp. capable of aniline degradation in the presence of secondary carbon sources. *Appl. Environ. Microbiol.* 55:385-389.

Kosono, S., Maeda, M., Fuji, F., Arai, H. and Kudo, T. (1997). Three of seven *pbhC* genes of *Rhodococcus erythropolis* TA421, isolated from termite ecosystem, are located on an indigenous plasmid associated with biphenyl degradation. *Appl. Environ. Microbiol.* 63:3282-3285

Kuhm, A. E., Schlomann, M., Knackmuss, H. J. and Pieper, D. H. (1990).Purification and characterization of dichloromuconate cycloisomerase from *Alcaligenes eutrophus* JM134. *Biochem.J.* 266:877-883.

Kulakova, A. N., Larkin, M.J. and Kulakov, L. A. (1997). The plasmid-located haloalkane dehalgenase gene from *Rhodococcus rhodochrous* NCIMB 13064. *Microbiology* 143:109-115.

Kulakova, A. N., Stafford, T. M., Larkin, M.J. and Kulakov, L. A. (1995). Plasmid pRTL1 controlling 1-haloalkane degradation by *Rhodococcus rhodochrous* NCIMB 13064. *Plasmid* 33:208-217.

Kruse, H. and Sorum, H. (1994). Transfer of multiple drug resistance plasmids between bacteria of diverse origins in natural microenvironments. *Appl. Environ. Microbiol.* 60:4015-4021.

Larkin, M. J., Allen, C.C. R., Kulakov, L. A. and Lipscomb, D. A. (1999). Purification and characterization of a novel naphthalene dioxygenase from *Rhodococcus* sp. strain NCIMB12038. *J. Bacteriol.* 181:6200-6204.

Larkin, M. J. and Day, M. J. (1986). The metabolism of carbaryl by three bacterial isolates, *Pseudomonas* sp. (NCIB12038), *Rhodococcus* sp. (NCIB 12038) from garden soil. *J. Appl. Bacteriol.* 60:233-242.

Laurie, A. D. and Lioyd, J. G. (1999). The *phn* genes of *Burkholderia* sp. strain RP007 constitute a divergent gene cluster for polycyclic aromatic hydrocarbon catabolism. *J. Bacteriol.* 181:1531-1540.

Larose, C. D., Labbe, D., Bergeron, H., Jones, A. M., Greer, C. W., Al-Hawari, J., Grossman, M. J., Saneky, B. M. and Lau, P. C. K. (1997). Conservation of plasmid dibenzothiophene desufurization genes in several *Rhodococci. Appl. Microbiol. Environ.* 63:2915-2919.

La Voie, E. J., Dolan, S. Dolan, S., Little, P., Wang, C.-X., Sugie, S. and Rivenson, A. (1988). Carcinogenicity of quinoline, 4-and-8-methylquinoline and benzoquinolines in newborn Mice and Rats. *Food Chem. Toxicol.* 26:625-629.

Layton, A. C., Lajoie, C. A., Easter, M., Muccini, M. and Sayler, G. S. (1994). Molecular diagnostic and chemical for assessing biodegradation of polychlorinated biphenyls in contaminated soils. *J. Ind. Microbiol.* 13:392-401.

Leahy, J. and Colwell, R. R. (1990). Microbial degradation of hydrocarbon in the environment. *Microbiol. Rev.* 54:305-315.

Lebenz, M R. and Williams, P. A. (1985). Complementation of deletion and insertion mutants of TOL plasmid pWWO: regulatory implications and location of *xylC* gene. *J. Gen. Microbiol.* 131:3261-3269.

Lella, D. (1988). Studies on the persistence of bromacil and diuron in different soil types in pot culture. *Pesticides* 22:30-32.

Lehrbach, P. R., Mcgregor, I., Ward, J. M. and Broda, P. (1983). Molecular relationship between *Pseudomonas* Inc P-9 degradative plasmids TOL, Nah, and SAL. *Plasmid* 10:164-174.

Lei, B. and Tu, S.C. (1996). Gene overexpression, purification, and identification of a desulfurization enzyme from *Rhodoccous* sp. Starin ITGS8 as a sulfide/sulfoxide monooxygenase. *J. Bacteriol.* 178:5699-5705.

Lewis, D. L., Kolling, H. P. and Hodson, R. E. (1986). Nutrient limitation and adaptation of microbial populations of chemical transformations. *Appl. Environm. Microbiol.* 51:598-603.

Loidl, M., Hinteregger, C., Ditzelmuller, G., Ferschl, A. and Streichsbier, F. (1990). Degradation of aniline and monochlorinated anilines by soil-born *Pseudomonas acidovorans* strains. *Arch. of Microbiol.* 155:56-61.

Madsen, E. L. (1997).Natural horizontal transfer of a naphthalene dioxygenase gene between native to a coal tar-contaminated field site. *Appl. Environ. Microbiol.* 63:330-2337.

Mae, I., Marits, R. O., Ausmees, N. R., Koiv, V. M. and Heinaru, A. L. (1993). Characterization of a new 2,4-dichlorophenoxyacetic acid degrading plasmid pEST4001: physical map and localization of catabolic genes. *J. Gen. Microbiol.* 139:3167-3170.

Mars, A. E., Kingma, J., Kaschabek, S R., Reineke, W. and Janssen, D. B. (1999). Conversion of 3-chlorocatechol by various catechol 2,3-dioxygenases and sequences analysis of the chlorocatechol dioxygenase region of *Pseudomonas putida* GJ31. *J. Bacteriol.* 181:1309-1318.

Masai, E., Yamada, A., Healy, J. M., Hatta, T., Kimbara, K., Fukuda, M. and Yano, K. (1995). Characterization of biphenyl catabolic genes of gram positive poly chlorinated degrader *Rhodococcus* sp. Strain RHA1. *Appl. Environ. Microbiol.* 61:2079-2085.

Mathur, A. and Verma, I. K. (1992). Tri (3-aminophenyl) phosphine oxid-based nadimide resins II. *J Appl. Polymer Sci.* 46:1749-1757.

McClure, N. C. and Venables, W. A. (1986). Adaptation of *Pseudomonas putida* mt-2 to growth on aromatic amines. *J. Gen. Microbiol.*132:2209-22118.

Mehlman, M. A. (1992). Dangerous and cancer-causing properties of products and chemicals in the oil refining and petrochemical industry. *Environ. Res.* 59:238-249.

Mulbry, W.W. and Eaton, R. W. (1991).Purification and characterization of the N-methylcarbamate hydrolase from *Pseudomonas* sp. strain CRL-OK. *Appl. Environ. Microbiol.* 57:3679-3682.

Mulbry, W. W., Kearney, P. C., Nelson, J. O. and Karns, J. S. (1987). Physical comparison of parathion hydrolase plasmids from *Pseudomonas diminuta* and *Flavobacterium* sp. *Plasmid* 18:173-177.

Munnecke, D.M. (1981). The use of microbial enzymes for pesticide detoxification,Pp.251-270. In T. Leisinger, A. Cook, R. Hutter, and J. Nuesch (ed.), Microbial degradation of xenobiotics and recalcitrant compounds. Academic Press, Inc., London.

Munnecke, D.M. and Hsieh, D.P.H. (1976). Pathway of microbial metabolism of parathion. *Appl. Environ. Microbiol.* 31:63-69.

Nakatsu, C., Ng, R., Singh, N., Straus, N. and Wyndham, C. (1991). Chlorobenzoate catabolic transposon Tn5271 is a composite class I element with flanking class II insertion sequence. *Proc. Natl. Acad. Sci.* USA 88:8321-8316.

Nakazawa, T., Inouye, S. and Nakazawa, A. (1985). Positive regulation and transcription initiation of xyl operon on TOL plasmid, In D. R. Heinski *et al.* (eds.) Plasmid in Bacteria. Plenum Press, New York , Pp. 415.

Nakazawa, T., Inouye, S. and Nakazawa, A. (1990). Regulatory systems for expression of *xyl* genes on the TOL plasmid.p133-141. *In* S. Silver, A. M. Chakrabarty, B.Iglewski and S. Kaplan (ed.), *Pseudomonas*: biotransformation, pathogenesis, and evolving biotechnology. American Society for microbiology, Washington, D.C.

Nakazawa , T. and Yokota (1973). Benzoate metabolism in *Pseudomonas putida* (arvilla) mt-2: demonstration of two benzoate pathways. *J. Bacteriol.* 115:262-267.

Negoro, S. (2000). Biodegradation of nylon oligomers. *APPl. Microbiol. Biotechnol.* 54:461-466.

Negoro, S., Kakuda, S., Urabe, I. and Okada, H. (1992). A new nylon oligomer degradation gene (*nylC*) on plasmid pOAD2 from *Flavobacterium* sp. *J. Bacteriol.* 174:7948-7953.

Neidle, E. L. and Ornston, L. N. (1986). Cloning and expression of *Acinetobacter calcoaceticus* catechol 1,2-dioxygenase structural gene *catA* in *E. coli*. *J. Bacteriol.*168:815-820.

Neilson, J. W, Josephson, K. L., Pepper, I. L., Arnold, R. B., Givanni, G. D. and Sinclair, N. A. (1994).Frequency of horizontal gene transfer of a large catabolic plasmid (pJP4) in soil. *Appl. Environ. Microbiol.* 60:4053-4058.

Newman, L. M. and Wackett, L. P. (1997).Trichloroethelene oxididation by purified toluene 2-monooxygenase: products, kinetics, and turnover-dependent inactivation. *J. Bacteriol.* 179:90-96.

Nishino, S. F. and Spain, J. C. (1993). Oxidative pathway for the biodegradation of nitrobenzene by *Pseudononas* sp. Strain JS765. *Appl. Environ. Microbiol.* 61:2308-2313.

Nordlund, I. and Shingler, V. (1990). Nucleotide sequence of the *meta*-cleavage pathway enzymes 2-hydroxymuconic semialdehyde dehydrogenase and 2-hydroxymuconic semialdehyde hydrolase from *Pseudomonas* CF600. *Biochim. Biophys. Acta* 1049:227-230.

O'connor, K., Buckley, C. M., Hartmans, S., and Dobson, D. W. (1995). Possible regulatory role for nonaromatic carbon source in styrene degradation by *Pseudomonas putida* CA-3. *Appl. Environ. Microbiol.* 61:544-548.

Ornston, L. N., Houghton, J., Neidle, E. L. and Gregg, L. A. (1990). Subtle selection and novel mutation during evolutionary divergence of the â-Ketoadipate pathway, Pp.207-225. *In* S. Silver, A. M. Chakrabarty, B.Iglewski, and S. Kaplan (ed.), *Pseudomonas*: biotransformation, pathogenesis, and evolving biotechnology. American Society for Microbiology, Washington, D.C.

Park, H. and Kim, H. (2000). Identification and characterization of the nitrobenzene catabolic plasmid pNB1 and pNB2 in *Pseudomonas putida* HS12. *J. Bacteriol.* 182:573-580.

Pereira, W. E., Rostad, C. E., Updegraff, D. M. and Bennett, J. L. (1987). Anaerobic microbial transformation of azaarenes in ground water at hazardous waste site, Pp. 111-123. *In* R. C. Averett and D. M. McKnight (ed.), chemical quality of water and the hydrologic cycle. Lewis publishers, Inc., Chelsea, Mich.

Perkins, E. J., Gordon, M.P, Caceres, O. and Lurquin, P. F. (1990). Organization and sequence analysis of the 2,4-dichlorophenol hydroxylase and dichlorocatechol oxidative operons of plasmid pJP4. *J. Bacteriol.* 172:2351-2359.

Peters, M., Heinaru, E., Talpsep, E., Wand, H., Stottmeister, U., Heinaru, A. and Nurk, A. (1997). Acquisition of a deliberately introduced phenol degradation operon, *pheBA*, by different indigenous *Pseudomonas* species. *Appl. Environ. Microbiol.* 63:4899-4906.

Piddington, C. S., Kovacevich, B. R. and Rambosek, J. (1995). Sequence and molecular characterization of a DNA region encoding the dibenzothiophene desulfurization operon of *Rhodococcus* sp. strain ITGS8. *Appl. Environ. Microbiol.* 61:468-475.

Pieper, D. H., Kuhm, A. E., Stadler-frietzsche, K., Fischer, P. and Knackmus, H. J. (1991). Metabolization of 3.5-dichlorocatechol by *Alcalgenes eutrophus* JM134. *Arch. Microbiol.* 156:218-222.

Pieper, D. H. and Reineke, W. (2000) Engineering bacteria for bioremediation. *Curr. Op. Bitechnol.* 11:262-270.

Radosevich, M., Traina, Hae, Y.L. and Tuovinen, O. H. (1995). Degradation and mineralization of atrazine by a soil bacterial isolate. *Appl. Rnviron. Microbiol.* 61:297-302.

Ramos, J. L., Stolz, A., Reineke, W. and Timmis, K.N. (1986). Altered effector specificities in regulators of gene expression: Tol plasmid *xylS* mutants and their use to engineer expansion of the rang of aromatics degraded by bacteria. *Proc. Natl. Acad. Sci.* USA 83: 8467-8471.

Ramos, J. L., Wasserfallen, A., Rose, K. and Timmis, K.N. (1989).Redesigning metabolic routs: manipulation of TOL plasmid pathway for catabolism of Aklylbenzoates. *Science* 235:593-596.

Rani, N. L. and Lalithakumari, D. (1994). Degradation of methylparathion by *Pseudomonas pudita. Can. J. Microbiol.* 40:1000-1006.

Rapp, P. and Timmis, K. N. (1999).Degradation of Chlorobenzenes at nanomolar concentration by *Burkholderia* sp. Strain PS14 in liquid cultures and in soil. *Appl. Environ. Microbiol.* 65:2547-2552.

Reineke, W. (1988). Microbial degradation of haloaromatics. *Annu. Rev. Microbiol.* 42:263-287.

Reineke, W. (1998) Development of hybride strains for the mineralization of chloroaromatics by patchwork assembly. *Annu. Rev. Microbiol.* 52:287-331.

Ripp, S., Ogunseitan, O. A. and Miller, R. V. (1994). Transduction of a freshwater microbiol community by a new *Pseudomonas aeruginosa* generalized transducing phage, UTI. *Mol.Ecol.* 3:121-126.

Romanowski, G., Lorenz, M. G. and Wackernagel, W. (1993). Plasmid DNA in a groundwater aquifer micrcosm-adsorption, DNAase resistance and natural genetic transformation of *Bacillus subtilis. Mol.Ecol.* 2:171-181.

Rothmel, R. K., Chakrabarty, A. M., Berry, A., and Darzins (1991). Genetic systems in *Pseudomonas, Methods Enzymol.* 204: 485-502.

Rusansky, S., Avigad, R., Michaeli, S., and Gutnick, D. (1987). Involvement of a plasmid in growth on and dispersion of crude oil by *Acinetobacter calcoaceticus* RA57. *Appl. Environ. Microbiol.* 53:1918-1923.

Sadowsky, M. J., Tong, Z., de Souza, M. L. and Wackett, L. P. (1998). AtzC is a new member of the amidohydrolase protein superfamily and is homologus to other atrazine-metabolizing enzymes. *J. Bacteriol.* 180:152-158.

Saeki, H., Akira, M., Furuhashi, K., Averhoff, B. and Gottschalk, G. (1999). Degradation of trichloroethene by a linear plasmid-encoded alkene monooxygenase in *Rhodococcus corallinus* (*Nocardia corallina*) B-276. *Microbiology* 145:1721-1730.

Saint, C. P., McClure, N. C. and Venables, W.A. (1990). Physical map of the aromatic amine and m-toluate catabolic plasmid pTDN1 in *Pseudomonas putida*: location of a unique *meta*-cleavage pathway. *J.Gen. Microbiol.* 163:615-625.

Salagovic, J., Kalian I. and Dudas, M. (2000). Genetic polymorphism of enzymes metabolizing xenobiotics as a factor in susceptibility to tumor development. *Bratisl. Lek. Listy.* 101:512-521.

Sallis, P. J., Armfield, S. J., Bull, A. F. and Hardman, D. H. (1990). Isolation and characterization of halidohydrolase from *Rhodococcus erythropolis* Y2. *J. Gen. Microbiol.* 136:115-120.

Sambrook, J., Fritsch, E. F. and Maniatis, T. (1989). Molecular cloning: A laboratory manual, (2nd ed.), Cold spring Harbor Laboratory, Cold spring Harbor, New York.

Santos, P. M., Blantny, J. M., Bartolo, I., Valla, S. and Zennaro, E. (2000). Physiological analysis of the expression of the styrene degradation gene cluster in *Pseudomonas fluorescens* ST. 66:1305-1310.

Sayler, G. S., Hooper, S. W., Layton, A. C. and King, J. M. (1990). Catabolic plasmids of environment and ecological significance. *Microbiol. Ecol.* 19:1-20.

Schell, M. A. (1985). Transcription control of the *nah* and *sal* hydrocarbon-degradation operons by the nahR gene product. *Gene* 36:301-309.

Schell, M. A. (1990). Regulation of naphthalene degradation of plasmid NAH7: example of a generalized positive control system in *Pseudomonas* and related bacteria, Pp165-177. *In* S. Siliver, A. M. Chakrabarty, B.Iglewski, and S. Kaplan (ed.), *Pseudomonas*: biotransformation, pathogenesis, and evolving biotechnology. American Society for microbiology, Washington, D.C.

Schell, M. A. and Poster, E. F. (1989) Cloning and expression in *E.coli* of the naphthalene degradation gene from plasmid NAH7. *J. Bacteriol.* 153:822-829.

Schneider, M. J. (1979). Persistent poisons Chemical Pollutants in the environment. New York. Academy of Science.

Scholtze, R., Leisinger, Suter, F. and Cook, A. M. (1987). Characterization of 1-haloalkane halidohydrolase, a dehalogenase of a wide substrate range from *Arthobacter* sp. *J. Bacteriol.* 169:5016-5021.

Schwartz, G., Senghas, Erben, A., Schaefer, B., Lingenes, F. and Hoeke, H. (1988). Microbial metabolism of quinoline and related compounds. 1. Isolation and characterization of Quinoline-degrading bacteria. *Syst. Appl. Microbiol.* 10:185-190.

Schwien, U., Schmidt, E., Knackmus, J. and Reineke, W. (1988). Degradation of chlorosubstituted aromatic compounds by *Pseudomonas* sp strain B13: fate of 3,5 dichlorocatechol. *Arch. Microbiol.* 150:78-84.

Serdar, C.M. and Gibson, D.T. (1985). Enzymatic hydrolysis of organophosphates: cloning expression of a parathion hydrolase gene from *Pseudomonas diminuta*. *Biotechnology* 3:567-571.

Serdar, C.M. and Gibson, D.T. (1989). Studies of nucleotide sequence homology between naphthalene-utilizing strains of bacteria. *Biochem. Biophys. Res. Commun.* 164:772-779.

Serdar, C.M., Gibson, D.T., Munnecke, D. M. and Lanccaster, J. H. (1982). Plasmid involvement in parathion hydrolase by *Pseudomonas diminuta*. *Appl. Environ. Microbiol.* 44:246-249.

Seto, M., Kimbara, K., shimura, M. Hatta, T. Fukuda, M. and Yano, K. (1995). A novel transformation of polychlorinated biphenyls by *Rhodococcus* sp. Strain RHA1. *Appl. Environ. Microbiol.* 61:3353-3358.

Sethunathan, N. and Yoshida, T. (1973). A *Flavobacterium* sp. That degrade diazinon and parathion. *Can. J.Microbiol.* 19:873-875.

Seiler, J. P. (1992). The genetic toxicity of phenoxy acids other than 2,4,5-T. *Mutation Res.*. 55:197-226.

Shingler, V., Powlowski, J. and Marklund, U. (1992). Nucleotide sequence and functional analysis of the complete phenol/3,4-dimethylphenol catabolic pathway of *Pseudomonas* strain CF600. *J. Bacteriol.* 174:711-724.

Shukla, O. P. (1987). Microbiological transformation and biodegradation of quinoline: isolation and characterization of quinoline-degrading bacteria and identification of early intermediates. *Biol.Mem.* 13:115-164.

Sorensen, S. J. (1993). Transfer of plasmid Rp4 from *E. coli*. K-12 to indigenous bacteria of seawater. *Microb. Releases.* 2:135-141.

Spain, J. C. and Gibson, D. T. (1991). Pathway for biodegradation of *p*-nitrophenol in *Moraxella* sp. Appl. *Environ. Microbiol.* 57:812-819.

Spain, J. C., and Nishino, S. F. (1987). Degradation of 1,4-dichlorobenzene by a *Pseudomonas* sp. *Appl. Environ. Microbiol.* 53:1010-1019.

Spooner, R. A., Lindsay, K. and Franklin, F.C.H. (1986). Genetic, functional and sequence analysis of the *xylR* and *xylS* regulatory genes of the Tol plasmid pWWO. *J. Gen. Microbiol.* 132:1347-1358.

Streber, W. R., Timmis, K. N. and Zenk, M. H. (1987). Analysis, cloning, and high-level expression of 2,4-dichlorophenoxyacetate monoxygenase gene *tfdA* of *Alcalgenes eutrophus. J. Bacteriol.* 169:2950-2955.

Sud, R. K., Sud, A. K. and Gupta, K. G. (1972).Degradation of sevin (1-naphthyl—N-methylcarbamate) by *Achromobacter* sp. *Arch. Microbiol.* 87:353-358.

Suflita, J. M. and Sewell, G. W. (1991). Anaerobic biotransformation of contaminates in the subsurface. US. EPA environmental Research brief. EPA./600/M-90/024. EPA, Washington, D. C.

Sutton, S. D., Pfaller, S. L., Ahann, J. R., Warshawsky, D., Kinkle, B. K. and Vestal, J. R. (1996). Aerobic biodegradation of 4-methylquinoline by a soil bacterium. *Appl. Environ. Microbiol.* 62:2910-2914.

Swinddoll, C. M., Aelion, C. M. and Pfaender, F. K. (1988). Influence of inorganic and organic nutrients on aerobic biodegradation and on the adaptation response of subsurface microbial communities. *Appl. Environ. Microbiol.* 54:212-217.

Taira, K., Hirose, J., Hayashida, S. and Furukawa, K. (1992). Analysis of *bph* operon from the polychlorinated biphenyl-degrading strain of *Pseudomonas pseudoalcaligenes* KF707. *J. Biol. Chem.* 267:4844-4853.

Timmis, K. N. (1985). Analysis and manipulation of plasmid-encoded pathway for the catabolism of aromatic compounds by soil bacteria. *In* D. Heliski *et al.* (eds.), *Plasmid in bacteria*. Plenum press, New york, Pp.719.

Tomasek, P. H. and Karns, J. S. (1989). Cloning of carbofuran hydrolase from *Achromobacter* sp strain WM111 and its expression in gram negative bacteria. *J. Bacteriol.* 171:4038-4044.

Tralau, T., Alasdair, M. C. and Juergen, R. (2001). Map of the IncP1â plasmid pTSA encoding the widespread genes (tsa) for p-Toluenesulfonate degradation in *Comamonas testosteroni* T-2. *Appl. Environ. Microbiol.* 67:1508-1516.

Tsuda, M. and Iino, T. (1988). Identification and characterization of Tn4653, a transposon covering the toluene transposon 4651 on TOL plasmid pWW0. *Mol Gen. Genet.* 213:72-77.

Tsuda, M. and Iino, T. (1990). Naphthalene degrading genes on plasmid NAH are on a defective transposon. *Mol.Gen. Genet.* 223:33-39.

Tsuda, M., Minegishi, K. I. and Iino, T. (1989). Toluene transposons Tn4651 are class II transposons *J. Bacteriol.* 171:1386-1393.

Tyler, J. E. and Finn, R. K. (1974). Growth rate of a *Pseudomond* on 2,4-dichlorophenoxyacetic acid and 2,4-dichlorophenol. *Appl. Environm. Microbiol.* 28:181-184.

Van der Meer, J. R., Eggen, R. I. L., Zehnder, A.J. B. and de Vos, W. M. (1991a). Sequence analysis of the *Pseudomonas* sp. strain P51 *tcb* gene cluster, which encodes metabolism of chlorinated catechol: evidence for specilization of catechol 1,2-dioxygenase for chlorinated substrates. *J. Bacteriol.* 173:2425-2434.

Van der Meer, J. R., van Neerven, A. R. W., de Vos, W. M. and Zehnder, A.J. B. (1991b). Cloning and characterization of plasmid-encoded genes for the degradation of 1,2-dichloro-1,4-dichloro, and 1,2,4, trichlorobenzene of *Pseudomonas* sp strain P51. *J. Bacteriol.* 173:6-15.

Van der Meer, J. R., Roelofsen, W., Schraa, G. and Zehnder, A.J. B. (1987) degradation of low concentration of dichlorobenzenes and 1,2,4-trichlorobenzene of *Pseudomonas* sp. strain P51. J. Bacteriol. 173: 6-15.

Van der Meer, Zehnder, A.J. B. and de Vos, W. M. (1991c). Identification of a novel composite transposable element, Tn5280, carrying chorobenzene dioxygenase genes of *Pseudomonas* sp. strain P51. *J. Bacteriol.* 173:7077-7083.

Velasco, A., Alonso, S., Garcia, J. L., Perera, J. and Diaz, E. (1998). Genetic and functional analysis of the styrene catabolic cluster of *Pseudomonas* sp. strain y2. *J.Bacteriol.* 180:1063-1071.

Ware , G. W. (1989). The pesticide book. Thomson Publications, Fresno, CA.

Warhurst, A. M., Clarke, K. F., Hill, R. A., Holt, R. A. and Fewson, C. A. (1994). Metabolism of styrene by *Rhodococcus rhodochrous* NCIMB 13259. *Appl. Environ. Microbiol.* 60:1137-1145.

Weisshaar, M. P., Franklin, F. C. H., and Reineke, W. (1987) Molecular cloning and expression of the 3-chlorobenzoate-degrading genes from *Pseudomonas* sp strain B13. *J. Bacteriol.* 169:394-402.

Wierenga, R. K., Tepstra, P. and Hol, W. G. L. (1986). Prediction of occurrence of the ADP-binding b}b-fold in protein, using an amino acid sequence fingerprint. *J. Mol. Biol.* 187:101-107.

Wiggins, B. A., Jones, H. S. and Alexander, M. (1987). Explanation for the acclimation period preceding the mineralization of organic chemicals in aquatic environments. *Appl. Environm. Microbiol.* 53:791-796.

Wong, C. L. and Dunn, N. W. (1976). Combined chromosomal and plasmid encoded control for the degradation of phenol in *Pseudomonas putida. Genet. Res.* 27:405-412.

Worsey, M.J. and Williams, A. P. (1975). Metabolism of toluene and xylenes by *Psudomonas putida* (arvilla) mt-2: evidence for a new function of the TOL plasmid . *J. Bacteriol.* 124:7-13.

Yanze-Kontchou, C., and Gschwind, N. (1994). Mineralization of the herbicide atrazine as a carbon source by a *Pseudomonas* strain. *Appl. Environ. Microbiol.* 60:4297-4302.

Yen, K. M., and Gunsalus, I.C. (1982).Plasmid gene organization: naphthalene/salicylate oxidation. Proc.Natl.Acad. Sci. USA. 79:874-978.

Yen, K. M. and Serdar, C. M. (1988). Genetics of naphthalene catabolism in *Pseudomonads. Crt. Rev. Microbiol.* 15:247-268.

Zeyer, J., Wasserfallen, A. and Timmis, K. N. (1985). Microbial mineralization of ring-substituted anilines through an *ortho*-cleavage pathway. *Appl. Environ. Microbiol.* 50:447-453.

Zillig, W., Kletzin, A., Schleper, C., Holz, I., Janekovick, D., Hain, J., Lanzendorfer, M. and Kristjansson, J. K. (1994). Screening for Sulfolobales, their plasmids and their viruses in Icelandic solfataras. *Syst. Appl. Microbiol.* 16:609-628.

Zuniga, M. C., Durham, D. R., and Welch, R. A. (1981). Plasmid and chromosome-mediated dissimilation of naphthalene and salicylate in Pseudomonas putida PMD-1. *J. Bacteriol.* 147:836-843.

Zylstra, G. J., Olsen, R. H. and Ballou, D. P. (1989). Genetic organization and sequence of the *Pseudomonas cepacia* genes for the alpha and beta subunits of protocatechuate 3,4-dioxygenase. *J. Bacteriol.* 171:5915-5921.

❑❑❑

CHAPTER **12**

Mechanisms Involved in Tolerance of Mycorrhizal Plants to Heavy Metals

K.S. Subramanian*, M. Thangaraju and R.A. Jegan

*Department of Soil Science & Agricultural Chemistry
Tamil Nadu, Agricultural University, Coimbatore 641 003, India

ABSTRACT

Heavy metal contaminated soils are biologically inactive and extremely poor in soil fertility status. These metals are highly resistant and cannot be easily degraded by either chemical reactions or amendments. Microorganisms are biological softwares capable of degrading the metals efficiently. Among microorganisms, mycorrhizas are known to circumvent metal toxicities through a series of processes that assist in immobilization in biological systems which converts toxic metals into less or non-toxic forms. There are five mechanisms such as biosorption, solubilization, hyphal transport, detoxification and dilution are involved in promoting tolerance of host plants to heavy metal toxicities. These processes are strategies that can be exploited to phytoremediate the contaminated soils. The phytoremediation strategies encompass tolerance of mycorrhizal fungi, phytostabilization (preventing entry of heavy metals into plants), phytoextraction (luxuriant uptake and transport to shoots) and intercropping or crop rotation with mycorrhizal dependent hyperaccumulators or non-hyperaccumulators. Despite mycorrhizas are natural, the indigenous population is not sufficient enough to induce tolerance of host plants to abiotic and biotic stress conditions. This necessitates adopting recommended agronomic practices to promote mycorrhizal colonization in natural agro-ecosystem or to derive full benefits from phytoremediation strategies. Thus mycorrhizas are important naturally occurring microorganisms capable of withstanding highly contaminated metal sites and offer protection against phytotoxicities caused by heavy metals.

Key words: mycorrhizal plant, heavy metals, mycorrhizal symbiosis, detoxification.

1. INTRODUCTION

The arable lands available for cultivation of crops is dwindling alarmingly as a direct consequence of desertification caused by contaminants derived

*Email: kssubra2001@rediffmail.com

from a wide range of industries. Metallic elements are intrinsic component of the environment and extremely difficult to eliminate them once they enter in it. Heavy metals are those elements measuring a density of more than 5 g cc^{-1}. The toxic heavy metals enter into the agro-ecosystems include Pb, Cd, Cu, Cr, Hg, As and Ni. Heavy metals are highly toxic and get accumulated as these metals do not easily degrade like organic pollutants. Physical and chemical methods of removing metals are extremely expensive and almost unrealistic. Since the metal contaminated soils are inactive, it is quite appropriate to introduce a microorganism that facilitates in promoting biological activities of soils thereby scavenging the toxic metals while keeping the rhizosphere free from contaminants. One of the biological means to remediate metal contaminated soils is by mycorrhizal symbiosis. Mycorrhizae are omnipresent soil borne fungi form symbiotic association with nearly 90% of the terrestrial plant species and provide nutritional benefits and tolerance to abiotic (drought, salinity, heavy metal toxicity) and biotic (diseases, pests and nematodes) stress conditions (Subramanian *et al.*, 2006; Subramanian and Charest, 2007). This book chapter reviews mechanisms involved in tolerance of host plants to heavy metal contaminated soils besides strategies to realize fullest benefits from the mycorrhizal symbiosis.

2. HEAVY METALS AND THEIR IMPACTS

Heavy metals constitute an important class of toxic substances which are encountered in occupational and environmental circumstances. Indeed, the impact of these toxic agents on human health is being extensively studied due its ubiquity nature of exposure. The toxicity of metal ions to mammalian systems is due to the reactivity of these ions with structural proteins, enzymes and membrane systems. The organ capable of accumulating heavy metals is vulnerable for its toxicity effects. Further, the toxic effects depend on the valence, volatility and lipid solubility etc. The sources and effects of toxic metals on plants and humans and their maximum allowable limits (MALs) in soils are given in Table. 1. Besides general toxicities of metals, certain metals such as Cr and Ni have been closely associated with the occurrence of cancers in human population. Plants grown in heavy metal accumulated soils are severely affected as a consequence of deterred enzyme activities, damaged cellular membrane, inhibited root and shoot development, reduced photosynthesis, decreased stomatal conductance and transpiration and uptake of essential nutrients. Heavy metals can also affect the microbial communities in soils by affecting growth, morphology and metabolism. Consequently, metal contamination drastically reduces the microbial biomass and soil enzyme activities (Chander and Brookes, 1991).

TABLE 1. Heavy metals, Sources, Impacts on humans, Critical concentrations in plants, and Maximum Allowable Limit (MAL) in soils.

Metals	Sources	Impacts on humans	Critical conc. in plants (mg/kg)	MAL in soils (mg/kg)
As	Fertilizer, paints and textile industries	Damages to intestine, liver and kidneys		
Cd	Fertilizer, electroplating, paints and pigments industries	Itai-itai disease (leg mylagia), skeletal deformation, pulmonary emphysema	5-10	3
Cr	Tanning, rubber and photographic industries	Ulcer, respiratory cancer, renal failure, chrome holes on skins	1-2	100
Pb	Painting and automobile industries	Encephalopathy, peripheral neuropathy, anemia	10-20	100
Mn	Fertilizer and pesticide industries	Central nervous disorders	50-100	200
Hg	Photo-film industry, chlor-alkali chemicals and instruments	Minamata disease, nervous disorders, Hunter-Russell syndrome	2-5	2
Ni	Electroplating, iron and steel industries, aerosols	Chronic rhinitis and sinusitis, respiratory cancer, dermatitis	20-30	50
Ti	Electroplating, iron and steel industries	Nervous system disorders, visual defects, pneumoconicosis	10-20	-

(Sauerbeck, 1982)

3. MYCORRHIZAL SYMBIOSIS

Arbuscular mycorrhizal (AM) fungi are soil borne fungi forming symbiotic relationship with majority of higher plants, providing a direct link between soil and plant roots. AM fungi undergo a multi-step colonization process before benefiting the host plant (Fig. 1). The AM fungi exist in soil as thick walled chlamydospores or as vegetative propagules. Spore germination is induced by root exudates carrying phenolic compounds (especially quercetin) that serve as signal molecules. After spore germination, the germ tube is attracted to grow towards the roots and offer mechanical pressure to split open the root epidermal cells and form appressoria indicating successful recognition between two symbionts. Fungal hyphae penetrate from the appressorium into the root cortex cells both intra- and intercellularly. At the deeper cortex region, the intercellular hyphae branched into the parenchymal host cells as haustorial structures called arbuscules. These structures are closely surrounded by plasmalemma of host cells and provide large surface area of contact. This is the site of exchange of bidirectional flow of minerals and metabolites. These

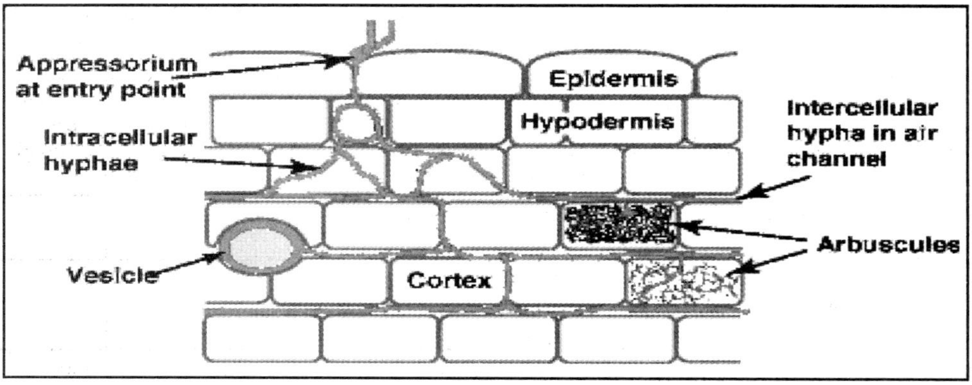

FIG. 1. Multi-step colonization process in mycorrhizal symbiosis (After M.C. Brundrett)

are ephemeral structures that live for 4-14 days and disappear due to autocatalytic activity of cells and believed to be a regulatory mechanism by the host plant to arrest carbon flow to VAM fungus. Further, vesicles are developed as apical or intercallery swellings of hyphae containing lipids. After the spread of the colonization in the roots, hyphae grow out of the root as extraradical mycelium which provides enormous surface area for the host plant roots to scavenge soil nutrients. The functional symbiosis is established within 4-6 weeks. The multi-step colonization process is dynamic as sporulation and spore germination may happen at the same time. After the establishment of functional symbiosis, the host plants derive nutritional benefits.

3.1 Heavy Metals and Mycorrhizal Symbiosis

Mycorrhizal association is known to improve the host plant tolerance to nutrient deficiencies (Smith and Read, 1997; Jeffries and Barea, 2003), drought (Nelson, 1987; Augé 2001; Subrmanian *et al.*, 2006), chilling (Charest *et al.*, 1993; Paradis *et al.*, 1995), salinity (Ruiz-Lozano *et al.*, 1996b; Azcon and El-Atrach, 1997), heavy metals (Weisenhorn *et al.*, 1995; Gaur and Adholeya, 2004; Audet and Charest, 2006) and root feeding pathogens (Linderman, 1994; Benhamou *et al.*, 1994). The role of AM fungi is utmost important in heavy metal contaminated soils where the biological activity is absolutely nil and nutritional status is extremely poor. Even under degraded soils, AM symbiosis can contribute to plant growth by increasing plant access to immobile mineral ions such as P (Vivas *et al.*, 2003; Yao *et al.*, 2003), improving physical conditions (Rillig and Steinberg, 2002; Steinberg and Rillig, 2003) and by binding heavy metals into roots that restricts their translocation into shoot tissues (Dehn and Schuepp, 1989; Kaldorff *et al.*, 1999). Thus there is an immense potential for mycorrhizal symbiosis in revegetation of contaminated soils. Even though AM fungi are omnipresent in terrestrial ecosystems, chemical pollutants in the soils can substantially reduce fungal population and their functionalities. The number of fungal spores and root colonization of plants are often reduced by soil disturbance (Waaland and Allen, 1987). However, some indigenous AM fungal species would have inherently developed built-in mechanisms that assist in adaptation to hostile environment (Gaur and Adholeya, 2004).

3.2 Mechanisms Involved in Tolerance of Mycorrhizal Plants to Heavy Metals

Arbuscular mycorrhizal (AM) fungal isolates derived from metal contaminated soils suggest that a potential adaptation of the indigenous AM populations. In turn, these fungi provide tolerance to their plants to sustain under harsh environment. There are five set of mechanisms involved in tolerance of mycorrhizal plants to heavy metals that include biosorption (adsorption, ion exchange, complexation, precipitation), solubilization, hyphal transport, detoxification (avoidance, compartmentalization) and dilution. These strategies reduce the metal toxicities either by preventing entry or diluting the toxicity of metals in mycorrhizal systems.

3.2.1 Biosorption of Heavy Metals

i. Adsorption

The AM fungal surface area serves as adsorptive sites for heavy metal cations and this process prevents entry of toxic metals into the host plants (Joner *et al.*, 2000). The binding of metals takes place on the fungal wall through electrostatic interactions. The heavy metal ions in soil solution interact with mycorrhizal cell wall which is rich in negatively charged ligands such as phosphoryl, carboxyl, hydroxyl and phenolic groups. Christie *et al.* (2004) indicated that the external mycelium of *Glomus mosseae* bound more than 3% of Zn on a dry weight basis which can be considered as a potential mechanism in alleviating Zn phytotoxicity in contaminated sites. Joner *et al.* (2000) reported that *Glomus mosseae* P2 isolated from contaminated sites accumulated four times of Cd (2000 μmol/g) in comparison to *Glomus mosseae* (450 μmol/g) in unpolluted soils (Fig. 2).

Further, the cell wall of the external mycelium carries bountiful of chitin which has greater metal binding capacity (Zhou, 1999). Binding of heavy metals in the fungal cell wall reduces its local concentration in soil. Passive adsorption of Cd on the hyphae has been quantified up to 0.5 mg Cd per mg of dry biomass (Joner *et al.*, 2000). The extensive surface area provided by the

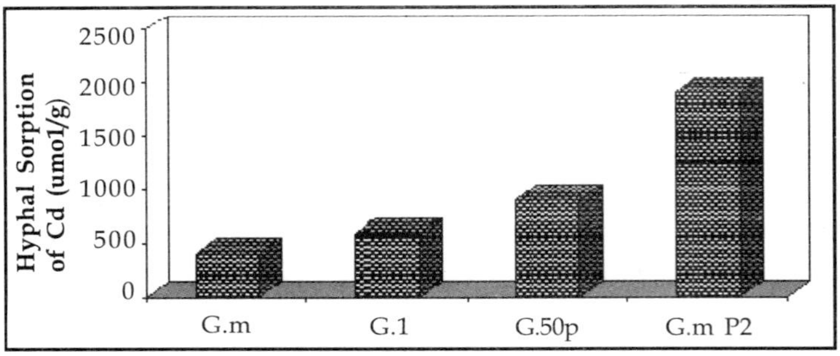

FIG. 2. Hyphal sorbtion of Cd in mycorrhizal species isolates from contaminated and *Glomus* sp. and *G. mossae* P2) and uncontaminated soils (*Glomus mossae (G.m)* and *Glomus lamellosum* (*G.l*) (Joner *et al.*, 2000).

external mycelium of AM fungi in the soil serves as a major sink for heavy metals. In addition, heavy metal tolerant fungi have higher affinity for heavy metals (2-4 times more than roots) and therefore particularly suitable for fixing heavy metals in soil (Joner *et al.*, 2000). Glomalin is a glycoprotein released by AM fungi binds heavy metals in soils (Gonzalez-Chavez *et al.*, 2004). Glomalin extracted from the contaminated soils carry considerable amounts of heavy metals (Khan, 2005). Gonzalez-Chavez *et al.* (2004) found that up to 4.3 mg Cu, 0.08 mg Cd and 1.12 mg Pb per gram of glomalin could be extracted from contaminated soils inoculated with AM fungi. The data suggest that AM fungal species secreting greater quantities of glomalin could be used for biostabilization efforts. The biosorption mechanism filters out the heavy metals from entering into the plant-mycorrhizal systems.

ii. Ion Exchange

The cell walls of mycorrhizal fungi contain polysaccharides and bivalent metal ions that can exchange with heavy metal ions resulting in the biosorptive uptake of heavy metals from the contaminated soil. Non-metabolic binding of metals to cell walls is usually rapid compared to metabolism-dependent absorption, and the former is quantitatively most important (Gadd, 1990). Joner *et al.* (2000) measured the Cation Exchange Capacity (CEC) of roots and mycorrhizal fungal biomass separately and found that the CEC of AM fungi is 10 times higher than that of roots suggesting enormous potential of AM fungi in ameliorating heavy metal toxicity through ion exchange mechanism.

iii. Complexation

Arbuscular mycorrhizal (AM) fungi excrete organic acids such as citric, malic and oxalic acids into rhizosphere (Landeweert *et al.*, 2001). The heavy metal removal from the soil solution takes place by complex formation on the fungal cell surface after the interaction between the heavy metal and the organic acids (the active groups). The organic acids produced by AM fungi may chelate toxic metals resulting in the formation of metallo-organic molecules. Heavy metals may be biosorbed or complexed by carboxyl groups found in mycorrhizal polysaccharides. The tolerance of AM fungi to heavy metals is also achieved by the release of siderophores and metallothionins and sequestration by plant derived compounds such as phytochelatins and phytates.

iv. Precipitation

The heavy metal uptake into AM fungal hyphae can precipitate the metallic cations with PO_4^- intracellularly (Kaldorf *et al* 1999). Turnau *et al.* (1993) have demonstrated greater accumulation of Cd, Ti and Ba in fungal structures than in the host plant cells. In addition, extracellular polysaccharide slime or mucilage is commonly produced by AM fungi, forming a coat around them that is capable of considerable metal immobilization. Turnau (1998) studied the localization of heavy metals within fungal mycelium and mycorrhizal roots of *Euphorbia cyparissias* from Zn contaminated wastes and found higher concentration of Zn as crystalloids within the fungal mycelium and cortical cells of mycorrhizal roots. These immobilization processes of heavy metals in AM fungal biomass act as a barrier to restrict the heavy metal transport from fungus to host plant.

3.2.2 Solubilization of minerals containing heavy metals

The mycorrhizal symbiosis acidifies the rhizosphere which facilitates weathering of rocks and minerals. Such events coincide with release of heavy metals to the soil solution. The AM fungal colonization has been shown to increase the mobilization of heavy metals from biotite (Leyval, *et al.*, 1990). These fungi can also increase the bioavailability of heavy metals in the rhizosphere by solubilizing minerals containing metals such as rock phosphates (Leyval *et al.*, 1993). Mycorrhizal fungi have been reported as solubilizers of water-insoluble inorganic phosphates and contribute to weathering of minerals (Paris *et al* 1994; Fomina *et al* 2004). This has been mainly attributed to the acidification of their environment and to the release of organic acids in the rhizosphere (Leyval *et al.*, 1993).

The availability of Cd from rock phosphates has been studied in the laboratory experiments with ectomycorrhizal fungi and three rock phosphates of different origin and with different chemical compositions and surface (Leyval *et al* 1993). Growing the fungi in flasks containing nutrient medium and rock phosphates in dialysis bags to avoid direct contact with the microorganisms showed an increase with the time of Cd concentration in the solution. The pH decreased in the medium and organic acids such as tartaric acid were released, which could explain the partial solubilization of the rock phosphates. Comparing the mycorrhizal fungi with a phosphate-solubilizing rhizobacterium showed that all the Cd released from the minerals remained in solution with the bacteria, but with the fungi, a large part of the Cd was bound to, or accumulated by the fungi. This suggested that both microorganisms contributed to the mobilization of Cd from rock phosphates by solubilization, but that the mycorrhizal fungi were able to further immobilize the metals.

3.2.3 Transport of heavy metals by AM fungi

Transport of elements by mycorrhizae is defined as the movement of an ion from soil solution into the plant through the fungus and consists of three separate steps or processes namely uptake, translocation and transfer (Cooper and Tinker, 1978). Uptake of heavy metals from soil solution to fungal cytoplasm is an energy-dependent active process against an ion concentration gradient. Translocation of heavy metals is accomplished by cytoplasmic streaming within the fungus from the site of uptake to the organ with an interface toward the host plant, a process that appears to be dependent on concentration gradients between the two sites. The transfer process between the fungus and the plant is carried out by intercellular or intracellular hyphae colonizing the root cortex that may be linked to symbiotic exchange of C compounds in the opposite direction.

Each of these steps has been optimized for transport of essential nutrients that are immobile in soil (P, NH_4, Zn, Cu and others) during mycorrhizal evolution and may also constitute a transport route for trace elements that are either non-essential or essential, but in excess of plant needs. To the extent that the symbiotic partners may distinguish between nutrients and non-essential trace elements or between adequate and excess amounts of an element, each step may also become a barrier against uptake of harmful elements. There are indications that AM restrict metal transport through fungal sequestration before transfer to the plants. This may be seen in experiments where enhanced metal concentrations in roots and reduced concentrations in shoots are due to mycorrhizal presence.

3.2.4 Detoxification of heavy metals by AM fungi

There are two different strategies used by arbuscular mycorrhizal (AM) fungi to detoxify toxic metal cations perhaps the most widely used is avoidance of uptake.

i. Avoidance mechanism

This mechanism ensures prevention of heavy metals entry into their cells. There is evidence that AM fungi can stop certain toxic heavy metals entering their cells (Meharg, 1993). This can be achieved through a range of biochemical adaptations such as suppression of toxicant (heavy metal) influx transporters, enhanced efflux of a toxicant, release of pollutant (heavy metal) complexing agents into the surrounding soil solution, precipitation or binding of toxicants on to cell surfaces and release of extra-cellular enzymes that convert the toxicant into non-toxic species.

ii. Compartmentalization strategies

In the compartmentalization strategies, the heavy metal is translocated to specific organs in the fungus where it is relatively non-toxic (due to these organs having limited metabolic function) or sub-cellular compartments where they can be stored away from the cytoplasm (Meharg, 2003). The compound that enters the cell may be altered to another species for either transport within the organ or for storage purposes. For example, metal chelators such as dicarboxylic organic acids, amino acids, metallothioneins released by arbuscular mycorrhizal fungi have the ability to sequestrate the heavy metal from the contaminated soils (Gadd, 1993).

iii. Dilution effect

It has been unequivocally demonstrated that mycorrhizal plants are nutritionally rich especially phosphorous and produce larger biomass than their counterpart non-mycorrhizal plants (Smith and Read, 1997). The external mycelium increase soil volume accessible for P acquisition (Li *et al.*, 1991) and the P is transported to plant roots (Smith and Smith, 1990). Consequently, both the P concentration and uptake of mycorrhizal plants are much higher than those of corresponding controls (Chen *et al.*, 2003). Under conditions of Zn contamination, improved P nutrition might help the plant tolerate toxicity through the resulting increase in plant growth. In addition, dilution of heavy metals is the possible metal tolerance mechanism takes place by increased root growth facilitated by arbuscular mycorrhizal fungi.

4. PHYTOREMEDIATION STRATEGIES

4.1 AM Fungi in Heavy Metal Contamination

Heavy metal contaminated soils which are nutritionally poor, low water holding capacity and biologically inactive. Under these circumstances, mycorrhizal colonization facilitates flare up of microorganisms in the rhizosphere and biogeocycling of nutrients. Survival of plants in contaminated soils depends on the ability of plants to colonize with AM fungi which have selective advantage over non-colonized plants. Mycorrhizal colonization of roots results in an increase in root surface area for nutrient acquisition. The external mycelium of AM fungi can extend beyond the rhizosphere and take up large amounts of nutrients and heavy metals to the host plant roots. Heavy metal contaminated soils have a wide range of mycorrhizal species with varying levels of abundance (Chaudhry *et al.*, 1999) suggesting that these AM fungi have evolved a heavy metal tolerance and that may play an important role in phytoremediation of contaminated sites. Studies have clearly shown that AM fungal ecotypes from heavy metal contaminated soils are more tolerant than reference strains from uncontaminated soils (Leyval *et al.*, 1995). Enhanced tolerance of specific heavy metals of fungi isolated from soils contaminated with Zn, Pb, Cd or Cu has been observed (Gaur and Adholeya, 2004). Specific AM species to various heavy metals is given in Table 2. The indigenous AM inocula from metal contaminated sites could be a promising tool for soil remediation efforts. If the toxic soil carries very low mycorrhizal population, inoculation with heavy metal resistant AM fungi from similar soils could be exploited to promote establishment of symbiosis and improve soil remediation. If indigenous AM fungi exist in the contaminated soil to be phytoremediated, management of the indigenous AM fungi and their associate rhizobial microflora would be an important strategy. During soil restoration, the evaluation of mycorrhizae development and other soil microflora could be used as an important indicator of ecosystem efficiency, in order to biomonitor the success of mycorrhizoremediation. If heavy metal contaminated soils carry metal content that exceeds the limit of plant tolerance, it may be possible to treat plants with AM fungi and thereby stabilization, revegetation and remediation of metal polluted soils.

4.2. Phytostabilization

Phytostabilization prevents spreading of heavy metals into the soil environment as well as their leakage from the soil due to erosion (Gohre and Paszkowski, 2006). Metal tolerant species with extensive root system is capable of resisting wind and water erosion. AM fungi contribute to the immobilization of heavy metals in the soil beyond the rhizosphere and thereby improve phytostabilization. The AM fungi prevent entry of metals into the host plant through precipitation in polyphosphate granules in the soil, adsorption to the fungal cell wall or chelation of metals within the fungus as explained in previous sections. These processes successfully immobilize the heavy metals in the rhizosphere while preventing their entry into the host plants and keep them free from phytotoxicity.

TABLE 2. Specificity of AM fungal species to heavy metal contamination

Heavy metals	Tolerant AM species	Host plant species	Reference
Cd	*Glomus mosseae*	*Trifolium repens*	Vivas *et al.*, 2003
	Glomus sp. *Gigaspora* sp.	*Hordeum vulgare*	Tullio *et al.*, 2003
	Glomus mosseae	*Trifolium subterraneum*	Joner and Leyval, 1997
	Glomus mosseae	*Allium porum*	Weissenhorn *et al.*, 1993
	Glomus intrardices *Gigaspora margarita*	*Daucus carota*	Janouskova and Vosatka, 2005
Cr	*Glomus* sp.	*Helianthus anuus*	Davies *et al.*, 2001
Ni	*Gigaspora* sp. *Glomus tenue*	*Berkheya coddii*	Turnau and Mesjasz-Przybylowic, 2003
Zn	*Glomus* sp. *Glomus* sp. *Glomus intrardices*	*Viola calaminaria* *Fragaria vesca* *Nicotiana tobaccum*	Tonin *et al.*, 2001 Turnau *et al.*, 2001 Audet and Charest, 2006
Pb	*Glomus intrardices*	*Agrostis capillaries* *Zea mays*	Malcova *et al.*, 2003
Cd, Cu	*Glomus caledonium*	*Zea mays*	Liao *et al.*, 2003
Zn, Pb	*Glomus mosseae* *Glomus macrocarpum*	*Lygeum spartum*	Diaz *et al.*, 1996
Cd, Zn	*Glomus mosseae*	*Trifolium subterraneum*	Joner *et al.*, 2000
Cd, Zn, Pb	*AM fungi*	*Plantago lanceolata*	Orlowska *et al.*, 2002
Cd, Zn, Cu	*Glomus mosseae* *Glomus mosseae*	*Trifolium subterraneum* *Phaseolus vulgaris*	Joner and Leyval, 2001 Guo *et al.*, 1996
Zn, Cd, Cu, Ni, Pb	*Glomus caledonium*	*Sorghum bicolor*	del Val *et al.*, 1999

4.3 Phytoextraction

Phytoextraction is fairly recent technology and represents the most effective strategy to clean up contaminated soils (Kramer, 2005). In this strategy, the plants are allowed to take heavy metals luxuriantly and transfer it from roots to shoots and accumulate in aerial parts. Consequently, heavy metals are removed upon harvest and can be recaptured (phytomining) or combusted for energy production (Peuke and Rennenberg, 2005). The efficiency of phytoextraction depends on biomass production by plants and their metal tolerance. The addition of AM fungi enhances uptake and accumulation of As in brake fern (*Pteris vittata*) to the tune of 88.1 g As per kg of dry mass (Leung *et al.*, 2006). Colonization of AM fungi diminished the phytotoxic effect of higher Zn levels thereby contributed overall to increased mobilization of Zn from soils (Chen *et al.*, 2003).

4.4 Cropping Systems

Intercropping and co-cropping systems are likely to improve the phytoremediation of heavy metal contamination. Wu *et al.* (2004) used an intercropping system to examine the interactions of mycorrhizosphere and rhizosphere on metal uptake by growing mycorrhizal non-hyperaccumulator *Zea mays* and non-mycorrhizal hyperaccumulator *Brassica juncea* in a split-pot experiment. The data showed that the intercropping system achieved higher phytoremediation efficiency in metal contaminated soil, especially with dual inoculation of beneficial rhizobacteria and AM fungi. Wu *et al.* (2004) used a co-cropping system with the metal hyperacumulator plants *Thlaspi caerulescens* and *Sedum alfredii* and the low accumulating but highly mycorrhiza dependent corn, *Zea mays L.* to increase phytoextraction efficiency from Zn-contaminated sewage sludge. The authors compared the co-crop systems with the mono-crop cultures and found that the hyperaccumulators when grown with maize as a co-crop extracted more Zn from the sewage sludge as compared with the monocultures of these plants. This is a step in the right direction to better understand the interactions between the non-mycorrhizal hyperaccumulators and the mycorrhizal plant species in order to increase the efficiency of the process and further experiments are needed using different combinations of co-crop systems.

Another possible strategy to exploit AM fungi for phytoremediation purposes is to use mycorrhiza-dependent plant species pre-inoculated with metal tolerant AM fungi on the site to be remediated followed by growing mycorrhiza-responsive plant species which can benefit from the AM fungal potential of soil left by the pre-crop. Panja and Chaudhuri (2004) applied this strategy and carried out a nursery experiment to study the effect of short season pre-cropping with different mycotropic herbaceous crops followed by growing AM-dependent mandarin orange. The authors recorded greater growth benefit to mandarine oranges by mycorrhizal pre-crops such as soyabean and onion than by the non-mycorrhizal crops such as mustard and ginger. Their results show that an AM-dependent crop grown even for a short season can substantially alter the inherent AM fungal potential of soils to significantly influence the performance of the succeeding mycorrhiza-dependent plants.

5. STRATEGIES TO PROMOTE MYCORRHIZAL SYMBIOSIS

Minimal Soil Disturbance: Heavy tillage has a major effect on the symbiosis, reducing root colonization primarily by disrupting the hyphal network (Evans, and Miller, 1988). No till system favours mycorrhizal symbiosis and therefore their adoption would enhance the functions of mycorrhiza and aid the sustainability of the system

Fertilizers: The literature on fertilizer impact on colonization of mycorrhiza is contradictory. Application of soluble P appears to reduce the germination of mycorrhizal spores and extent of colonization (Miranda and Harris, 1994). However, some fungal species (Glomus intraradices) are insensitive to fertilizer application

Crop Rotations: Mycorrhizas are obligate symbionts and thus highly dependent on living host plants. Fallowing also results in reduction of inoculum potential but the effect is smaller than the inclusion of non-mycotrophic plants in the cropping sequence (Wu *et al.*, 2004; Panja and Chaudhuri, 2004).

Pesticides: Mycorrhizas are highly sensitive to pesticides such as methyl bromide and if used, it impedes the development of mycorrhizal symbiosis in the crop. Most fungicides have shown to be detrimental to either colonization or functions of mycorrhizas (O'Connor et al., 2002). Rationalization of their use by selection of fungicides that do not interact negatively with mycorrhiza could bring significant benefits.

Crop Breeding: Some crop varieties are more responsive to mycorrhizal colonization and can be exploited for low-input production system (Subramanian and Charest, 1997).

6. CONCLUSION

The occurrence of arbuscular mycorrhizal (AM) fungi is ubiquitous even in harsh environment such as metal contaminated soils and mine spoils. Since the metal contaminated soils are biologically inactive it is imperative to induce the biological activity of the soil utilizing mycorrhizal fungi. Mycorrhizal plants have developed various tolerance mechanisms such as biosorption, solubilization, hyphal transport, detoxification and dilution that can be fully exploited in order to greening the grey areas. The indigenous mycorrhizal fungi may have adapted to contaminated soils that can be screened, multiplied and used for revegetation programmes. Microbial consortia may be evolved to target a wide range of metal contaminants instead of single metal approach. Mechanisms involved in the tolerance of these AM species to metal contamination has to be studied at the molecular level so as to knock out the genes responsible for tolerance that can be transferred to plants to confer resistance. The indigenous population can be favoured by adopting agronomic strategies such as minimum tillage, reduced fertilizers, proper crop rotation and selective pesticide use. Further, utilization of AM fungal inoculation to alleviate heavy metal toxicities in vegetable farming in peri-urban areas should be studied extensively. Overall, the data suggest that mycorrhizal fungi are naturally occurring microorganisms capable of alleviating a wide array of metal contaminants by orchestrating biological activities in the rhizosphere besides detoxify them through a series of bioprocesses.

REFERENCES

Audet, P and Charest, C (2006). Effects of AM colonization on "Wild tobacco" Plants grown in zinc- contaminated soil. *Mycorrhiza* 16: 277-283.

Augé, R.M.(2001). Water relations, drought and vesicular-arbuscular mycorrhizal symbiosis, *Mycorrhiza* 11:3-42.

Azon, R and El-Atrach, F (1997) Influence of arbuscular mycorrhizae and phosphorus fertilization on growth, nodulation and N_2 in *Medicago fativa* at four salinity levels. *Biol. Fertility Soil* 24:81-86.

Benhamou, N., Fortin, J.A., Hamel, C., St.Arnand, M and Shatilla, A (1994). Resistance responses of mycorrhizal Ri T-DNA-transformed carrot roots to infection by *Fusarium oxysporum* f..Sp. chrysantheni. *Phytopathol.* 84: 958-968.

Chander, K and Brookes, P.C (1991) Effects of heavy metals from pest applications on microbial biomass and organic matter accumulation in a sandy loam U.K,Soil. *Soil Biol. Biochem.* 23:927-932.

Charest, C., Dalpé, Y and Brown, A (1993).The effect of vesicular-arbuscular mycorrhizae and chilling on two hybrids of Maize. *Mycorrhiza* 4:89-92.

Chaudhry,T.M., Hill,L., Khan,A.G and Kuek,C (1999). Colonization of iron and zinc contaminated dumped filter cake waste by microbes, plants and associated mycorrhizae, In: Wong, M.H., Wong, J.W.C., Baker, A.J.M.(Eds), *Remediation and management of degraded land*, CRC Press, Baca Roton, Chap. 27: 275-283.

Chen, B.D., Li, X.L., Tao, H.Q., Christie, P and Wong, M.H. (2003). The role of arbuscular mycorrhiza in zinc uptake by red clover growing in a calcareous soil spiked with various quantities of zinc. *Chemosphere* 50: 839-846.

Christie, P., Li, X. and Chen, B (2004). Arbuscular mycorrhizae can depress translocation of zinc to shoots of host plants in soils moderately polluted with zinc. *Plant Soil* 261: 209-217.

Cooper, K. M. and Tinker, P. B. (1978).Translocation and transfer of nutrients in Vesicular-arbuscular mycorrhizas.II. uptake and translocation of Phosphorus zinc and sulphur. *New Phytol.*81:43-52.

Davies, F.T., Puryear, J.D., Newton, R.J., Egilla, J.N. and Saraiva-Grossi, J.A. (2001). Mycorrhizal fungi enhance accumulation and tolerance of chromium in sunflower (*Helianthus annuus*) *J. Plant Physiol.*158: 777-786.

Dehn,B. and Schuepp, H. (1989). Influence of VA mycorrhizae on the uptake and distribution of heavy metal in plants. *Agric. Ecosyst. Environ.* 29: 79-83.

del Val, C., Barea, J.M. and Azon-Aguilar, C. (1999). Assessing the tolerance to heavy metals of arbuscular mycorrhizal fungi isolated from sewage sludge – contaminated soils, *Appl. Soil Eco.* 11:261-269.

Diaz,G., Azcon-Aguilar, C and Honrubia, M. (1996). Influence of arbuscular mycorrhizae on heavy metal (Zn and pb) uptake and growth of Lygedum spartum and *Anthyllis cytisoides. Plant Soil* 180:241-249.

Evans, D.G. and Miller, M.H. (1988). Vesicular arbuscular mycorrhizas and the soil induced reduction of nutrient absorption in maize, 1 casual relation. *New Phytol* 110:67-74.

Fliback, A., Martens, R and Reber, H. H (1994). Soil microbial biomass and Microbial activity in soils treated with heavy metal contaminated sewage sludge. *Soil Biol. Biochem.* 26:1201-1205.

Fomina, M., Alexander, I. J., Colpaert, J. V. and Gadd, G.M (2004). Solubilization of toxic metal minerals and metal tolerance of Mycorrhizal Fungi. *Soil Biol. Biochem.* 37:851-866.

Gadd, G.M. (1990). Heavy metal accumulation by bacteria and other microorganisms. *Experimentia* 46: 834-840.

Gadd, G. M.(1993). Interactions of fungi with toxic metal. *New Phytol.* 124:25-60.

Gaur, A. and Adholeya, A. (2004) Prospects of arbuscular mycorrhizal fungi in phytoremediation of heavy metal contaminated soils. *Curr. Sci.* 86:528-534.

Gohre, V. and Paszkowski, U. (2006). Contribution of arbuscular mycorrhizal symbiosis to heavy metal phytoremediation. *Planta* 223: 1115-1122.

Gonzalez –chavez, M.C., Carrillo-Gonzalez, R., Wright, S.F. and Nichols, K.A. (2004). The role of glomalin, a protein produced by arbuscular mycorrhizal fungi in sequestrating potentially toxic elements. *Environ. Pollut* 130: 317-323.

Guo,Y., and Geroge, E and Marschner, H (1996). Contribution of an arbuscular mycorrhizal fungus to the uptake of cadmium and nickel in bean and maize plants. *Plant Soil,* 184:195-205.

Janouskova,M. and Vosatka, M. (2005). Response to cadmium of Daucus cartoa hairy roots dual cultures with Glomus intraradices or *Gigaspora margarita. Mycorrhiza* 15:217-224.

Jeffries, P. and Barea, J. M. (2003). Arbuscular mycorrhiza- a key component in plant soil Ecosystems. *The Mycota* 9: 95-113.

Joner, E. J., Briones, R. and Leyval, C. (2000) Metal binding capacity of arbuscular mycorrhizal mycelium. *Plant Soil.* 226: 227-234.

Joner, E. J and Leyval, C (1997). Uptake of ^{109}Cd by roots and hyphae of a *Glomus mosseae/ Trifolium subterraneum* mycorrhiza from soil amended with high and low concentrations of cadmium. *New Phytol* 135: 353-360.

Joner, E.J. and Leyval, C (2001). Time course of heavy metal uptake in maize and clover as affected by root density and different mycorrhizal inoculation regimes *Biol. Fertil. Soils* 33:351-357.

Kaldorf, M., Kuhn, A.J., Schroder, W.H., Hildebrandt, U and Bothe, H (1999). Selective element deposits in maize colonized by a heavy metal tolerance conferring arbuscular mycorrhizal fungus, *J Plant Physiol.* 154:718-728.

Khan, A.G. (2005). Role of soil microbes in the rhizospheres of plants growing on trace metal contaminated soil in phytoremediation – A Review. *J. Trace elements in Medicine & Biol* 18: 355-364.

Kramer, U (2005).Phytoremediation: novel approaches to cleaning up polluted soils. *Curr. Opin Biotechnol.* 16: 33-141.

Landeweert, R., Hoffland, E., Finlay, R.D., Kuyper, T.W. and Van Breemen, N (2001). Linking plants to rocks: ectomycorrhizal fungi mobilize nutrients from minerals. *Trends in Ecol.& Evol.*16: 248-254.

Leung, H.M., Ye, Z. H and Wong, M.H. (2006). Interactions of mycorrhizal fungi with Pteris vittata (as hyperaccumulator) in as contaminated soils, *Environ. Pollut.* 139:1-8.

Leyval, C., Singh, B.R. and Joner, E.J (1995). Occurrence and infectivity of arbuscular mycorrhizal fungi in some Norwegian soils influenced by heavy metals and soil properties *Water Air Soil Pollut.* 84: 203-216.

Leyval, C., Surtiningsih, T. and Berthelin, J. (1993). Mobilization of Cd from rock phosphates by rhizospheric microorganisms (phosphate-issolving bacteria and ectomycorrhizal fungi) Phosphorus, Sulfur and Silicon. *New Phytol.* 77:133-136.

Leyval, C; Laheurte, F; Belgy, G. and Berthelin, J. (1990). Weathering of micas in the rhizospheres of maize, pine and beech seedlings influenced by mycorrhizal and bacterial inoculation. *Symbiosis*.9: 105-109.

Li, X. L., George, E. and Marschner, H (1991). Phosphorus depletion and pH decrease at the root-soil and hyphae - soil interfaces of VA mycorrhizal white clover fertilized with ammonium. *New Phytol.* 119:397-404.

Liao, J, P., Lin, X. G., Cao, Z.H., Shi, Y.Q and Wong, M.H (2003). Interactions between arbuscular mycorrhizae and heavy metals under sand culture experiment. *Chemosphere* 50: 847-853.

Linderman, R.G. (1994). Role of AM fungi in biocontrol. *In*: ds. Pfleger, F.L. and Linderman, R.G., Eds. Mycorrhizae and Plant Health, EAPS Press, The American society of Phytopathology, Minnesota, USA.

Malcova, R., Vosatka, M. and Gryndler, M. (2003). Effects of inoculation with *Glomus intraradices* on lead uptake by *Zea mays* L and *Agrostis Capillaris*. L *Appl. Soil Eco.* 23:55-67.

Meharg, A. (2003). The Mechanistic basis of interactions between mycorrhizal associations & toxic metal cations. *Mycol. Res.* 11:1253-1265.

Meharg, A.A. (1993). The role of the plasmalemma in metal tolerance in angiosperms. *Physiologia plantarum*. 88:191-198.

Miranda, J.C.C. and Harris, P.J. (1994). Effects of soil P on spore germination and hyphal growth of fungi. *New Phytol.* 128:103-108.

Nelsen, C.E (1987). The water relations of VAM systems. *In*: Safir, G.R.., Eds. Ecophysiology of Mycorrhizal Plants, CRC Press, Boca Raton, Fla. pp. 71-92.

O'Connor PJ, Smith SE, Smith EA (2002). Arbuscular mycorrhizas influence plant diversity and community structure in semi-arid herbland. *New Phytol.* 154:209-218.

Orlowska,E., Zubek, S., Jurkiewicz, A., Szarek-ukaszewska, G and Turnau, K. (2002). Influence of restoration on arbuscular mycorrhiza of *Biscutella laevigata* L (Brassicaceae) and *Plantago lanceolata* (Plantaginceae) from calamine spoil mounds. *Mycorrhiza,* 12:153-160.

Panja, B.N and Chandhuri, S (2004). Exploitation of soil arbuscular mycorrhizal potential for AM-dependent mandarin orange plants by pre-cropping with mycotrophic crops. *Appl. Soil Eco.* 26:249-255.

Paradis, R., Dalpé, Y. and Charest, C (1995). The combined effect of arbuscular mycorrhizas and short-term cold exposure on wheat. *New Phytol.* 129. 637-642.

Paris, F., Bonnaud, P., Ranger, J. and Lapeyrie, F. (1994). Phyllosilicate alteration by ecto-mycorrhizal fungi in vitro, *Acta Bot Gallica.* 141:529-532.

Peuka, A.D, and Rennenberg, H. (2005). *Phytoremediation EMBO Rep.* 6: 497-501.

Rillig, M.C. and Steinberg, P.D.(2002). Glomalin production by an arbuscular mycorrhizal fungus: a mechanism of habitat modification. *Soil Biol. Biochem.* 34:1371-1374.

Ruiz-Lozano, J. M., Azcon, R. and Plama, J.M. (1996b). Superoxide dismutase activity in arbuscular mycorrhizal Lactuca sativa plants subjected to drought stress. *New Phytol.*134 327-333.

Sauerbeck, D. (1982). Which heavy metal concentrations in plants should not be exceeded inorder to avoid detrimental effects on their growth. *Landw. Forsch. Sonderh.* 39:108-129.

Smith, S.E and Smith, F.A. (1990). Structure and function of the interfaces in biotrophic symbioses as they relate to nutrient transport. *New Phytol* 114:1-38.

Smith, S.E. and Read, D.J. (1997). *Mycorrhizal symbiosis*. Academic press, London, England.

Steinberg,, P.D and Rillig, M.C.(2003). Differential decomposition of arbuscular mycorrhizal fungal hyphae and glomalin. *Soil Biol. Biochem*. 35:191-194.

Subramanian, K.S. Charest, C. 2007. Arbuscular mycorrhiza and drought management of crops *In: Arbuscular mycorrhizae in arid and semi-arid environments*. Montaño NM, Camargo- Ricalde SL, García-Sánchez R, A. Monroy-Ata (Eds.) (pages 78-109).Mundi-Prensa SA of CV,Instituto Nacional de Ecología-SEMARNAT, UAM-Iztapalapa, FES-Zaragoza-UNAM. Distrito Federal, México. 460 pp.

Subramanian, K.S., and Charest, C. (1997). Nutritional, growth and reproductive responses of maize (Zea mays L) to arbuscular mycorrhizal inoculation during and after drought stress at tasselling, *Mycorrhiza* 7:25-32.

Subramanian, K.S. Santhanakrishnan, P. and Balasubramanian, P. (2006). Responses of field grown tomato plants to arbuscular mycorrhizal fungal colonization under varying intensities of drought stress, *Scientia Hort*. 107:245-253.

Tonin, C., Vandenkoornhuyse, P., Joner, E.J., Straczeb, J and Leyval, C (2001).Assessment of arbuscular mycorrhizal fungi diversity in the rhizophere of viola calaminaria and effect of these fungi on heavy metal uptake by clover, *Mycorrhiza* 10: 161-168.

Tullio, M., Pierandrei, F., Salerno, A and Rea, E (2003). Tolerance to cadmium of vesicular arbuscular mycorrhizae spores isolated from cadmium polluted and unpolluted. *Soil. Biol. Fertil, Soils* 37: 211-214.

Turnau, K., Kottke, I and Oberwinkler, F (1993). Element localization in mycorrhizal roots of pteridium aquilinum L Kuhh collected from experimental plots treated with cadmium dust. *New Phytol* 123: 313-324.

Turnau, K., Ryszka,P., Gianinazzi-Pearson, V., and Van Tuinen,D (2001). Identification of arbuscular mycorrhizal fungi in soils and roots of plants colonizing zinc wastes in Southern Poland. *Mycorrhiza* 10:169-174.

Turnau, K (1998). Heavy metal content and localization in mycorrhizal Euphorbia cyparissias from zinc wastes in southern Poland, *Acta Soci. Bot. Poloniae*, 67:105-113.

Turnau, K and Mesjasz-przybylowicz, J (2003). Arbuscular mycorrhiza of Berkheya Coddii and other Ni-hyper accumulating members of asteraceae from ultramafic soils in South Africa, *Mycorrhiza* 13:185-190.

Vivas, A., Marulanda, A., Gomez, M., Barea, J.M and Azcon, R (2003). Physiological characteristics (SDH and ALP activities) of arbuscular mycorrhizal colonization as affected by *Bacillus thuringiensis* inoculation under two phosphorus levels. *Soil Biol, Biochem*. 35:987-996.

Waaland, M.E and Allen, E. B.(1987). Relationships between VA mycorrhizal fungi and plant cover following surface mining in wyoming. *J Range Manage* 40: 271-276.

Weissenhorn, I. Leyval, C, Belgy, G and Berthelin, J (1995). Arbuscular mycorrhizal contribution to heavy metal uptake by maize (*Zea mays* L) in pot culture with contaminated soil. *Mycorrhiza* 5:245-251.

Weissenhorn, I., Leyval, C and Berthelin, J (1993). Cd-tolerant arbuscular mycorrhizal (AM) fungi from heavy metal polluted soils. *Plant Soil* 157:247-256.

Wu, Q.T., Liu, X.M., Merg, Q.Q, Long, X.X., Schwartz, C and Morel, J. L. (2004). Co-crop hyperaccumulator and low-accumulating plants for treating heavy metal contaminated soil and sludge *In*: proceedings of the fifth International conference on environmental geochemistry in the tropics, March 21-26, Haiko, Haimen, china. Nanjing, P R china, Institute Soil Science, Chinese academy of Science, pp. 71.

Yao, Q., Li, X., Weidang, A and Christie, P. (2003). Bi-directional transfer of phosphorus between red clover and perennial ryegrass via arbuscular mycorrhizal hyphal links. *Eur. J. Soil Biol.* 39:47-54.

Zhou, J.L. (1999). Zn biosorption by *Rhizopus arrhizus* and other fungi. *Appl. Microbiol. Biotechnol* 51:686-693.

❑❑❑

CHAPTER **13**

Impact of Ultraviolet-B Radiation on Microbes and Human Health : The Bitechnological Prespectives

R. P. Sinha[1], M. B. Tyagi[2] and Ashok Kumar[3*]

[1]Centre of Advanced Study in Botany, [2]Department of Botany, Mahila Mahavidyalaya, [3]School of Biotechnology, Banaras Hindu University, Varanasi-221005, India

ABSTRACT

The continuing depletion of the stratospheric ozone layer mainly due to release of man-made atmospheric pollutants such as chlorofluorocarbons (CFCs) have resulted in the increase of solar ultraviolet-B radiation (UV-B; 280-315 nm) reaching to the Earth's surface. UV-B is a small (less than 1% of total energy) but highly active component of the solar spectrum that can penetrate to biologically significant depths in lakes, ponds, rivers and oceans. It has been demonstrated that UV-B can cause wide ranging effects including mutagenesis, destruction of proteins/enzymes, inhibition of a number of vital metabolic processes and formation of thymine dimers in DNA leading to death of a number of organisms. A number of microbes including strict photoautotrophs always grow in lighted habitats and thereby always sense and face the stress of solar UV-B radiation. Similarly, humans are also exposed to varying levels of UV-B radiation depending upon the altitudes and latitudes of the location. A wide ranging effect of UV-B such as cataracts, skin cancer, immunosupression etc. has been reported in human. Damaging effects of increased UV-B radiation has also been shown in a number of microbes. These effects include inhibition of growth and survival, decrease in photosynthetic pigments, loss of PS II and RuBISCO activity, inactivation of nitrogenase enzyme and inhibitory effects on several other metabolic processes. Among photosynthetic pigments, phycobiliproteins seem to be the primary target of action since phycobilisomes are completely disintegrated after UV-B exposure. In this contribution an attempt has been made to put in perspective the main effects of UV-B radiation on microbes and human health.

Key words : microbes, human health, UV-B radiation, DNA damage, sun-screens

1. INTRODUCTION

Ultraviolet radiation is an important component of solar radiation having many deleterious effects on biological systems (Häder *et al.,* 1998; Vincent

*E.mail: kasokbt@rediffmail.com

and Roy, 1993). Based on wavelength, it has been arbitrarily divided into UV-A (315 - 400 nm), UV-B (280 - 315 nm) and UV-C (<280 nm). Prior to their attenuation by the Earth's atmosphere these wavelengths represent 6.3, 1.5 and 0.5 % of the total solar irradiance corresponding to an intensity of 85.7, 21.7 and 6.4 W/m^2 respectively (Frederick *et al.*, 1989). Among different kinds of UV radiations UV-C is strongly mutagenic and lethal whereas UV-A is biologically least reactive component. Ozone present in the stratosphere acts as an efficient screen for short wavelength UV radiation absorbing all UV-C radiation, but allowing all UV-A to reach the Earth's surface. On the other hand it allows some UV-B radiation to reach the ground surface. Ozone column affects both the intensity and quality of UV-B radiation; with decrease in ozone column, the intensity increases and the wavelength composition is shifted to proportionately shorter wavelength (Frederick *et al.*, 1998; Madronich *et al.*, 1998).

Continuing depletion of stratospheric ozone caused by the anthropogenic emission of chlorofluorocarbons and other gaseous pollutants into the atmosphere has resulted in a specific increase in UV-B radiation (Crutzen, 1992; Madronich *et al.*, 1998; Weatherhead and Anderson 2006). Such increases in UV-B radiation are most pronounced in Antarctica, but similar trends have been reported at other latitudes also (Kerr and McElroy, 1993).

The continued increase in UV-B radiation has been demonstrated to affect both primary production and higher trophic levels including human beings in biological food webs causing instabilities in ecosystems (Smith *et al.*, 1992; Häder *et al.*, 1998). Besides exerting a variety of adverse effects on terrestrial vegetation, UV-B radiation is also known to penetrate fairly deep in freshwater and marine environments (Smith *et al.*, 1992) where it affects the plankton in the euphotic zone as well as benthic organisms (Häder *et al.*, 1998). The damaging effects of UV-B radiation on cyanobacteria are of special concern because they form a dominant flora of nitrogen-fixing microorganisms in tropical rice-fields and enrich the fertility of soil (Singh, 1961). Similarly bacterioplankton and viruses have been reported to be adversely affected by UV radiation. The UV-induced damage to DNA leading to mutations, and damage to physiological and biochemical processes have been extensively studied. Following absorption of UV radiation by DNA, formation of a variety of photoproducts may occur. These photoproducts include: cyclobutane-type dimers of thymine, cytosine and uracil, pyrimidine adducts, photohydrates, and DNA- protein cross-links. On the other hand physiological changes after increased UV-B radiation exposure may include, growth inhibition followed by loss of survival, inhibition of photosynthetic processes, loss of enzymatic activity, increased synthesis of UV-B absorbing pigments, induction of stress protein, and others (Quesada *et al.*, 1999). These changes could result from a number of primary UV-B mediated events such as direct photosynthetic damage, loss of permeability/membrane changes, pigment destruction, protein/enzyme inactivation, reduced DNA and protein synthesis, reduced uptake of nutrients, hormone inactivation and signal transduction through phycochrome or signal transduction via a specific UV-B photoreceptor (Vincent and Roy, 1993; Sinha and Häder, 2002; Kumar *et al.*, 2003).

Besides exhibiting drastic effects on microbes, solar UV-B radiation also causes acute and chronic effects on human health. The most common effects include inflammation of cornea, malignancy of cornea and uveal tract and snow blindness. Additionally there are pronounced effects on skin leading to sun burn, tanning, photoaging and skin cancer. There is also report of immuno-suppression by exposure of UV-B radiation. This chapter deals with the effects of UV-B radiation on microbes and human health.

2. UV-B-MEDIATED EFFECTS ON MICROBES

UV-B radiation has been reported to affects the growth, survival, pigmentation, orientation, general metabolism, photosynthesis, nitrogen fixation and assimilation of nitrogen in diverse microbial species. There is still considerable discussion about the primary targets of UV-B radiation damage but it is generally felt that the deleterious effects are in part due to the direct effects of UV-B radiation on membrane component (Sinha and Häder, 2002), DNA (Quesada *et al.*,1999), enzymes (Döhler *et al.*, 1986), growth regulators and the photosynthetic apparatus, particularly the reaction center of photosystem II and RuBISCO (Wilson et al., 1995; Sinha and Häder, 2002) or to its indirect effect through the formation of reactive species such as super oxide (O_2^-), hydroxyl radicals (OH^-) or hydrogen peroxide (H_2O_2) (He and Häder, 2002). Although DNA has an absorption spectrum that peaks at 260 nm, a significant absorption also occurs in the UV-B range (280 - 315 nm). This results in production of cytotoxic lesions such as pyrimidine base pairs. Such lesions, if left uncorrected, interfere with DNA replication and transcription ultimately leading to mutation or death. Proteins also show considerable absorption in the UV-B region, which is primarily due to the presence of aromatic amino acids such as tryptophan and tyrosine. Such UV-B-induced damage/denaturation of various proteins/enzymes may also cause changes in the structure and function ultimately leading to changes in membrane permeability and functioning of various enzymes. The known effects of UV-B radiation on several metabolic events of microbes are briefly discussed below:

2.1 Motility

As such cyanobacteria employ motility and orientation mechanisms to escape harmful UV conditions and high light intensities but it has been reported that these avoidance mechanisms themselves are severely affected by UV-B radiation, resulting in a decrease or loss of their ability to escape from harmful UV conditions (Donkor *et al.*, 1993). Filamentous cyanobacteria such as *Anabaena variabilis, Oscillatoria tenuis* and *Phormidium uncinatum* use gliding motility to optimize their position in their microhabitat. Donkor *et al.* (1993) found a drastic reduction in the percentage of motile filaments of the above cyanobacteria within 10-30 min of exposure to unfiltered solar radiation at Ghana (4.30^0 N). Removal of UV-B radiation by an appropriate long-pass or band- pass filters resulted in higher percentage of motility in most cases suggesting the inhibitory effect of UV-B radiation on gliding motility of these cyanobacteria. Likewise, inhibition of phototactic orientation and photophobic responses reduce the ability of the organisms to orient themselves in their photo-environment (Bebout and Garcia-Pichel 1995). In the event of failure to quickly respond to changes in the photo-environment their death may occur (Häder and Worrest, 1991).

2.2 Cell differentiation

Differentiation of vegetative cells into heterocysts or akinetes is also affected by UV-B radiation. Blakefield and Harris (1994) observed delayed heterocyst differentiation and akinete formation in Anabaena aqualis following exposure to UV-B radiation although increase in akinete production was noted. It has been suggested that UV-B exposure alters the C:N ratio which in turn affects the spacing pattern of heterocyst in a filament (Sinha *et al.*, 1996). A decrease in the level of major heterocyst polypeptides (26, 54 and 55 kDa) following UV-B treatment has been noted which is presumed to be responsible for the disruption of multi-layered heterocyst wall, the essential component for maintaining the active form of nitrogenase. Role of UV-B radiation in the opening of specific calcium channels in the heterocyst of Anabaena sp. has also been demonstrated (Richter *et al.*, 1999). Whether this opening is mediated through signal transduction pathway and/or has any significance in heterocyst formation and function is yet to be proved.

2.3 Growth and survival

UV-B radiation has been shown to severely affect the growth and survival of several rice-field and Antarctica cyanobacteria (Tyagi *et al.*, 1992; Sinha *et al.*, 1995, 1996; Quesada and Vincent, 1997). Most of these studies have been performed under laboratory conditions using artificial lamps specifically emitting UV-B wavelengths. Long-term studies in natural habitats under ambient levels of UV-B radiation are lacking. Various species differ with respect to their tolerance to UV-B radiation and even closely related strains show differential sensitivity. Complete killing and loss of survival within 120 - 180 min of UV-B exposure has been reported for several cyanobacteria (Tyagi *et al.*, 1992; Sinha *et al.*, 1995). Strains such as Nostoc commune and *Scytonema* sp. whose filaments are embedded in a mucilage sheath are more tolerant to UV-B radiation than the species which do not have such mucilage covering (Sinha *et al.*, 1996). A phycoerythrin rich (brown) strain of *Nostoc* sp. was found to be more tolerant to UV-B radiation than the phycocyanin rich (blue-green) form of this strain (Tyagi *et al.*, 1992). Two Antarctica cyanobacteria, Phormidium murrayi and Oscillatoria priestleyi, isolated from a common habitat showed differential sensitivity to UV-B radiation. In contrast to complete suppression of growth in *Oscillatoria priestleyi*, only a 62 % inhibition of growth of *Phormidium murrayi* was observed following exposure to the same quantum of UV-B intensity (Quesada and Vincent, 1997).

2.4 Pigmentation

Several workers have demonstrated that UV-B radiation causes photobleaching of photosynthetic pigments, a decrease in phycocyanin/ chlorophyll a ratio and disintegration of phycobilisome complex leading to the release of phycobiliprotein pigment complex (Häder and Worrest 1991; Tyagi *et al.*, 1992; Quesada and Vincent 1997). Phycobiliprotein seems to be the most sensitive pigment most probably due to its water soluble property. Sinha *et al.* (1995) have reported a decrease in the phycobiliprotein content and disassembly of phycobilisomal complex following UV-B irradiation in a number of cyanobacteria. The pattern of fluorescence emission spectra of phycobiliproteins observed after UV-B irradiation suggests an impairment of energy transfer from the accessory pigments to the photosynthetic reaction centers (Sinha *et al.*, 1995). Exposure of cyanobacterial cultures to intense UV-B radiation causes bleaching of phycocyanin and phycoerythrin which is in accordance with the disintegration of phycobilisomes.

2.5 Photosynthesis

Photosynthetic parameters such as $^{14}CO_2$ uptake, O_2 evolution and ribulose-1,5 bisphosphate carboxylose (RuBISCO) activity are greatly inhibited following exposure to UV-B radiation (Wilson et al., 1995; Quesada and Vincent, 1997). Photosynthetic O_2 evolution of a *Phormidium* strain from Lake Baikal declined within minutes of exposure to solar irradiation (Häder and Worrest, 1991). Several workers have reported an inhibition of $^{14}CO_2$ uptake by a number of cyanobacteria following exposure to UV-B irradiation (Newton *et al.*, 1979; Tyagi et al., 1992; Sinha *et al.*, 1996). In a study of diurnal variability of the effect of solar UV-radiation on photosynthesis in an algal mat comprising of Lyngbya aestuarii, Cockell and Rothschild (1999) found a mean increase in the total primary productivity (measured in terms of $^{14}CO_2$ uptake) over the day when solar UV-B was screened out. It is felt that RuBISCO may undergo a number of modifications such as photo-degradation, fragmentation and denaturation of polypeptide chain, change in active site and increased solubility of membrane proteins. In addition supply of ATP and $NADPH_2$ may also be inhibited which may cause inhibition of CO_2- fixing ability. The D_1 and D_2 polypeptides which are the major constituents of PS-II reaction centers are also degraded, even after exposure to intermediate levels of UV-B radiation (Sass *et al.*, 1997; Campbell *et al.*, 1998). UV-B radiation may also induce modifications and photo-degradation of membrane proteins, which may subsequently affect the integrity of thylakoid membrane thereby causing the damage of photosynthetic components which are otherwise essential for active CO_2 fixation.

2.6 Nitrogen metabolism

Among all the key enzymes involved in nitrogen metabolism, the nitrogen-fixing enzyme nitrogenase has been found to be extremely sensitive to UV-B radiation (Newton *et al.*, 1979; Tyagi et al., 1992; Kumar et al., 1996; Kumar *et al.*, 2003). In a *Nostoc* sp, nitrogenase activity was lost within 45 min of UV-B exposure whereas nitrate reductase and glutamine synthetase activity remained more or less unaffected (Tyagi *et al.*, 1992). Kumar *et al.* (2003) reported complete inactivation of nitrogenase activity within 25-40 min of UV-B exposure in a number of cyanobacteria including Anabaena sp., *Nostoc* sp., *Calothrix* sp. and *Scytonema* sp. Employing inhibitors of protein synthesis and PS-II activity they have demonstrated that restoration of nitrogenase activity in UV-B-treated cultures occurs only after fresh synthesis of nitrogenase polypeptide. The inactivation of activity most probably occurs due to complete damage to the nitrogenase polypeptide. It has been suggested that the aromatic amino acids present or the native structure of nitrogenase are responsible for extreme susceptibility of nitrogenase to UV-B radiation. They have further suggested that the nitrogenase may be used as a marker enzyme for assessing the impact of UV-B radiation in diazotrophic species (Kumar *et al.*, 2003).

2.7 Protein profile

Exposure to UV-B radiation is known to causes a subtle change in the protein profile of several cyanobacteria (Sinha et al., 1995; Sinha and Häder,

2002). Increased duration of UV-B radiation exposure brings about a linear decrease in the number and quantity of protein bands of a number of cyanobacteria (Fig. 1). In a *Nostoc* sp., the low molecular weight proteins of

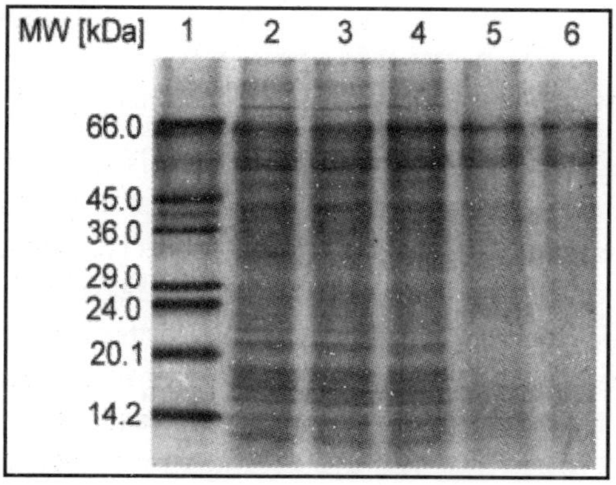

FIG. 1. Effects of UV-B radiation on total protein profileof the cyanobacterium, Anabaena sp. Lane 1- marker; lane 2- control; lane 3- 30 min; lane 4- 60 min; lane 5- 90 min and lane 6- 120 min exposure of UV-B radiation

around 20 kDa (ab monomers of phycocyanin) were the most affected one. It is interesting to find that UV-B radiation has a drastic effect specifically on low molecular weight proteins. In a comparative study on total protein profile of several cyanobacteria, Sinha *et al.* (1995), reported complete loss of protein bands between 14.5 and 45 kDa after 90 and 120 min of UV-B exposure in *Nostoc carmium* and *Anabaena* sp. However protein bands of relatively higher molecular weight viz., 55 and 66 kDa were unaffected even after 120 min. In species such as *N. commune* and *Scytonema* sp, which contain a mucilage covering/sheath, complete elimination of bands occurred only after 150 min of UV-B exposure (Sinha *et al.*, 1995). The increased tolerance to UV-B radiation in certain species may be ascribed to the presence of mucilage covering or photoprotectants such as scytonemin and mycosporine-like amino acids (MAAs) in the cell. Appearance of certain new proteins may be due to UV-B shock and may be similar to the hsp group of proteins (Shibata *et al.*, 1991).

2.8 DNA damage

A variety of lesions in DNA after UV-B radiation treatment have been identified. These lesions result either due to direct absorption of UV-B radiation by DNA or from free radicals and reactive oxygen species formed by various photochemical processes. The main class of UV-induced lesions consist of dimeric photoproducts such as cis-syn cyclobutane pyrimidine dimers (CPDs) and pyrimidine (6-4) pyrimidone photoproducts (6-4 PPs) (Mitchell and Karentz, 1993; Friedberg *et al.*,1995). Dimers have been reported to inhibit the progress of DNA polymerase, and mammalian RNA polymerase is known to stall at both CPDs and 6-4 PPs (Britt, 1995; Kumar *et al.* 2004). Photoproducts formation interferes with DNA synthesis and RNA transcription, ultimately leading to mutation or death of the organism.

The damaging effects of UV-B radiation on DNA and formation of various photoproducts are well known in a number of organisms including the Antarctic marine planktonic diatom communities (Karentz *et al.*, 1991a; Buma *et al.*, 1995). However little information is available regarding the specific effects of UV-B radiation on DNA of cyanobacteria, although few studies have been conducted with UV-C radiation (Levine and Thiel, 1987; Eker *et al.*, 1990). Sinha *et al.* (2001a) have demonstrated that exposure to UV-B radiation (10 W/m^2) could induce the formation of thymine dimers in three rice-field cyanobacteria, *Nostoc*, *Anabaena* and *Scytonema* sp. The frequency of dimer formation increased with increase in exposure time and a 120 min exposure of these cyanobacteria to UV-B radiation led to dimer frequency of 35 - 40%. By using blotting and chemiluminescence method these workers successfully developed a relatively quick, simple and efficient method for the quantitative analysis of thymine dimers in aquatic primary producers including cyanobacteria (Sinha *et al.*, 2001a).

3. EFFECTS ON HUMAN HEALTH

Exposure to elevated levels of UV-B radiation may impose diverse health related hazards to human beings, the magnitude of which is considerably modified by human behaviour and life style. UV-B is strongly absorbed by the tissues and because of its low penetrance it directly affects the exposed organs such as eye and skin (Young, 2006). The absorption of light energy by various chromophores present in UV-B exposed tissues cause a variety of diverse biological effects on eye and skin and can also lead to some systemic affects including the suppression of immunological responses (Schwarz, 2005). The Langerhan's cells which reside in or migrate through the epidermis can mediate such effects on immune system. Although it is difficult to obtain quantitative risk estimates for these effects however some quantitative studies have been made with cataract and skin cancer, but comparable data for immuno suppression effects are not available.

3.1 Eye

Both acute and chronic (occurring due to extended or repeated exposure to levels of UV radiation which are below those required to cause acute affects) effects of UV-B radiation on eye have been observed. The ocular damage caused by UV- radiation includes damage to cornea, lens, retina, iris, and associated epithelial and conjunctival tissues (Table 1). Since most of the UV

TABLE 1. Possible effects of UV-B radiation on eye

Structure affected	Disease	Symptoms
Cornea/Conjunctiva	Photokeratitis (Snow blindness)	Inflammation (reddening) of the eyeball, severe pain, tearing, photophobia, twitching
	Climâte Droplet Keratopathy (CDK)	Degeneration of the fibrous layer of the cornea and the accumulation of droplet shaped deposits
	Pterygium	Out growth of the conjunctiva over the cornea resulting in loss of transparency
	Pinguecula	Appearance of a raised opaque mass just adjacent to the cornea
	Squamous cell carcinoma (SCC)	Malignant neoplasm.
Uveal tract (Iris, ciliary body and choroid)	Malignant melanoma	Malignancy
Lens	Increased incidence of cortical and posterior subcapsular cataract	Clouding of the eye lens causing impaired vision and sometimes blindness

radiation is absorbed by cornea and the lens, only little amount reaches to retina. However, lens removal (in case of cataract surgery) may put retina at risk for UV- damage.

3.1.1 Cornea/Conjunctiva

The principal corneal damage linked to UV-B exposures is photokeratitis commonly known as snowblindness, which is a painful but usually transient condition of the eyes and is most frequently seen in skiers and other outdoor recreationists. The inflammation accompanying photokeratitis has been attributed to the increased production of biologically active chemicals such as cytokines. In vitro and in vivo exposure of human corneal stromatal cells to low doses of UV radiation (10-100 mJ/cm^2) results in significant increase in production of cytokines such as interleukin (IL)-1, IL-6, IL-8 and tumors necrosis factor alpha (TNF-á). (Kennedy *et al.,* 1997).

Epidemiological studies involving relationship between solar exposure and other ocular effects on cornea/conjunctiva are strongest with climatic droplet keratopathy (CDK) and pterygium. The incidence of both diseases is frequently associated with persons working in environments such as water, sand and concrete showing high surface reflectance. In a study of pterygium approximately 36 fold increased risk in disease occurrence was found in people with early intense exposure for the first five years. A study of squamous cell carcinoma (SCC) of eye conducted to test the risk associated with solar UVR exposure across the world showed that an increase in incidence of SCC of eye with UV-B exposure that is equivalent to almost 40 % decline in incidence of the disease for 10 ^{0} increase in the latitude.

3.1.2 Uveal tract

Uveal tissues constitute the middle vascular coat of the eyeball. The most commonly occurring primary ocular malignancy is malignant melanoma of the uveal tract. Malignant melanoma of choroids, observed between 40-70 years of age, is rare in blacks but is more common is whites. Holly *et al.,* (1996) found an increased risk of intraocular melanomas associated with sensitivity to UV radiation.

3.1.3 Lens

The development of opacity in lens is known as cataract and such age related changes are attributed to side-effects of metabolic processes and formation of reactive oxygen species. Since exposure to UV radiation is also known to generate reactive oxygen species, it is also suspected to be involved in the deterioration of the lens by oxidation (Dillon, 1994).

Of the three main types of cataracts viz cortical cataract (CC), nuclear cataract (NC) and posterior sub-capsular cataract (PSC) UV-B has been identified as a risk factor for pure and mixed cortical cataracts and PSC with the exception of cortical/nuclear combination. Several studies have indicated that in case of cortical cataract there is between 1-and-3 fold relative risks associated with increased sun exposure (West *et al.,* 1998). In a study conducted on watermen in Maryland a highly significant correlation was found between CC and solar UV exposure.

Although NC is more common in people residing in tropical climates, but evidence relating NC to solar UV exposure are lacking (Saski *et al.,* 1999). The available evidence of cortical opacities associated with increasing UV exposure suggests the necessity of warning people and to advise them to avoid solar exposure to their eyes (West, 1999).

3.2 Skin

Several types of molecules/chromophores present in the epidermal and dermal layers absorb UV-radiation which in turn damage or modify the primary targets to cause acute and chronic effects on the skin. The outermost epidermal layer is the major absorber of UV-B however only < 10% of incident UV-B wavebands penetrates to the dermis. The acute effects of UV exposures are sunburn and DNA damage whereas the chronic effects include photoaging (wrinkling, thinning, and loss of elasticity of the skin) and incidence of skin cancers.

3.2.1 Sunburn

Sunburn, characterized by a reddening of the skin (erythema) and appearance of blisters in severe cases, is the most frequently observed acute symptom observed on the skin, which has been related to solar UV exposure. Sensitivity to sunburn is often related to skin pigmentation and light skinned individuals are more prone to such effects whereas deeply pigmented individuals are most resistant and often become more deeply pigmented on the other hand upon exposure. Sunburn is initially characterized by appearance of sunburn cells and infiltration of the dermis and epidermis by a variety of inflammatory cells such as mast cells, monocytes and lymphocytes (Garmyn *et al.*, 1995). Abarca *et al.*, (2002) in a study conducted at a city located at the southern Chile observed a correlation between episodes of ozone depletion, increased UV-B radiation and sunburn during the spring months. A dramatic rise in the number of sunburn cases also coincided with Sunday outdoor recreational exposure after sudden ozone depletion.

3.2.2 Polymorphic Light Eruption (PMLE)

It is a common type of skin reaction attributed to solar UV radiation and is characterized by recurrent, abnormal, delayed reaction resulting in the appearance of small raised spots and blisters on the skin which develop 30 minutes to several hours after sun exposure. The eruptions may persist for a few hours to as long as 2 weeks. Although reported in all races, PMLE is more common in light-skinned individuals (Fitzpatrick skin typé I-IV). The symptoms are more common on arms but the legs and chest may also be affected. Interestingly females are affected two to three times more often than males since the female hormone 17-estradiol prevents the UV radiation induced suppression of the contact hypersensitivity response caused by the release of immunosuppressive cytokine (IL-10) from keratinocytes.

PMLE appears to be an immune-mediated type IV delayed hypersensitivity reaction (Tutrone et al., 2003). A reduced expression of tumour necrosis factor, IL-4 and IL-10 in the UV exposed skin of patients with PMLE may be important in the pathogenesis of PMLE.

3.2.3 Photoaging

The photoaged skin is characterized by wrinkling, altered pigmentation and loss of elasticity. Skin aging is the result of the loss of elasticity of connective tissue fibers and is, itself, the consequence of solar effects. Several studies indicate that reactive oxygen species (ROS) are involved in photoaging of the skin. UV-A is considered more effective in generation of ROS but a combination of UV-B and UV-A was found to be less effective than UV-B alone in the induction of photoaging suggesting the crucial role of UV-B radiation in causing photoaging effects.

3.2.4 Skin cancer

Skin cancer is the most prominent cancer in light skinned populations and is very common in people living in sub-tropical areas of the USA and Australia. Carcinogenesis is a complex series of events which involve three stages: initiation, promotion and progression (Fig. 2). UV radiation can act both as initiator and promoter in carcinogenesis. Widely studied skin cancers are: i) basal cell carcinoma (BCC), ii) squamous cell carcinoma (SCC), and iii) cutaneous melanoma (CM). BCC and SCC are the malignant transformation of the major structural cell of the skin called keratinocytes whereas CM is characterized by the malignancy of skin's pigment producing cells termed melanocytes. BCC and SCC are collectively referred to as non-melanoma skin cancer (NMSC). Severe DNA damage caused by UV-radiation can lead to abnormal cell proliferation and tumour development through inactivation of one or more tumour suppressor genes or the over-activation of the growth stimulatory protooncogenes (Fig.3). p53 is the most commonly mutated tumour suppressor gene and is very important for other types of cancers besides skin cancer. UV-B specific mutations have been found in ras protooncogens in some cases of human BCCs and SCCs. Amplification of ras gene has also been observed in individuals with xeroderma pigmentosum (XP).

Epidemiological studies have revealed that incidence of all three types of skin cancers is related to solar (UV) exposure although their biological behaviour and relationship to solar exposure differ qualitatively and

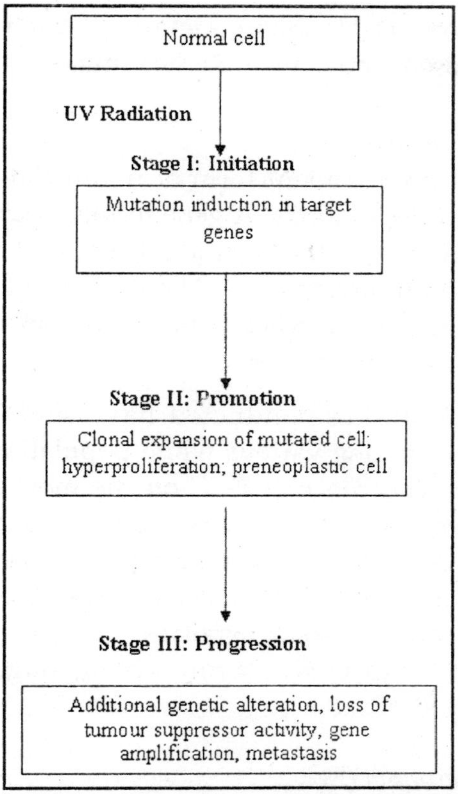

FIG. 2. Stages in tumour development following UV radition exposure

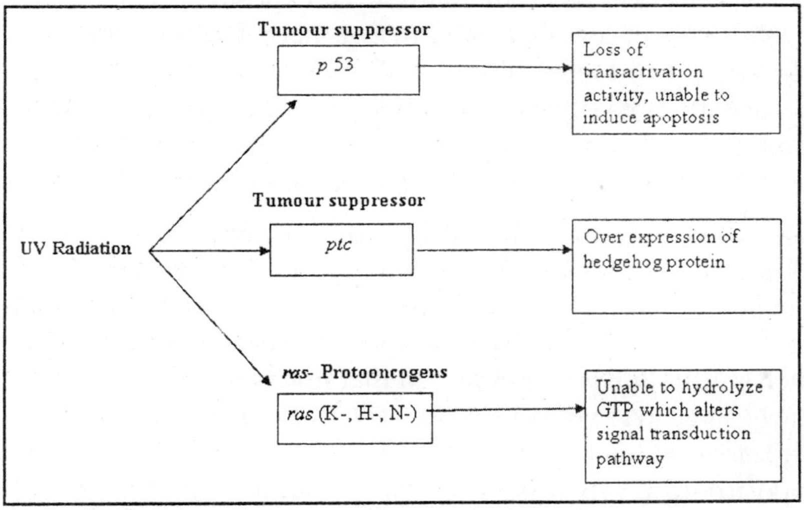

FIG. 3. Effects of UV radiation on target genes.

quantitatively (Savona *et al.*, 2005). Details of each type of carcinoma/ melanoma are discussed under separate heading.

i. Basal cell carcinoma (BCC)

This is the most predominant form of non-melanoma cancer observed among white skinned populations. A gene termed 'patched' (ptc) associated with nevoid BCC syndrome has been identified and its presence has been correlated to the susceptibility to BCC. About 50 % of alterations in this gene are caused by UV-B radiation indicating a direct molecular evidence for the role of UV-B radiation in causing BCC.

Epidemiological studies conducted have shown an increase in the incidence rate of BCC among various white populations. Studies conducted in USA showed the occurrence of BCC on the most heavily exposed sites such as head, neck and extremities and was directly related to cumulative life time solar exposure. However risk of BCC in intermittently exposed sites such as trunk showed an increasing trend with increasing exposure. A plateau in BCC risk at higher level of solar exposure has been observed. Moreover intermittent exposure especially during young stage (age 15-19) is more important in causing BCC.

ii. Squamous cell carcinoma (SCC)

Squamous cell carcinoma is less common than BCC, but more common than CM in USA. Lifetime cumulative UV exposure has been identified as critical risk factor in causing SCC in Southern Europe. On the contrary a Canadian study conducted on 180 SCC patients did not find any increased risk related to life-time exposure. Indirect evidence has indicated that persistent UV exposure is important in the final stages of tumor development. The increased risk of SCC has also been correlated to childhood sunburn.

UV related mutations in the p53 tumor suppressor gene have been reported in majority of SCC. It has been found that about 60 % of precursor lesions of AK (actinic keratosis) bear such mutations. Such mutations have also been found in UV-B induced murine skin cancers in experimental mice. From such studies it has been inferred that mutations in the p53 gene occur at a very early stage in the development of SCC. Human skin grafted on immune deficient mice can also develop AK and SCC within 1 to 2 years following daily exposure to UV-B, and such kind of studies may provide new models for studying induction of skin cancer in response to UV-B exposure.

iii. Cutaneous melanoma (CM)

Cutaneous melanoma accounts for only 2-3 % of the skin cancers associated with solar exposure in USA nevertheless they are more dangerous than other skin cancers as their mortality rate amounts to about 25%, due to a high risk of metastasis. The incidence of CM is rising among white-skinned races throughout the world. In terms of its relationship to UV-B exposure, its biological behavior and in incidence it significantly differs prom NMSC. As with BCC the increased risk of CM is generally associated with an increased intermittent exposure rather than continuous exposure. Several epidemiological studies have indicated an increased melanoma risk in individuals who lived in exposed sunny areas during their child hood. Moreover childhood and adult exposures act interdependently and in individuals having childhood exposure there is a significant risk for CM only if there is substantial sun exposure during adult life also. The presence of one or more forms of pigmented lesions on the skin either freckles or moles (often known as melanocytic nevi) is an additional risk factor for CM. The observation of frequent occurrence of nevus in pathological examinations of melanomas suggests that some nevi may be precursors for melanomas. Studies in some animal systems have suggested that etiology of melanoma probably involves a number of steps governed by UV-B and UV-A- induced changes involving several different molecules.

3.2.5 Epidemiological trends in skin cancer

BCC, SCC and CM show an increasing order of malignancy and decreasing order of incidence and all three types are related to sun exposure Mortality rates are highest for CM. Increasing trends with time in incidence of BCC and SCC have been reported although the magnitude of increase varies substantially. Studies conducted in different countries have revealed an increase of 20 -70 % for BCC and 15 - 90 % for SCC during 1985-1996. Large increase (up to 2- fold) during 10 years has also been shown for CM. However studies conduced at Arizona, USA did not show any consistent increase either in BCC or SCC during 1985-1996 (Harris *et al.*, 2001).

In human skin cells UV-B specific point mutation in the p53 tumor suppressor gene has been found in large majority of skin carcinomas (BCC and SCC). In animal systems causal link between the p53 mutant clones and SCC have also been established (Rebel *et al.*, 2001). Both epidemiological data and experimental proof in mice suggest the correlation between UV-B exposure and occurrence of SCC and CM.

In vast majority of BCC a particular proliferative pathway is activated after mutations in the ptc gene and some of these mutations are characteristically produced by UV-B exposure. Such evidences indicate the role of UV-B radiation in causing and enhancing the development of BCC by activating the proliferative pathway and by making the P53 tumors suppressor protein dysnctional.

Likewise studies conducted on animal systems have indicated that CM arise from a defective tumor suppressive pathway involving P16/ INK4a protein and activation of a cell proliferative pathway involving RAS/BRAF. However sporadic data showing specific involvement of UV-B radiation in causing such charges are available. In opossums (*Monodelphis domestica*) a 3-weeks moderate UV-B exposure to sucklings lead to occurrence of CM in some of these animals later in life and some of these CM were found to have UV-B specific mutations in the InK4a gene (Chan *et al.*, 2001). However the role of UV-B radiation in disrupting the P16/InK4a or RAS/BRAF pathway in human and the genesis of CM are not clear.

3.3 Immunosuppression

Although UV radiation mediated skin cancer effects occur mainly in the white-skinned races, its effects on the immune system occur irrespective of skin colour. Exposure to UV radiation can alter the immune system (Akaraphanth and Lim, 1999, Schwarz, 2005) thereby causing diminished immune response to various infectious agents and skin cancers (Fig. 4). This can occur in three ways; i) Langerhan or antigen presenting cells (APC) residing in the skin, whose function is to capture and carry the invading foreign intruders such as antigens and pathogen to the lymph nodes where initiation of immune response occur, are directly damaged by UV-radiation. This results in production of either aberrant immune response or no response. ii) Some cells in the skin produce chemical mediators that lead either to immunity development or suppression of immunity. Following exposure to UV-B radiation, these cells release chemical mediators which favour immune suppression, and iii) urocanic acid present in the keratin layer of skin, undergoes chemical change following UV exposure, bringing about triggering of the release of chemical mediators from skin mast cells, which also leads to immune suppression.

Suppression of both local and systemic immune responses to a number of antigens has been shown in both humans and animals following UV-B exposure. Several factors such as genetic constitution, behavioral choices, diet and medical treatment, age of exposure modify the above mentioned effects of solar UV-exposure. The genetic differences known to modify the susceptibility to UV-radiation include skin pigmentation, ability to repair damaged lesions in DNA, expression of oncogenes and cytokines and ability to make an immune response to certain antigen. Individuals with xeroderma pigmentosum (defective in repair of UV-induced DNA damage) have a 2000 -fold increased risk of developing skin cancer.

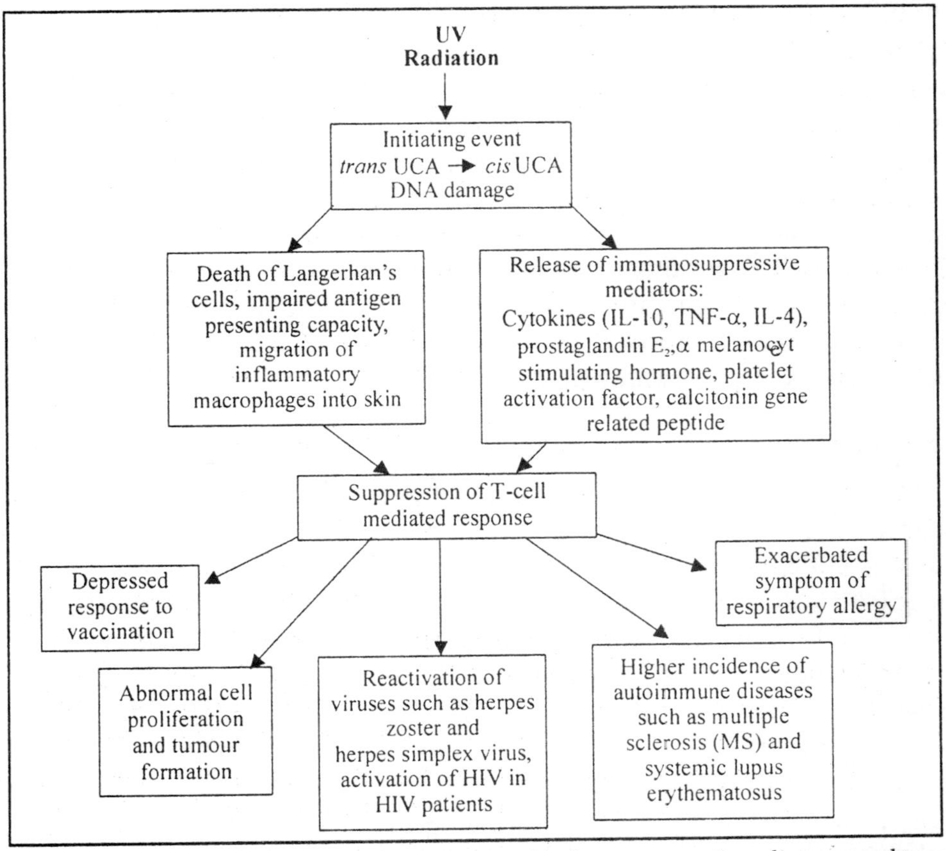

FIG 4. UV radiation-induced immunosuppression and its impact on various diseases and tumour formation

3.4 Vitamin D

Exposure to solar UV radiation has many adverse health effects in humans, but it has at least one very beneficial effect since it contributes to the vitamin D status. The vitamin D status of an individual is based on measuring the serum or plasma concentration of 25-hydroxyvitamin D[25(OH)D], and the active form of vitamin D is 1,25-dihydroxyvitamin D3 [1,25(OH)2D3], which is synthesized in the final step of the metabolic pathway. Currently the plasma levels of 25(OH) D considered excessive, sufficient, insufficient and deficient are >250, 50-250, 25-50 and <25 nmol/L (Wolpowitz and Gilchrest, 2006; Norval *et al.*, 2007). For most people, more than 90% of their vitamin D requirement is acquired from exposure to solar UVR. Because of its ability to absorb UV-B, melanin in the skin can also affect vitamin D status although, with sufficient UVB, adequate vitamin D status can be achieved (Matsuoka *et al.*, 1991). Vitamin D is a very important hormone for many aspects of general health. Vitamin D status is now implicated in the prevention of an increasing list of non-skeletal disorders including several internal cancers and autoimmune diseases, and hypertension. 1,25(OH)2D3 acts as an anti-apoptotic factor and a stimulant to differentiation. In addition, following the discovery of vitamin D receptors (VDRs) on several populations of immune cells, it is now known that vitamin D status can affect the immune system by suppressing T cell proliferation, down-regulating antigen presentation, stimulating the generation of Treg cells and Th2 cells, and activating macrophage function (Mathieu *et al.*, 2004).

4. BIOTECHNOLOGICAL PERSPECTIVES

Since cyanobacteria are believed to have originated in the Precambrian era at a time when ozone shield was absent, they presumably faced high fluxes of UV radiation which must have acted as an evolutionary pressure leading to selection for efficient UV radiation protecting mechanisms. Tolerance of cyanobacteria to intense sunlight as well as UV radiation might have contributed to their success during early stages of colonization. As a consequence they have developed effective mechanisms to counteract the damaging effects of UV-B radiation. In case of microbes behavioral responses as well as external and internal mechanisms of protection against UV-induced damages have been reported and it has been proposed that several of these mechanisms for UV-protection and/or repair of UV-induced damage might have worked singly or in various combinations (Pierson *et al.*, 1993). Further differential tolerance level against UV-B radiation in human beings has been reported from different geographical regions. Role of genetic constitution has been implicated in tolerance mechanism. Evidence generated on the role of protective mechanisms, chemicals/components involved and molecular mechanisms operating in different systems give ample opportunity to explore biotechnological potentials of above processes in mitigating the anticipated health hazards of UV-B radiation. Details of certain UV protecting/sunscreen pigments are given below.

4.1 Sunscreens

An important mechanism to prevent UV-induced photodamage, is the synthesis of UV-absorbing compounds. Phenylpropanoids, mainly flavonoid derivatives and melanins are known to protect higher plants and animals, respectively, from harmful UV radiation. In cyanobacteria photoprotectants such as scytonemin and mycosporine-like amino acids (MAAs), having strong absorption in the UV-A and/or UV-B region of the spectrum play an important role in allowing these organisms to grow and survive in habitats exposed to strong irradiation. In a terrestrial Nostoc commune strain, subjected to regular cycles of desiccation and rewetting, UV-absorbing compounds play a key role as UV-protectants (Ehling-Schulz *et al.,* 1997). Role of different screening compounds in UV-B protection is discussed below.

4.1.1 Scytonemin

Scytonemin is a yellow-brown colored pigment occurring in the extracellular sheath of a number of terrestrial cyanobacteria especially in those species growing in habitats exposed to intense solar radiation (Garcia-Pichel and Castenholz, 1991; 1993). It is a lipid soluble compound having in vivo absorption maximum at 370 nm in the UV-A region but shows considerable absorption in the UV-B region also. Purified scytonemin shows absorption maximum at 386 nm although there being appreciable absorbance at 252, 278 and 300 nm. It is a novel dimeric pigment composed of indole and phenolic subunits with a molecular mass of 544 Da. Microspectrophotometric measurements of the transmittance of pigmented sheath and the quenching of ultraviolet excitation of phycocyanin fluorescence suggest that the pigment is effective in shielding the cells from incoming UV radiation and is able to reduce the incident UV-A radiation entering the cells to around 90 % (Garcia-Pichel and Castenholz, 1991; Brenowitz and Castenholz, 1997).

Scytonemin may constitute as much as 5 % of the total cellular dry weight or even higher in some naturally occurring cyanobacteria (Castenholz, 1997). Since rapid photodegradation of scytonemin does not occur, it is highly stable and able to persist very long in terrestrial cyanobacterial crusts or dried mats (Brenowitz and Castenholz, 1997; Quesada *et al.,* 1999). It has been proposed that appearance of scytonemin during the Precambrian era must have helped the cyanobacteria in the colonization of exposed, shallow water and terrestrial habitats (Quesda *et al.,* 1999). Besides having a role as UV protectant, scytonemin may have additional roles such as protection against pathogenic attack, bacterial decomposition or herbivore grazing (Sinha and Häder, 2002).

4.1.2 *Mycosporine-like amino acids (MAAs)*

The other class of photoprotectants is colorless, water-soluble MAAs having absorption maxima in the UV range between 310 - 360 nm (Torres *et al.*, 2006; Callone *et al.*, 2006). First identified in fungi, these compounds are now reported to be present in a number of taxonomically diverse groups of organisms from eukaryotic and prokaryotic classes. They are present in cyanobacteria (Garcia-Pichel *et al.*, 1993; Sinha et al., 1999, 2001b), eukaryotic algae (Carreto *et al.*, 1990; Karentz *et al.*, 1991b), marine heterotrophic bacteria and marine invertebrates (Karentz *et al.*, 1991b), fish (Dunlap *et al.*, 1989) and in a wide variety of other marine organisms (Karentz *et al.*, 1991b; McClintock and Karentz, 1997; Dunlap and Shick, 1998). Among cyanobacteria, MAAs has been reported from species growing in freshwater, marine or terrestrial habitats (Garcia-Pichel and Castenholz, 1993; Garcia-Pichel *et al.*, 1993; Zhang *et al.*, 2007). Their presence has also been reported from strains growing in Antarctica (Quesada and Vincent, 1997) as well as from a community of halophilic species (Oren, 1997).

Structurally these compounds contain cyclohexenone or cyclohexenime chromophore conjugated with the nitrogen substituent of an amino acid or its imino alcohol and have an average molecular weight of 300 (Cockell and Knowland, 1999). The specific MAA content in several isolates has been found to vary between 0.16 to 0.84 % of the total dry weight. Sinha *et al.* (2001b) have demonstrated the induction of MAAs by solar ultraviolet-B radiation in three cyanobacteria namely, *Anabaena* sp., *Scytonema* sp. and Nostoc commune. They reported that PAR and UV-A were ineffective in causing induction of MAAs. All these cyanobacteria possessed only a single type of MAA which was identified as shinorine.

It has been suggested that the content of MAAs in the cell is regulated by osmotic mechanisms (Oren, 1997), may transfer energy to the photosynthetic reaction centers (Sivalingam *et al.*, 1976; Sinha et al., 1999), may have a role in osmotic regulation (Oren, 1997) and, they may act as antioxidants and prevent cellular damage resulting from UV-induced production of highly reactive oxygen species. Interestingly, in laboratory-grown cultures of Chlorogloeopsis sp., both UV and osmotic stress are able to induce and regulate the synthesis of MAAs (Portwich and Garcia-Pichel, 1999, 2000). In this species, a specific UV-B photoreceptor, probably pterin may mediate MAA induction (Portwich and Garcia-Pichel 2000).

A terrestrial species of Nostoc commune when exposed to high solar radiation was found to possess a UV-A/B absorbing pigment with absorption maxima at 312 and 325 nm (Scherer *et al.*, 1988). In this species, an oligosaccharide MAA (OS-MAA), has been shown to be located in the extracellular glycan sheath (Böhm et al., 1995; Ehling-Schulz *et al.*, 1997). It is the first mycosporine so far known to be covalently linked to the oligosaccharide and located outside the cell (Böhm *et al.*, 1995). This pigment besides providing protection by absorbing UV radiation may also provide additional protection by quenching (Dunlap and Yamamoto, 1995).

4.1.3 Skin type (melanin) and sunscreens

Skin type affects the degree to which some people burn and the time it takes them to burn. The Food and Drug Administration (FDA) classifies skin type on a scale from 1 to 6. Individuals with lower number skin types (1 and 2) have fair skin and tend to burn rapidly and more severely. Individuals with higher number skin types (5 and 6), though capable of burning, have darker skin and do not burn as easily. Sunscreens contain active chemical ingredients that provide protection from the sun through absorption, reflection, and scattering of UV radiation (Nicol, 1989). They are usually divided into two categories: physical and chemical. Physical sunscreens are thick, opaque substances that prevent UV radiation from reaching the skin by scattering the sunlight. They contain titanium dioxide, zinc oxide, or talc as their active ingredients (Lewis and Fischer, 1987). These sunscreens are very effective in blocking both UV-A and UV-B radiation. Chemical sunscreens are the most commonly used and are available in a variety of formulations (Nicol, 1989). They contain one or more of the UV-absorbing chemicals. Some of the most widely used chemical groups that block UV-B radiation are PABA (p-aminobenzoic acid), PABA esters (padimate O), cinnamates (cinoxate, ethylhexyl-p-methoxycinnamate), salicylates (octylsalicylate, homosalate), and anthranilates (methyl anthranilate). While these all block UVB radiation, the chemical group known as benzophenone (oxybenzone and dioxybenzone) provides protection against both UVA and UVB (Lester, 1983).

4.1.4 Sun protection factor (SPF)

Sunscreens have been assigned Sun Protection Factor (SPF) values by the U.S. Food and Drug Administration (FDA) since 1978 (Warshauer and Steinbaugh, 1983). SPF is a number that refers to the product's (sunscreen) ability to block UVB radiation. This number does not show the blockade against UV-A radiation. Sunscreen products with SPFs of 2 to 50 are currently available (Nicol, 1989). A product with a SPF of 15 can protect the skin 15 times longer from UV-B exposure than without sunscreen. The exact amount of time will vary from person to person, the altitude, and proximity to the equator. SPF 15 will block 95 % of the UV-B wavelengths. However SPF 30 does not work twice as well, it provides another 3 % of protection.

4.1.5 Others

Besides scytonemin and MAAs, occurrence of a few other UV-absorbing compounds, conferring resistance to UV radiation to the organisms, have been reported (Kumar *et al.*, 1996). The shielding role of a brown-coloured pigment (having major absorption peak at 314 nm) from Scytonema hofmanii and a phycoerythrin rich pink extract fom Nostoc spongiaeforme, against UV-B-induced damage, has been reported (Kumar *et al.*, 1996). A strain of *Nostoc* sp. showing resistance to high light intensity and UV radiation has been reported to produce three different UV-absorbing compounds having absorption maxima at 256, 314 and 400 nm (de Chazal and Smith, 1994).

During the last two decades vast information on UV-B absorbing/sun-screening pigments from diverse organisms have been gathered and this has led to the development of a database on photoprotective compounds (Gröniger *et al.*, 2000). Readers may find detailed information on the absorption maxima and extinction coefficients of different compounds together with experimental protocols on the web site (see http://www. biologie.uni-erlangen.de/ botanikl/index.html). There is need to develop protocol for large scale industrial production of these compounds so that they can find application as UV-protecting cosmetics.

5. CONCLUSIONS

Increased intensity of ultraviolet-B (UV-B) radiation has now been unequivocally demonstrated to show detrimental effect on a number of vital physiological and biochemical processes of cyanobacteria. UV-B affects pigmentation, disintegrates phycobilisomes, severely inhibits photosynthetic processes and causes inactivation of nitrogenase. Major alterations in general protein profile have also been reported. Formation of dimers between two adjacent pyrimidine bases, cis-syn cyclobutane dimers and pyrimidine (6-4) pyrimidone products following UV-B exposure in certain cyanobacteria has also been demonstrated. Till date most of the studies pertaining to UV-B radiation effects on cyanobacteria are based on short term laboratory experiments employing high UV-B flux. Only a few short-term experiments that utilize solar UV-B radiation obtained by using various long-pass filters are available (Donkor *et al.*, 1993). Conclusions about the long-term ecological impacts of ambient natural UV-B radiation cannot be accurately forecast from results obtained through short-term incubation studies.

It is felt that other environmental and biological variables should be taken into account while attempting a realistic assessment of the effects of enhanced levels of UV-B radiation. The existence of various protective/repair strategies provide a high level of natural defense against the potential effects of elevated UV-B radiation. Information related to the ecological significance of sun-screening substances such as scytonemin and MAAs and spatial distribution of MAAs from a wide range of microbes are added every day and data available suggest their great potential. Diverse types of effects on human following enhanced level of UV-B exposure have become major concern for health risks. Increase in the number of eye and skin cancers from various parts of world reflect the exposure level of solar UV-B radiation. It is a challenge to the scientific community to fight against the deadly environmental problem in coming years. There is need to explore in detail the protective role of various sunscreen compounds against UV-B radiation. Efforts are also needed to reveal the genetic level of tolerance in microbes as well as human beings. In conclusion it appears that during the last 25 years vast knowledge covering diverse aspects of UV-B effects on living organisms have accumulated but there are a number of gaps left in our understanding which need immediate attention. This fertile area of research has vast biotechnological potentials.

Acknowledgements

Research works in the laboratory of Prof. A. Kumar was supported by research grants received from the Department of Biotechnology, Government of India (BT/PR 1239/AGR/02/065/98).

REFERENCES

Abarca, J.F., Casiccia, C.C. and Zamorano, F.D. (2002). Increase in sunburns and photosensitiviry disorders at the edge of the Anartic ozone hole, Southern Chile, 1986-2000. *J. Am. Acad. Dermatol.* 46:193-199.

Akaraphanth R. and Lim, H.W. (1999). UV and immunosuppression, *Photodermatol. Photoimmunol. Photomed.* 15:28-31.

Bebout, B.M. and Garcia-Pichel, F. (1995). UV-B induced vertical migrations of cyanobacteria in a microbial mat. *App. Environ. Microbiol.* 61:4215-4222.

Blakefield, M.K. and Harris, D.O. (1994). Delay of cell differentiation in Anabaena aequalis caused by UV-B radiation and the role of photo-reactivation and excision repair. *Photochem. Photobiol.* 59:204-208.

Böhm, G.A., Pfleiderer, W., Böger, P. and Scherer, S. (1995). Structure of a novel oligosaccharide-mycosporine-amino acid ultraviolet-A/B sunscreen pigment from the terrestrial cyanobacterium Nostoc commune. *J. Biol. Chem.* 270:8536-8539.

Brenowitz, S. and Castenholz, R.W. (1997). Long-term effects of UV and visible irradiance on natural populations of scytonemin-containing cyanobacterium (Calothrix sp.). *FEMS Microbiol. Ecol.* 24:343-352.

Britt, A.B. (1995). Repair of DNA damage induced by ultraviolet radiation. *Plant Physiol.* 108:891-896.

Buma, A.G.J., Van Hannen, E.J., Roza, L., Veldhuis, M. J. W. and Gieskes, W. W. C. (1995). Monitoring ultraviolet-B induced DNA damage in individual diatom cells by immunofluorescent thymine dimer detection. *J. Phycol.* 51:314-321.

Callone, A.I., Carignan, M., Montoya, N.G. and Carreto, J.I. (2006) Biotransformation of mycosporine like amino acids (MAAs) in the toxic dinoflagellate Alaxandrium tamarense. J. Photochem. *Photobiol. B: Biol.* 84:204-212.

Campbell, D., Eriksson, M.-J., Öquist, G., Gustafsson, P. and Clarke, A.K. (1998). The cyanobacterium Synechococcus resists UV-B by exchanging photosystem II reaction-centre D1 proteins. *Proc. Nat. Acad. Sci.* (USA) 95:364-369.

Carreto, J.I., Carignan, M.O., Daleo, G. and de Marco, S.G. (1990). Occurrence of mycosporine-like amino acids in the red tide dinoflagellate Alexandrium excavatum: UV-protective compounds? J. *Plankton Res.* 12:909-921.

Castenholz, R.W. (1997). Multiple strategies for UV tolerance in cyanobacteria. *The Spectrum* 10: 10-16.

Chan, J., Robinson, E.S., Atencio, J., Wang, Z., Kazianis, S., Coletta, L.D., Nairn, R.S. and McCarrey, J.R. (2001). Characterization of the CDKN2A and ARF genes in UV-induced melanocytic hyperplasias and melanomas of an opossum (Monodelphis domestica). *Mol. Carcinog.* 31: 16-26.

Cockell, C. S. and Knowland, J. (1999). Ultraviolet radiation screening compounds. *Biol. Rev.* 74:311-345.

Cockell, C.S. and Rothschild, L.J. (1999). The effects of UV radiation A and B on diurnal variation in photosynthesis in three taxonomically and ecologically diverse microbial mats. Photochem. *Photobiol.* 69:203-210.

Crutzen, P.J. (1992). Ultraviolet on the increase. *Nature* 356:104-105.

de Chazal, N.M. and Smith, G.D. (1994). Characterization of a brown *Nostoc* sp. from Java that is resistant to high light intensity and UV. *Microbiology* 140:3183-3189.

Dillon, J. (1994). UV-B as a pro-aging and pro-cataract factor Doc. *Ophthalmol.* 88:339-344.

Döhler, G., Biermann, I. and Zink, J. (1986). Impact of UV-B radiation on photosynthetic assimilation of ^{14}C-bicarbonate and inorganic ^{15}N-compounds by cyanobacteria. Z. *Nature.* 41: 426-432.

Donkor, V.A., Damian, H.A.K. and Häder, D. P. (1993). Effects of tropical solar radiation on the motility of filamentous cyanobacteria. *FEMS Microbiol. Ecol.* 12:143-148.

Dunlap, W.C., Williams, D.M., Chalker, B.E. and Banaszak, A.T. (1989). Biochemical photoadaptations in vision: UV-absorbing pigments in fish eye tissues. *Comp. Biochem. Physiol.* 93: 601-607.

Dunlap, W.C. and Shick, J.M. (1998). Ultraviolet radiation-absorbing mycosporine-like amino acids in coral reef organisms: a biochemical and environmental perspective. *J. Phycol.* 34:418-430.

Dunlap, W.C. and Yamamoto, Y. (1995). Small-molecule antioxidants in marine organisms: antioxidant activity of mycosporine-glycine. Comp. *Biochem. Physiol.* 112:105-114.

Ehling-Schulz, M., Bilger, W. and Scherer, S. (1997). UV-B-induced synthesis of photoprotective pigments and extracellular polysaccharides in the terrestrial cyanobacterium Nostoc commune. J. Bacteriol. 179: 1940-1945.

Eker, A.P., Kooiman, P., Hessels, J.K.C. and Yasui, A. (1990). DNA photoreactivating enzyme from the cyanobacterium Anacystis nidulans. *J. Biol. Chem.* 14:8009-8015.

Frederick, J.E., Snell, H.E. and Haywood, E.K. (1989). Solar ultraviolet radiation at the earth's surface. *Photochem. Photobiol.* 50:443-450.

Friedberg, E.C., Walker, G.C. and Siede, W. (1995). DNA repair and mutagenesis. American Society of Microbiology Press, Washington, DC.

Garcia-Pichel, F. and Castenholz, R.W. (1991). Characterization and biological implications of scytonemin, a cyanobacterial sheath pigment. *J. Phycol.* 27:395-409.

Garcia-Pichel, F. and Castenholz, R.W. (1993). Occurrence of UV-absorbing mycosporine-like compounds among cyanobacterial isolates and an estimation of their screening capacity. *Appl. Environ. Microbiol.* 59:163-169.

Garcia-Pichel, F., Wingard, C.E. and Castenholz, R.W. (1993). Evidence regarding the UV sunscreen role of a mycosporine-like compound in the cyanobacterium *Gloeocapsa* sp. *Appl. Environ. Microbiol.* 59:170-176.

Garmyn, M., Degreef, H. and Gilcrest, B.A. (1995). The effect of acute and chronic photodamage on gene expression in human keratinocytes. *Dermatology* 190: 305-308.

Gröniger, A., Sinha, R.P., Klisch, M. and Häder, D.P. (2000).Photoprotective compounds in cyanobacteria, phytoplankton and macroalgae- a database. *J. Photochem. Photobiol. B: Biol.* 58: 115-122.

Häder, D.P. and Worrest, R.C. (1991). Effects of enhanced solar radiation on aquatic ecosystems. *Photochem. Photobiol.* 53:717-725 .

Häder, D.P., Kumar, H.D., Smith, R.C. and Worrest, R.C. (1998). Effects of solar ultraviolet radiation on aquatic ecosystems. *J. Photochem. Photobiol. B: Biol.* 46:53-68.

Harris, R.B., Griffith, K. and Moon, T.E. (2001). Trends in the incidence of nonmelanoma skin cancers in Southeastern Arizona, 1985-1996. *J. Am. Acad. Dermatol.* 45: 528-536.

He, Y.Y. and Häder, D.-P. (2002). Involvement of reactive oxygen species in the UV-B damage of the cyanobacterium, *Anabaena* sp. *Photochem. Photobiol.* 66:73-80.

Holly, E.A., Aston, D.A., Ahn, D.K. and Smith, A.H. (1996). Intraocular melanoma linked to occupations and chemical exposures. *Epidemiology* 7:55-61.

Karentz, D., Cleaver, J.E. and Mitchell, D.L. (1991a). Cell survival characteristics and molecular responses of Antarctic phytoplankton to ultraviolet-B radiation. *J. Phycol.* 27:326-341.

Karentz, D., Mceuen, F.S., Land, M.C. and Dunlap, W.C. (1991b). Survey of mycosporine-like amino acid compounds in Antarctic marine organism: potential protection from ultraviolet exposure. *Mar. Biol.* 108:157-166.

Kennedy, M., Kim, K.H., Harten, B., Brown, J., Planck, S., Meshul, C., Edelhauser, H., Rosenbaum, J.T., Armstrong; C.A. and Ansel, J.C. (1997). Ultraviolet irradiation induces the production of multiple cytokines by human corneal cells. Invest. Ophthalm. *Visual Sci.* 38:2483-2491.

Kerr, J.B. and Mcelroy, C.T. (1993). Evidence for large upward trends of ultraviolet-B radiation linked to ozone depletion. *Science* 262: 1032-1034.

Kumar, A., Tyagi, M.B., Srinivas, G., Singh, N., Kumar, H.D., Sinha, R.P. and Häder, D.-P. (1996). UV-B shielding role of $FeCl_3$ and certain cyanobacterial pigments. Photochem. *Photobiol.* 64:321-325.

Kumar, A., Tyagi, M.B., Jha, P.N., Srinivas, G. and Singh, A. (2003). Inactivation of cyanobacterial nitrogenase after exposure to ultraviolet-B radiation. *Curr. Microbiol.* 46:380-384.

Kumar A., Tyagi, M.B. and Jha, P.N. 2004. Evidences showing ultraviolet-B radiation-induced damage of DNA in cyanobacteria and its detection by PCR assay. *Biochem. Biophys. Res. Commun.* 318:1025-1030.

Lester, R.S. (1983). Topical formulary for the pediatrician. Ped. Clin. North Am. 30: 749-750.

Levine, E. and Thiel, T. (1987). UV-inducible DNA repair in the cyanobacteria Anabaena sp. *J. Bacteriol.* 169:3988-3993.

Lewis, R.M. and Fischer, R.G. (1987). Sunscreen agents. *Pediatr. Nurs.* 13:200-201.

Madronich, S., Mckenzie, R.L., Björn, L.O. and Caldwell, M.M. (1998). Changes in biologically active ultraviolet radiation reaching the Earth's surface. *J. Photochem. Photobiol. B: Biol.* 46:5-19.

Mathieu, C., van Etten, E., Decallonne, B., Guilietti, A., Gysemans, C., Bouillon, R. and Overbergh, L. (2004). Vitamin D and 1, 25-dihydroxyvitamin D3 as modulators in the immune system. *J. Steroid Biochem. Mol. Biol.* 89-90:449-452.

Matsuoka, L.Y., Wortsman, J., Haddad, J.G., Kolm, P. and Hollis, B.W. (1991). Racial pigmentation and the cutaneous synthesis of vitamin *D. Arch. Dermato.* 127:536-538.

McClintock, J.B. and Karentz, D. (1997). Mycosporine-like amino acids in 38 species of subtidal marine organisms from McMurdo Sound, Antarctica. *Antarc. Sci.* 9:392-398.

Mitchell, D.L. and Karentz, D. (1993). The induction and repair of DNA photodamage in the environment. In Environmental UV Photobiology (eds. A. R. Young, L. Björn, J. Moan and W. Nultsch), pp 345-377. Plenum Press, London.

Newton, J.W., Tyler, D.D. and Slodki, M.E. (1979). Effect of ultraviolet-B (280-320 nm) radiation on blue-green algae (cyanobacteria): possible biological indicators of stratospheric ozone depletion. *Appl. Environ. Microbiol.* 37:1137-1141.

Nicol, N.H. (1989). What's new with sunscreens? choices-choices-choices. Pediatr. Nurs. 15:417-418.

Norval, M., Cullen, A.P., de Gruijl, F.R., Longstreth, J., Takizawa, W., Lucas, R.M., Noonan, P. and van der Leun, J.C. (2007). The effects on human health from stratospheric ozone depletion and its interactions with climate change. *Photochem. Photobiol. Sci.* 6:232-251.

Oren, A. (1997). Mycosporine-like amino acids as osmotic solutes in a community of halophilic cyanobacteria. *Geomicro. J.* 14:231-240.

Pierson, B.K, Mitchell, H.K. and Ruff-Roberts, A.L. (1993). Chloroflexus aurantiacus and ultraviolet radiation: implications for archean shallow water stromatolites. *Ori. Life Evolu. Biosph.* 23:243-260.

Portwich, A. and Garcia-Pichel, F. (1999). Ultraviolet and osmotic stresses induce and regulate the synthesis of mycosporines in the cyanobacterium Chlorogloeopsis PCC 6912. *Arch. Microbiol.* 172:187-192.

Portwich, A. and Garcia-Pichel, F. (2000).A novel prokaryotic UV-B photoreceptor in the cyanobacterium Chlorogloeopsis PCC 6912. *Photochem. Photobiol.* 71:493-498.

Quesada, A. and Vincent, W.F. (1997). Strategies of adaptation by antarctic cyanobacteria to ultraviolet radiation. *Eur. J. Phycol.* 32:335-342.

Quesada, A., Vincent, W.F. and Lean-David, R.S. (1999). Community and pigment structure of arctic cyanobacterial assemblages: the occurrence and distribution of UV-absorbing compounds. *FEMS Microbiol. Ecol.* 28:315-323.

Rebel, H., Mosnier, L.O., Berg, R.J., Westerman-de Vries, A., Van Steeg, H., Van Kranen, H.J. and De Gruijl, F.R. (2001). Early p^{53} -positive foci as indicators of tumor risk in ultraviolet-exposed hairless mice: kinetics of induction, effects of DNA repair deficiency, and p^{53} heterozygosity. *Cancer Res.* 61:977-983.

Richter, P., Krywult, M., Sinha, R.P. and Häder, D.-P. (1999). Calcium signals from heterocysts of Anabaena sp. after UV irradiation. *J. Plant Physiol.* 154:137-139.

Sasaki, K., Sasaki, H., Kojima, M., Shui, Y.B., Hockwin, O., Jonasson, F., Cheng, H.M., Ono, M. and Katoh, N. (1999). Epidemiological studies on UV-related cataract in climatically different countries. *J. Epidemiol.* 9:S33-38.

Sass, L., Spetea, C., Mate, Z., Nagy, F. and Vass, I. (1997). Repair of UV-B induced damage of photosystem II via de novo synthesis of the D1 and D2 reaction centre subunits in Synechocystis sp. PCC6803. *Photosyn. Res.* 54:55-62.

Savona, M.R., Jacobsen, M.D., James, R. and Owen, M.D. (2005). Ultraviolet radiation and the risk of cutaneous malignant melanoma and non- melanoma skin cancer: perceptions and behaviours of Danish and American adolescents. *Eur. J. Cancer. Prev.* 14:57-62.

Scherer, S., Chen, T.W. and Böger, P. (1988). A new UV-A/B protecting pigment in the terrestrial cyanobacterium Nostoc commune. *Plant Physiol.* 88: 1055-1057.

Schwarz, T. (2005). Mechanisms of UV-induced immunosuppression. *Keio.J.Med.* 54:165-171.

Shibata, H., Baba, K. and Ochiai, H. (1991). Near UV irradiation induces shock proteins in Anacystis nidulans R-2; possible role of oxygen. *Plant Cell Physiol.* 32:771-776.

Singh, R.N. (1961). Role of blue-green algae in nitrogen economy of Indian agriculture, Indian Council of Agricultural Research, New Delhi, India.

Sinha, R.P. and Häder, D.-P. (2002). Life under solar UV radiation in aquatic organisms. *Adv. Space Res.* 30:1547-1556.

Sinha, R.P., Kumar, H.D., Kumar, A. and Häder, D.P. (1995). Effects of UV-B irradiation on growth, survival, pigmentation and nitrogen metabolism enzymes in cyanobacteria. *Acta Protozool.* 34:187-192.

Sinha, R.P., Singh, N., Kumar, A., Kumar, H.D., Häder, M. and Häder, D.P. (1996). Effect of UV radiation on certain physiological and biochemical processes in cyanobacteria. *J. Photochem. Photobiol. B: Biol.* 32:107-113.

Sinha, R.P., Klisch, M. and Häder, D.P. (1999). Induction of a mycosporine-like amino acid (MAA) in the rice-field cyanobacterium Anabaena sp. by UV irradiation. J. Photochem. *Photobiol. B. Biol.* 52:59-64.

Sinha, R.P., Dautz, M. and Häder, D.-P. (2001a). A simple and efficient method for the quantitative analysis of thymine dimers in cyanobacteria, phytoplankton and macroalgae. *Acta Protozool.* 40: 187-195.

Sinha, R.P., Klisch, M., Helbling, E.W. and Häder, D.-P. (2001b). Induction of mycosporine-like amino acids (MAAs) in cyanobacteria by solar ultraviolet-B radiation. J. Photochem. *Photobiol. B: Biol.* 60:129-135.

Sivalingam, P.M., Ikawa, T. and Nisizawa, K. (1976). Physiological roles of a substance 334 in algae. *Bot. Mar.* 19:9-21.

Smith, R.C., Prezelin, B.B., Baker, K.S., Bidigare, R.R., Boucher, N.P., Coley, T., Karentz, D., MacIntyre, S., Matlick, H.A., Menzies, D., Ondrusek, M., Wan, Z. and Waters, K.J. (1992). Ozone depletion: ultraviolet radiation and phytoplankton biology in Antarctic waters. *Science* 255:952-959.

Torres, A., Enk, C.D., Hochberg, M. and Srebnik, M. (2006) Porphyra-334, a potential natural source for UVA protective sunscreens. *Photochem. Photobiol. Sci.* 5:432-43.

Tutrone, W.D., Spann, C.T., Scheinfeld, N. And Deleo, VA (2003). Polymorphic light eruption. *Dermatol. Ther.* 16:28-39.

Tyagi, R., Srinivas, G., Vyas, D., Kumar, A. and Kumar, H.D. (1992). Differential effect of ultraviolet-B radiation on certain metabolic processes in a chromatically adapting Nostoc. Photochem. *Photobiol.* 55:401-407.

Vincent, W.F. and Roy, S. (1993). Solar ultraviolet-B radiation and aquatic primary production: damage, protection and recovery. *Environ. Rev.* 1:1-12.

Warshauer, D.M., Steinbaugh, J.R. (1983). Sunlight and protection of the skin. *AFP* 27:09-115.

Weatherhead, E.C. and Anderson, S.B. (2006) The search for signs of recovery of the ozone layer. *Nature* 444:39-45.

West, S.K. (1999). Ocular ultraviolet B exposure and lens opacities: a review. *J. Epidemiol.* 9:S97-101.

West, S.K., Duncan, D.D., Munoz, B., Rubin, G.S., Fried. L.P., Bandeen-Roche, K. and Schein, O.D. (1998). Sunlight exposure and risk of lens opacities in a population-based study: the Salisbury Eye Evaluation project. *JAMA* 280:714-718.

Wilson, M.I., Ghosh, S., Gerhardt, K.E., Holland, N., Sudhakar Babu, T., Edeeman, M., Dumbroff, E.B. and Greenberg, B.M. (1995). In vivo photomodification of ribulose-1,5-bisphosphate carboxylase/oxygenase holoenzyme by ultraviolet-B radiation. *Plant Physiol.* 109:221-229.

Wolpowitz, D. and Gilchrest, B.A. (2006). The vitamin D questions: how much do you need and how should you get it? *J. Am. Acad. Dermatol.* 54:301-317.

Young, A.R. (2006). Acute effects of UVR on human eyes and skin. *Prog. Biophys. Mol. Biol.* 92:80-85.

Zhang, L., Li, L. and Wu, Q. (2007) Protective effects of mycosporine-like amino acids of Synechocystis sp. PCC 6803 and their partial characterization. *J. Photochem. Photobiol. B. Biol.* 86:240-245.

□□□

CHAPTER 14

Antimicrobial Resistance in Gram Positive and Gram Negative Bacteria Progress and Challenges

Debajit Thakur*, Minali Baishya, Barnali Sarma, Tarun Ch. Bora and Ratul Saikia
Biotechnology Division, North East Institute of Science & Technology
(Formerly Regional Research Laboratory) (CSIR), Jorhat 785 006, Assam, India

ABSTRACT

Serious infections caused by bacteria that have become resistant to commonly used antibiotics have become a major global healthcare problem in the 21st century. Antibiotic resistance, initially a problem of the hospital setting associated with an increased number of hospitals acquired infections usually in critically ill and immunosuppressed patients, has now extended into the community causing severe infections difficult to diagnose and treat. In hospitals, most common resistant bacteria include methicillin-resistant *Staphylococcus aureus*, vancomycin-resistant enterococci and gram-negative rods including Enterobacteriaceae and *Pseudomonas aeruginosa*. Vancomycin intermediate and resistant *S. aureus*, represent a new treatment challenge. In the community, penicillin and macrolide-resistant pneumococci developed several decades ago and are now present all over the world. More recently, community-acquired methicillin-resistant *S. aureus* has become a problem in several countries causing skin infections but also severe diseases. The molecular mechanisms by which bacteria have become resistant to antibiotics are diverse and complex. Bacteria have developed resistance to all different classes of antibiotics discovered to date. The most frequent type of resistance is acquired and transmitted horizontally via the conjugation of a plasmid. In recent times new mechanisms of resistance have resulted in the simultaneous development of resistance to several antibiotic classes creating very dangerous multidrug-resistant (MDR) bacterial strains, some also known as "superbugs". In many cases the use of antibiotics is unnecessary or questionable. The indiscriminate and inappropriate use of antibiotics in outpatient clinics, hospitalized patients and in the food industry is the single largest factor leading to

*Email: debajitthakur@yahoo.co.uk

antibiotic resistance. In recent years, the number of new antibiotics licensed for human use in different parts of the world has been lower than in the recent past. In addition, there has been less innovation in the field of antimicrobial discovery research and development. The pharmaceutical industry, large academic institutions or the government are not investing the necessary resources to produce the next generation of newer safe and effective antimicrobial drugs. In many cases, large pharmaceutical companies have terminated their anti-infective research programs altogether due to economic reasons. The potential negative consequences of all these events are relevant because they put society at risk for the spread of potentially serious MDR bacterial infections.

Key words: antimicrobial resistance, antibacterial agents, antibiotic, antimicrobial activity.

1. INTRODUCTION

The term "antibiotic" originally referred to a natural compound produced by a fungus or another microorganism that kills bacteria, which cause disease in humans offective against most gram-negative pathogens (e.g. Salmonella, Shigella, *Bordetella pertussis, Yersinia pestis,* Pseudomonas) with the notable exception of *Neisseria gonorrhoeae*.

The list of bacteria developing resistance is impressive, from sulfonamide and penicillin-resistant *Staphylococcus aureus* in the 1930s and 1940s (Levy, 2002; Rammelkamp, 1942) to penicillin-resistant *Neisseria gonorrhoeae* (PPNG), and β-lactamase-producing *Haemophilus influenzae* in the 1970s, (Jaffe *et al.*, 1981; Lind, 1990; Williams and Moosdeen, 1986; Jorgensen, 1993) methicillin resistant *Staphylococcus aureus* (MRSA) and the resurgence of multi-drug resistant (MDR) *Mycobacterium tuberculosis* in the late 1970s and 1980s, (Lowy, 1998; Deresinski, 2005; Lowy, 2003; Foster, 2004; Pablos-Mendez *et al.*, 1998; Espinal *et al.*, 2001) and several resistant strains of common enteric and non-enteric gram-negative bacteria such as *Shigella* sp., *Salmonella* sp., *Vibrio cholerae, E. coli, Klebsiella pneumoniae, Acinetobacter baumanii, Pseudomonas aeruginosa,* some of these associated with the use of antimicrobials in animals grown for human food consumption in the 1980s and 1990s (Waterer and Wunderink, 2001; Rupp and Fey 2003; White *et al.*, 2001; Smith *et al.*, 1999; Wegener, 1999; Fey *et al.*, 2000). Recently we have also witnessed the report of very worrisome cases of previously unthinkable resistance as well as the spread of resistant bacteria outside the hospital causing community-acquired infections. Such is the case or strains of Group A, Streptococcus becoming resistant to macrolide antibiotics (Seppala *et al.*, 1995; Cizman *et al.*, 2001; Martin *et al.*, 2002; Huovinen, 2002), *Streptococcus pneumoniae* developing resistance to different antibiotic classes, including penicillin, and causing serious infections (Nuorti *et al.*, 1998; Whitney *et al.*, 2000; Hofmann *et al.*, 1995; Tomasz, 1995; Amsden, 2004; Vanderkooi *et al.*, 2005; File, 2004; Jacobs, 2004) and more virulent strains of MRSA (due to the expression of certain toxins such as the so-called Panton-Valentine leukocidin) spreading to the community (Saravolatz *et al.*, 1982; Robinson *et al.*, 2005; Kaplan *et al.*, 2005), as well as *Staphylococcus aureus* and Enterococci becoming resistant to vancomycin (Smith *et al.*, 1999; Sieradzki *et al.*, 1999; Waldvogel, 1999).

The period of the late 1940s and early 1950s saw the discovery and introduction of streptomycin, chloramphenicol, and tetracycline, and the age of antibiotic chemotherapy came into full being. These antibiotics were effective against the full array of bacterial pathogens including gram-positive and gram-negative bacteria, intracellular parasites, and the tuberculosis bacillus. However, by 1953, during a *Shigella* outbreak in Japan, a strain of the dysentery bacillus was isolated, exhibiting resistances to chloramphenicol, tetracycline, streptomycin, and the sulfanilamides. There was also evidence mounting that bacteria could pass genes for multiple drug resistance between strains and even between species.

Antibiotic resistance, initially a problem of the hospital setting associated with an increased number of hospital acquired infections usually in critically ill and immunosuppressed patients, has now extended into the community causing severe infections difficult to diagnose and treat. In hospitals, most common resistant bacteria include methicillin-resistant *Staphylococcus aureus*, vancomycin-resistant enterococci and gram-negative rods including Enterobacteriaceae and *Pseudomonas aeruginosa*. Gram-positive bacteria are common causes of bloodstream and other infections in hospitalized patients, and the percentage of nosocomial bloodstream infections caused by antibiotic-resistant gram-positive bacteria is increasing. Methicillin-resistant *Staphylococcus aureus* (MRSA) and vancomycin-resistant *enterococci* (VRE) are of particular concern. Vancomycin is the standard treatment for serious MRSA infections, but a few cases of vancomycin-resistant *S. aureus* (VRSA) have recently emerged, represent a new treatment challenge. More recently, community-acquired methicillin-resistant *S. aureus* (CA-MRSA) has become a problem in several countries causing skin infections but also severe diseases. Resistance to co-trimoxazole in *Escherichia coli* has changed empirical treatment of urinary tract infections.

It was also apparent that *Mycobacterium tuberculosis* was capable of rapid development of resistance to streptomycin, which had become a mainstay in tuberculosis therapy. It is estimated that one third of the world's population is infected with TB, with about 8 million new cases reported annually (National Institute of Allergy and Infectious Diseases (NIAID), Fact sheet on tuberculosis. www.niaid.nih.go.factsheets/ tb.htm). India is one of the developing countries having TB and HIV infection on rise in an alarming speed. Still worse is the situation due to the emergence of multi-drug-resistant (MDR) strains of *M. tuberculosis* along with human immunodeficiency virus (HIV) infection. Moreover, secondary fugal infections that occur towards the later part of medication, among the TB patients aggravate the situation. In July 2003 report, the WHO has called TB (not AIDS) is the biggest health concern of India (Gupta, 2003). This reflects the magnitude and the seriousness of the problem.

In recent times new mechanisms of resistance have resulted in the simultaneous development of resistance to several antibiotic classes creating very dangerous multidrug-resistant (MDR) bacterial strains, some also known as "superbugs". The indiscriminate and inappropriate use of antibiotics in outpatient clinics, hospitalized patients and in the food industry is the single largest factor leading to antibiotic resistance. In recent years, the number of new antibiotics licensed for human use in different parts of the world has been lower than in the recent past. In addition, there has been less innovation in the field of antimicrobial discovery research and development. The pharmaceutical industry, large academic institutions or the government are not investing the necessary resources to produce the next generation of newer safe and effective antimicrobial drugs. In many cases, large pharmaceutical companies have terminated their anti-infective research programs altogether due to economic reasons. The potential negative consequences of all these events are relevant because they put society at risk for the spread of potentially serious MDR bacterial infections.

2. MECHANISMS OF RESISTANCE TO ANTIBACTERIAL AGENTS

Bacteria may manifest resistance to antibacterial drugs through a variety of mechanisms (Table 1). Of greater concern are cases of acquired resistance, where initially susceptible populations of bacteria become resistant to an antibacterial agent and proliferate and spread under the selective pressure of use of that agent. Several mechanisms of antimicrobial resistance are readily spread to a variety of bacterial general

(1) The organism may acquire genes encoding enzymes, such as β-lactamases, that destroy the antibacterial agent before it can have an effect.

(2) Bacteria may acquire efflux pumps that extrude the antibacterial agent from the cell before it can reach its target site and exert its effect.

(3) Bacteria may acquire several genes for a metabolic pathway, which ultimately produces altered bacterial cell walls that no longer contain the binding site of the antimicrobial agent, or bacteria may acquire

mutations that limit access of antimicrobial agents to the intracellular target site via down regulation of porin genes.

Thus, normally susceptible populations of bacteria may become resistant to antimicrobial agents through mutation and selection, or by acquiring from other bacteria the genetic information that encodes resistance. The last event may occur through one of several genetic mechanisms, including transformation, conjugation, or transduction. Through genetic exchange mechanisms, many bacteria have become resistant to multiple classes of antibacterial agents, and these bacteria with multidrug resistance have become a cause for serious concern, particularly in hospitals and other healthcare institutions where they tend to occur most commonly. As noted above, susceptible bacteria can acquire resistance to an antimicrobial agent via new mutations. Such spontaneous mutations may cause resistance by-

- altering the target protein to which the antibacterial agent binds by modifying or eliminating the binding site (e.g., change in penicillin-binding protein 2b in pneumococci, which results in penicillin resistance),
- upregulating the production of enzymes that inactivate the antimicrobial agent (e.g., erythromycin ribosomal methylase in staphylococci),
- down regulating or altering an outer membrane protein channel that the drug requires for cell entry (e.g., OmpF in *E. coli*), or upregulating pumps that expel the drug from the cell [efflux of fluoroquinolones in *S. aureus* (McManus, 1997)]. In all of these cases, strains of bacteria carrying resistance-conferring mutations are selected by antimicrobial use, which kills the susceptible strains but allows the newly resistant strains to survive and grow.

Acquired resistance that develops due to chromosomal mutation and selection is termed vertical evolution. Vertical evolution is strictly a matter of Darwinian evolution driven by principles of natural selection: a spontaneous mutation in the bacterial chromosome imparts resistance to a member of the bacterial population. Bacteria also develop resistance through the acquisition of new genetic material from other resistant organisms. This is termed horizontal evolution, and may occur between strains of the same species or between different bacterial species or genera.

Bacteria are able to exchange genes in nature by three processes: conjugation, transduction and transformation (McManus, 1997). For each of these processes, transposons may facilitate the transfer and incorporation of the acquired resistance genes into the host's genome or into plasmids.

Conjugation: Conjugation is the most important and the most common mechanism of transmission of resistance in bacteria. During conjugation, a gram-negative bacterium transfers plasmid-containing resistance genes to an adjacent bacterium, often via an elongated proteinaceous structure termed as pilus, which joins the two organisms. Conjugation among gram-positive bacteria is usually initiated by production of sex pheromones by the mating pair, which facilitate the clumping of donor and recipient organisms, allowing the exchange of DNA.

Transformation: Transformation is another form of transmission of bacterial resistance genes and takes place when there is direct passage of free DNA (also known as "naked DNA") from one cell to another. The "naked DNA" usually originates from other bacteria that have died and broken apart close to the receiving bacteria. The receiving bacteria then simply introduce the free DNA into their cytoplasm and incorporate it into their own DNA.

Transduction: Transduction is a third mechanism of genetic transfer and occurs via the use of a "vector", most often viruses capable of infecting bacteria also known as "bacteriophages" (or simply "phages"). The virus containing the bacterial gene that codifies antibiotic resistance (the "resistant DNA") infects the new bacterial cell and introduces this genetic material into the receiving bacteria. Most times, the infecting bacteriophage also introduces to the receiving bacteria its own viral DNA, which then takes over the bacterial replication system forcing the cell to produce more copies of the infecting virus until the bacterial cell dies and liberates these new bacteriophages, which then go on to infect other cells.

Mutation and selection, together with the mechanisms of genetic exchange, enable many bacterial species to adapt quickly to the introduction of antibacterial agents into their environment. Genetic recombination can follow the transfer of DNA from one cell to another leading to the emergence of a new genotype (recombinant). It is common for DNA to be transferred as plasmids between mating bacteria. Since bacteria usually develop their genes for drug resistance on plasmids (called **resistance transfer factors or RTFs**), they are able to spread drug resistance to other strains and species during genetic exchange processes.

2.1 Biological Mechanisms of Resistance

Which ever way a gene is transferred to a bacterium, the development of antibiotic resistance occurs when the gene is able to express itself and produce a tangible biological effect resulting in the loss of activity of the antibiotic. These biological mechanisms are many and varied but they can be summarized as follows.

2.1.1 Antibiotic Destruction or Antibiotic Transformation

This destruction or transformation occurs when the bacteria produces one or more enzymes that chemically degrade or modify the antimicrobial making them inactive against the bacteria. This is a common mechanism of resistance and probably one of the oldest ones affecting several antibiotics but especially β-lactam antibiotics via the bacterial production of β-lactamases (Jacoby and Munoz-Price, 2005).

2.1.2 Antibiotic Active Efflux

Efflux was first described for tetracycline and macrolide antibiotics (Roberts, 1996; Leclercq, 2002) but is now common for many other antibiotics such as fluoroquinolones (Sefton, 2002; Hooper, 2005). Antibiotic active efflux is relevant for antibiotics that act inside the bacteria and takes place when the microorganism is capable of developing an active transport mechanism that pumps the antibiotic molecules that penetrated into the cell to the outside milieu until it reaches a concentration below that necessary for the antibiotic to have antibacterial activity. This means that the efflux transport mechanism must be stronger than the influx mechanism in order to be effective (Hooper, 2005).

2.2 Receptor Modification

Receptor modification occurs when the intracellular target or receptor of the antibiotic drug is altered by the bacteria, resulting in the lack of binding and consequently the lack of antibacterial effect. Examples of this mechanism include modifications in the structural conformation of penicillin-binding proteins (PBPs) observed in certain types of penicillin resistance, ribosomal alterations that can render aminoglycosides, macrolides or tetracyclines inactive, and DNA-gyrase modifications resulting in resistance to

TABLE 1. Mechanisms of resistance

Mechanism	**Class of antibiotic**
1. Reduced uptake into cell	Chloramphenicol, β-lactams
2. Active efflux from cell	Tetracycline, Quinolones Sulfonamides, Trimethoprim
3. Modification of target to eliminate or reduce binding of antibiotic	β-lactams, Erythromycin Rifampicin, minoglycosides
4. Detoxification of antibiotic by enzyme modification: • Hydrolysis • derivatization	β-lactams, Erythromycin Aminoglycosides, Chloramphenicol Fosfomycin, Lincomycin
5. Sequestration of antibiotic by protein binding	β-lactams, Fusidic acid
6. Metabolic bypass of inhibited reaction	Sulfonamides, Trimethoprim
7. Binding of specific immunity protein to antibiotic	Bleomycin
8. Overproduction of antibiotic target (titration)	Sulfonamides, Trimethoprim
9. Failure to active prodrug	Isoniazid

fluoroquinolones (Levy and Marshall, 2004; Sefton, 2002). It is likely that more and newer biological mechanisms of resistance will develop in the future.

3. MECHANISMS OF ACTION OF ANTIBACTERIAL AGENTS

Most antimicrobial agents used for the treatment of bacterial infections may be categorized according to their principal mechanism of action. There are 4 major modes of action (Table 2): (1) interference with cell wall synthesis, (2) inhibition of protein synthesis, (3) interference with nucleic acid synthesis, and (4) inhibition of a metabolic pathway (Neu, 1992).

Antibacterial drugs that work by inhibiting bacterial cell wall synthesis include the β-lactams, such as the penicillins, cephalosporins, carbapenems, and monobactams, and the glycopeptides, including vancomycin and teicoplanin (Neu, 1992; McManus, 1997). β-Lactam agents inhibit synthesis of the bacterial cell wall by interfering with the enzymes required for the synthesis of the peptidoglycan layer (McManus, 1997). Vancomycin and teicoplanin also interfere with cell wall synthesis, but do so by binding to the terminal D-alanine residues of the nascent peptidoglycan chain, thereby preventing the cross-linking steps required for stable cell wall synthesis (McManus, 1997). Macrolides, aminoglycosides, tetracyclines, chloramphenicol, streptogramins, and oxazolidinones produce their antibacterial effects by inhibiting protein synthesis (McManus, 1997). Bacterial ribosomes differ in structure from their counterparts in eukaryotic cells. Antibacterial agents take advantage of these differences to selectively inhibit bacterial growth. Macrolides, aminoglycosides, and tetracyclines bind to the 30S subunit of the ribosome, whereas chloramphenicol binds to the 50S subunit. Fluoroquinolones exert their antibacterial effects by disrupting DNA synthesis and causing lethal double-strand DNA breaks during DNA replication, (Drlica and Zhao, 1997) whereas sulfonamides and trimethoprim (TMP) block the pathway for folic acid synthesis, which ultimately inhibits DNA synthesis (Yao and Moellering, 2003; Petri, 2006). The common antibacterial drug combination of TMP, a folic acid analogue, plus sulfamethoxazole (SMX) (a sulfonamide) inhibits 2 steps in the enzymatic pathway for bacterial folate synthesis. Disruption of bacterial membrane structure may be a fifth, although less well characterized, mechanism of action. It is postulated that polymyxins exert their inhibitory effects by increasing bacterial membrane permeability, causing leakage of bacterial contents (Storm, 1977). The cyclic lipopeptide daptomycin apparently inserts its lipid tail into the bacterial cell membrane, (Carpenter and Chambers, 2004) causing membrane depolarization and eventual death of the bacterium.

TABLE 2. Mechanisms of action of antibacterial agents

Mechanisms	Antibiotic class
Interference with cell wall synthesis	β-Lactams: penicillins, cephalosporins, carbapenems, monobactams Glycopeptides: vancomycin, teicoplanin
Protein synthesis inhibition Bind to 50S ribosomal subunit Bind to 30S ribosomal subunit Bind to bacterial isoleucyl-tRNA synthetase	macrolides, chloramphenicol, clindamycin, quinupristin-dalfopristin, linezolid aminoglycosides, tetracyclines mupirocin
Interference with nucleic acid synthesis i. Inhibit DNA synthesis ii. Inhibit RNA synthesis	 fluoroquinolones rifampin
Inhibition of metabolic pathway	sulfonamides, folic acid analogues
Disruption of bacterial membrane structure	polymyxins, daptomycin

4. ANTIMICROBIAL RESISTANCE IN GRAM-POSITIVE BACTERIA

Gram-positive bacteria—particularly gram-positive cocci like coagulase-negative staphylococci, *Staphylococcus aureus*, and *Enterococcus* spp—are extremely important pathogens in the hospital environment. Data from the Surveillance and Control of Pathogens of Epidemiologic Importance (SCOPE) project, which monitors significant bloodstream infections in hospitalized patients in the United States, indicated that 60% of nosocomial bloodstream infections for the 3-year period from April 1995 through April 1998 involved gram-positive bacteria (Edmond *et al.*, 1999). Coagulase-negative staphylococci were the causes of 31.9% of monomicrobial nosocomial bloodstream infections, followed by *S. aureus* in 15.7%, *enterococci* species in 11.1%, and viridans streptococci in 1%. Updated figures from the SCOPE project covering March, 1995 through September, 2002 revealed similar findings (Wisplinghoff *et al.*, 2004). Increasing resistance in gram-positive bacteria can be expected to complicate treatment and potentially lead to increased morbidity and mortality.

4.1 Methicillin-resistant *S. aureus*

The first of these problematic bacteria was *S. aureus*. Virtually all *S. aureus* strains were susceptible to penicillin G when the latter was initially introduced in the early 1940s, but by 1944 the first reports of penicillin-resistant *S. aureus* had already appeared, and today virtually all strains of *S. aureus* are resistant to natural penicillins, aminopenicillins, and antipseudomonal penicillins (Chambers, 2001; Neu, 1992). Resistance to these drugs occurs because of the acquisition of genes that encode drug-inactivating enzymes, initially known as penicillinases and now called β-lactamases. At first, cases of penicillin resistant *S. aureus* were limited and appeared only in healthcare settings, but over time resistant species were observed and became increasingly prevalent in the wider community as well (Chambers, 2001). Methicillin and other penicillinase-resistant penicillins were developed to treat infections caused by penicillin-resistant *S. aureus* and met with initial success. MRSA is currently recognized as a major problem in hospitals and the broader community in the United States and throughout the world (Diekema *et al.*, 2001). The most recent SCOPE project report showed that methicillin resistance was present in 44% of *S. aureus* bloodstream isolates from ICU infections (Wisplinghoff *et al.*, 2004).

Methicillin was introduced in 1959 to treat infections caused by penicillin-resistant *Staphylococcus aureus*. In 1961, there were reports from the United Kingdom of *S. aureus* isolates that had acquired resistance to methicillin (methicillin-resistant *S. aureus*, MRSA) (Jevons, 1961), and MRSA isolates were soon recovered from other European countries, and later from Japan, Australia (Palavecino, 2004). The acquisition of methicillin resistance gave these bacteria the name methicillin-resistant *S. aureus* (MRSA), but in reality these were multidrug-resistant *S. aureus* (Wadsworth *et al.*, 1992). MRSA is now a problem in hospitals worldwide and is increasingly recovered from nursing homes and the community (Hussain *et al.*, 2000). Periodic outbreaks of MRSA were observed in various countries throughout the 1970s and were typically associated with high methicillin use in intensive care units (Palavecino, 2004), but it was not until the 1980s that MRSA began to become a really significant problem in United States hospitals—first in hospitals with a large number of beds and then in community hospitals (Panlilio *et al.*, 1992).

The methicillin resistance gene (mecA) encodes a methicillin-resistant penicillin-binding protein that is not present in susceptible strains and is believed to have been acquired from a distantly related species (Hiramatsu *et al.*, 2001). Expression of the mecA gene encoding low-affinity penicillin-binding protein PBP2a confers resistance to other β-lactams in addition to methicillin (Utsui and Yokota, 1985), but the resistance pattern of MRSA typically includes other classes of antibiotics as well.

Nosocomial MRSA is remarkable for its clonal pattern of spread. A recent study looking at 359 MRSA isolates collected from 20 countries from 1961 to 1999 identified 11 major MRSA clones within 5 groups of related genotypes (Enright *et al.*, 2002). Results from a recent study in the United Kingdom examining a new epidemic strain of MRSA, designated EMRSA-17, illustrate multidrug resistance (Aucken *et al.*, 2002). EMRSA-17 expressed resistance to methicillin, fluoroquinolones (ciprofloxacin), macrolides (erythromycin), aminoglycosides (gentamicin, kanamycin, neomycin, and streptomycin), tetracycline, rifampin, and fusidic acid. Virtually all variants of this strain were multidrug resistant. In particular, fluoroquinolone resistance is a hallmark of nosocomial MRSA, although this was not always the case. When ciprofloxacin was first introduced, it was recommended as the first orally administered treatment effective against MRSA. However, within one year many hospitals observed dramatic increases in the rate of ciprofloxacin resistance in MRSA. High-level ciprofloxacin resistance was observed within 3 months of ciprofloxacin introduction, and within one year, 79% of all MRSA from hospitalized patients exhibited resistance (Blumberg *et al.*, 1991). For many years, vancomycin was the only effective treatment for serious MRSA infections. But in the past 4 years, 4 new agents with anti-MRSA activity have been introduced (quinupristin-dalfopristin, linezolid, daptomycin, and tigecycline). These new agents are most welcome, since the past decade has seen the emergence of vancomycin resistance in *S. aureus. S. aureus* with intermediate resistance to vancomycin (vancomycin-intermediate *S. aureus* [VISA]; minimum inhibitory concentration [MIC], 8 to 16 mg/L) was first observed in 1996 in a strain isolated from a hospitalized patient in Japan (Hiramatsu *et al.*, 1997). The mechanism of resistance in VISA has been linked to cell wall thickening, which may cause vancomycin molecules to become trapped in the outer layers of the cell wall, thereby limiting access to the cytoplasmic membrane where the functional targets of vancomycin are located (Cui *et al.*, 2003). The mechanism for this high-level vancomycin resistance involves the horizontal transfer of a transposon containing vanA and associated genes from VRE (Rice, 2006).

Many studies have characterized MRSA isolates from individual hospitals or countries and have identified strains that appear to be well adapted to the hospital environment, are established in several hospitals within a country, or have spread internationally (epidemic MRSA, EMRSA). MRSA isolates are generally characterized by pulsed-field gel electrophoresis, a powerful technique for identifying the relatedness of isolates from recent outbreaks within a hospital, but are not well suited to long-term global epidemiology, which requires a procedure that is highly discriminatory but that indexes variation that accumulates slowly. Multilocus sequence typing (MLST) provides such a procedure and characterizes isolates of bacteria unambiguously by using the sequences of internal fragments of seven housekeeping genes (Enright and Spratt, 1999). MLST has been developed and validated for *S. aureus* (Enright *et al.*, 2000), and provides a discriminatory method that allows related strains recovered in different countries to be readily identified. The origins of the major MRSA clones are still poorly understood. Kreiswirth *et al.* (1993) proposed that all MRSAs were descended from a single ancestral *S. aureus* strain that acquired mecA, but more recent studies (Musser and Kapur, 1992) show that some MRSAs are very divergent, implying that mecA has been transferred between *S. aureus* lineages.

4.2 Multidrug-Resistant (MRD) Enterococcal Species

Enterococcus faecalis and E. faecium are species normally found in human gut flora and, apart from their rare occurrence as the cause of urinary tract infections, they were not usually considered as pathogenic. However, when patients were given immunosuppressive therapy, enterococci were increasingly found as the causative organism of infection. Like MRSA, the final defense against these gram-positive bacteria was considered to be vancomycin, but, by the end of the 1980s, they had acquired the ability to resist high levels of vancomycin (Woodford *et al.*, 1993). Despite the fact that vancomycin has been in clinical use since the late 1950s, VRE were not observed until the mid-1980s, and in the United States, VRE were virtually non-existent as recently as 1989 (Martone, 1998). During the 1990s, however, a dramatic rise in VRE occurred—first in Intensive Care Units (ICUs), then essentially throughout hospitals. The latest NNIS report indicated that nearly 30% of all enterococci isolated from patients infected in ICUs are now resistant to vancomycin. Perhaps an even more remarkable aspect of this outbreak is the fact that an overwhelming majority of VRE are *Enterococcus faecium*. The Surveillance Network Database–USA showed that resistance to both vancomycin and ampicillin was much more prevalent among *E. faecium* than among *Enterococcus faecalis* in patients with nosocomial bloodstream infections in 1995 and 1997 (Sahm *et al.*, 1999).

The mechanism of vancomycin resistance is particularly sophisticated and, although there are different variations of it, they all involve the reversal of the bacterium's normal manufacture of the cell wall. This is accompanied by the synthesis of a modified cell wall, which includes D-alanyl-D-lactate, or D-alanyl-D-serine terminal dipeptides, which are less capable of binding vancomycin than the normal bacterial dipeptide D-alanyl-D-alanine (Arthur *et al.*, 1996). Vancomycin resistance is mediated by either of 2 classes of related gene clusters: one class contains vanA and other class contains vanB (Arthur *et al.*, 1996). Both produce resistance by altering the target for vancomycin from D-alanine-D-alanine to D-alanine-D-lactate.

5. ANTIMICROBIAL RESISTANCE IN GRAM-NEGATIVE BACTERIA

Pseudomonas aeruginosa, Acinetobacter baumannii and *Klebsiella pneumoniae* are among the bacteria that readily develop multiple resistance mechanisms to various classes of antibiotics (Hanlon, 2005; Helfand and Bonomo, 2005; McGowan Jr, 2006; Ramphal and Ambrose, 2006; Rossolini and Mantengoli, 2005). In addition, they are important nosocomial pathogens affecting both immunocompetent and immunocompromised patients and are responsible for a considerable proportion of infections in patients in ICUs worldwide. Thus, infections by multidrug-resistant (MDR) *P. aeruginosa, A. baumannii* and *K. pneumoniae* strains have become common in healthcare institutions. The continuously evolving resistance to antibiotics of *P. aeruginosa, A. baumannii* and *K. pneumoniae* has led to the emergence of clinical isolates susceptible to only one class of antimicrobial agents and eventually to pandrug-resistant (PDR) isolates, i.e. resistant to all available antibiotics (Kuo *et al.*, 2004; Marais *et al.*, 2004; Yoon *et al.*, 2004). Polymyxins, an old class of polypeptide cationic antibiotic that was abandoned during the 1980s and 1990s in most parts of the world, have been used as the last class of available antibiotics to which some of these bacterial isolates were susceptible *in vitro*. Several recent clinical studies have reported on the therapeutic use of polymyxins in patients with infections by organisms with such a phenotype (i.e. susceptible only to polymyxins) Garnacho-Montero *et al.*, 2003; Michalopoulos *et al.*, 2005; Falagas *et al.*, 2006; Levin *et al.*, 1999). In addition, there have been some recent clinical studies that reported on infections with isolates of *P. aeruginosa, A. baumannii* and *K. pneumoniae* that were characterized as PDR. However, in some of these studies either the isolates were not tested in vitro against polymyxins or they were tested and found to be susceptible to them (Kuo *et al.*, 2004; Marais *et al.*, 2004). PDR infections due to the aforementioned gram-negative bacteria represent a fearful clinical situation with tremendous public health implications in which the clinician is left with practically no rational choice of antibiotic treatment. Thus, data regarding the frequency of PDR clinical isolates, the morbidity and mortality related to infections by these isolates and the therapeutic options, if any, are of great clinical and public health importance.

6. STRATEGIES FOR REDUCING THE IMPACT OF RESISTANCE

Antibiotic resistance is driven by antibiotic use. When antibiotics are superseded and therefore used less, strains resistant to these tend to disappear.

6.1. In the Community

In the UK, more than 80% of human use of antibiotics occurs in the community, mostly for respiratory tract infections. The Standing Medical Advisory Committee, in its report The Path of Least Resistance, made recommendations to reduce inappropriate prescribing (Standing Medical Advisory Committee, Department of Health. The path of least resistance. London: HMSO, 1998).

- No antibiotics should be given for simple coughs and colds.

- Antibiotics should not be routinely prescribed for sore throats, unless there is evidence of streptococcal infection.
- Antibiotics are not routinely required for acute otitis media and sinusitis-like symptoms; if given, courses can be limited to 3 days. In addition, 3 days' treatment should suffice in otherwise healthy women with uncomplicated cystitis. This strategy has been useful in reducing prescribing by GPs, but anxiety has resulted from anecdotal reports of an increase in bacterial respiratory infections. In hospital, antibiotic use can be reduced by several means.
- Routine use of antibiotics for surgical prophylaxis should be reduced to a minimum.
- Antibiotics should not be started immediately in all suspected infections. Certain patients (e.g. those with febrile neutropenia or evidence of septicaemia) require urgent antibiotic therapy, but in many other cases (e.g. mild pyrexia postoperatively), it is safe and ultimately preferable to withhold antibiotics until culture results are known or there is clear evidence of bacterial infection.
- An alternative is to discontinue antibiotics as soon as information is available suggesting that the problem is not bacterial or has resolved itself. This is termed an 'antibiotic-stop' policy, and the aim is to encourage doctors to actively review the need for antibiotics after, say, the second day, given information on cultures and surrogate markers that has by then become available.
- Certain antibiotics can be withheld from the hospital formulary. This is the main benefit of an agreed antibiotic policy. The choice of restricted antibiotics may be decided on the basis of cost rather than likely selection of resistance (these antibiotics must be used occasionally, however, when resistance to all other available drugs has been selected).
- In theory, antibiotics can be rotated such that, for example, predominantly penicillins are used at some times, and cephalosporins or quinolones at others. There is little clear scientific evidence that this has any effect, and major, complicated studies would be needed to determine the effect of change in use on both normal and infecting flora. However, it has been shown that, in a setting of heavy cephalosporin use, discontinuation of use of this class of drugs leads to a reduction in the risk of colonization with glycopeptide-resistant enterococci and 1difficile associated diarrhoea.

6.2 Other Strategies

Development of resistance in human pathogens might be delayed if antibiotics were not used so widely in animal husbandry, particularly for growth promotion.

7. THE FUTURE

Resistance to antibiotics is one of the greatest threats to the success of modern medicine. It has recently become more serious because we can no longer be sure that any antibiotic chosen empirically will work, and because of the emergence of totally resistant bacteria. How to reduce resistance without simply discontinuing use of all antibiotics is a dilemma. We do not know to what degree antibiotic use must be reduced to decrease the selective pressure and allow reversion to sensitivity, nor do we know how to protect the few remaining drugs that can be used to treat resistant infections. Doctors and vets should overcome the view that they have an inalienable right to prescribe empirical antibiotics, and should set targets for reduction of their personal prescribing.

8. NOVEL AGENTS FOR RESISTANT BACTERIAL INFECTIONS

Emerging bacterial resistance to many available antimicrobial agents has led to a recent surge of pharmaceutical development, which has resulted in production of several new antibiotics. Novel agents, almost all derivatives of old compounds, have recently been introduced into clinical practice.

8.1 Linezolid

The oxazolidinones are a unique family of antimicrobial agents, first developed in the late 1970s for agricultural use (Brickner *et al.*, 1996). The agents are effective against a wide range of gram-positive bacteria, anaerobes and *Mycobacterium tuberculosis*. Linezolid was the first derivative with acceptable tolerability in man to advance to clinical trials, and is the first Federal Drug Administration approved oxazolidinone for the treatment of pneumonia and skin and soft tissue infections caused by susceptible organisms, and infections caused by VRE. It has not yet been approved for the treatment of penicillin-resistant *Streptococcus pneumoniae* (Cammarata *et al.*, 2000).

Linezolid possesses activity against staphylococci, including *Staphylococcus aureus*, irrespective of its oxacillin susceptibility (Cercenado, 2001; Gemmell, 2001), and against the recently isolated glycopeptide-intermediately-resistant *S. aureus* (GISA) (Rybak, 2000). Linezolid is active against glycopeptide resistant coagulase-negative staphylococci (Cercenado, 2001). In addition, activity against other major multidrug-resistant Gram positive pathogens, including VRE and penicillin resistant *Streptococcus pneumoniae* has also been demonstrated, independent of their resistance profile to other antibacterials (Jones et al., 1999; Noskin *et al.*, 1999).

8.2 Daptomycin

Daptomycin is the first in a new class of cyclic lipopeptide drugs derived from the fermentation of *Streptomyces roseosporus*. It was first discovered in the 1980s and was shown to be effective in the treatment of skin and soft tissue infection (SSTI). However, because of muscle toxicity, which appeared at higher doses as used in the treatment of endovascular infections and endocarditis, and because of clinical failures (Garrison *et al.*, 1989) clinical trials were suspended. The emergence of infections caused by resistant gram positive bacteria led in 1997 to renewed interest in daptomycin, with new data obtained in the laboratory and in animal and human studies.

Daptomycin is active solely against gram-positive bacteria. Daptomycin was also active against betahaemolytic streptococci, viridans streptococci, as well as against gram-positive rods such as *Leuconostoc*, which are characteristically resistant to glycopeptides (Jones and Barry 1987; Barry, 2001; King and Phillips 2001). The effect of calcium in the medium was especially evident when enterococcal susceptibility was evaluated. Daptomycin was active against most strains, including VRE. However, added calcium increased the activity of daptomycin two-to four fold (Barry, 2001; King and Phillips 2001). Unlike most other antibiotics, which are only bacteriostatic against enterococci, daptomycin exhibits rapid, concentration dependent bactericidal activity against these pathogens, as well as against other bacteria (Snydman, 2001).

8.3 GAR-936

Tetracyclines have been important broad-spectrum antimicrobial agents for the last 40 years. In order to overcome resistance to these drugs, modification of their structures was undertaken. Glycylcyclines are obtained by modification of the 9-position of chlortertracycline, minocycline, or doxycycline. GAR-936 is a novel derivative of minocycline, active against tetracycline-resistant organisms (Sum and Petersen, 1999)

GAR-936 is generally less active than its parent compound against minocycline sensitive, oxacillin susceptible and resistant staphylococci, but it does show in vitro activity against minocycline-resistant strains of *Staphylococcus aureus* and CONS. However, mean MIC against these isolates is higher than the MIC against the susceptible strains (Boucher, 2000; Petersen, 1999).

The activity against aerobic gram-negative rods is fair. Many enterobacteriaceae, including *Escherichia coli* were inhibited by GAR-936 (Proteus spp. excluded). However, other strains were resistant, with the potency of other antibiotics, including ciprofloxacin, ceftazidime and imipenem, being significantly better (Petersen, 1999; van Ogtrop, 2000). In addition, GAR-936 retains its activity against some nonfermentative bacteria such as *Stenotrophomonas maltophilia,* although it is not active against *P. aeruginosa* (van Ogtrop, 2000; Gales and Jones, 2000).

Most strains of *Neisseria gonorrhoeae* are susceptible to GAR-936, including penicillin- or tetracycline resistant strains (Whittington, 1995). In addition, GAR-936 is highly active against other sexually transmitted diseases pathogens, including *Chlamydia trachomatis* (Roblin and Hammerschlag, 2000), as well as against *Mycoplasma* spp. and *Ureaplasma urealyticum* (Kenny GE, Cartwright, 2001).

8.4 LY-333328 (Oritavancin)

Vancomycin was introduced in 1956 because of its effectiveness against penicillin-resistant staphylococci. Approximately 30-years later, acquired glycopeptide resistance in *Enterococcus* spp. was reported followed by clinical isolates of GISA in 1997 (Fekety, 2000). However, to overcome resistance, vancomycin derivatives that has hydrophobic substituents on the vancosamine nitrogen have been developed. LY-264826 is a naturally occurring glycopeptide with the same core structure as vancomycin but with differences in associated sugars present. Alkyl modification of the disaccharide amino led to several compounds, the most active of which is LY-333328 (Nicas *et al.,* 1996).

LY-333328 proved effective against the major multi-drug-resistant gram-positive bacteria (Zeckel *et al.,* 2000). LY-333328 was equally effective against oxacillin susceptible and methicillin-resistant *Staphylococcus aureus* (MRSA). However, its activity for most strains was either identical or less efficient than other glycopeptides (vancomycin and teicoplanin) (Zeckel *et al.,* 2000; Biavasco, 1997). LY-333328 possesses good in vitro activity against *Enterococcal* spp., including glycopeptide-resistant strains of the three major phenotypes (VanA, VanB, and VanC), with MICs significantly lower than the other glycopeptides (Zeckel *et al.,* 2000; Garcia-Garrote *et al.,* 1998; Biavasco *et al.,* 1997; Fraise *et al.,* 1997).

9. CONCLUSION

Development of new antibiotics has not kept pace with the emergence of antibiotic resistance. In the past few decades, pharmaceutical companies have shifted their developmental efforts to drugs for chronic diseases such as heart diseases and high blood pressure and anti-viral instead of antibiotics. The steady increase in resistant organisms is related to the widespread use of antibiotics in community and hospital settings. New therapeutic options are needed, including treatments for infections caused by antibiotic-resistant gram-positive organisms. Moreover, resurgence of TB in an epidemic form due to HIV infection especially in both developed and underdeveloped countries together with the development of multi-drug resistant strains among the patients have assumed a serious situation in health care sector. Now, the only answer is to search for novel and effective drug regime to fight this menace.

Today, intensive programmes seeking for antibiotics are running worldwide. In our laboratory (NEIST, formerly RRL (CSIR), Jorhat, India), mainly, by screening soils and water, and testing the activity of the isolated organisms against specific pathogenic bacteria or fungi that are resistant to several known antibiotics. In our laboratory, a bacterial strain, namely, *Streptomyces* sp. 201 was isolated from the tea garden soil of Jorhat, Assam, India, (26.44 N and 94.10 E) exhibited promising anti-fungal and anti-bacterial activity against a wide range of pathogens including *M. tuberculosis* (Bordoloi *et al.*, 2001; Bordoloi *et al.*, 2002). Absolute stereochemistry of the new antifungal and antibacterial antibiotic produced by *Streptomyces* sp.201 has been established by achieving the total synthesis of the product (Boruwa *et al.*, 2004).

The vast resources of microorganisms have received increasing attention from chemists and pharmacologists and played an important role in the explosive growth of biomedical science during the past two decades. The recent advent of sophisticated chromatographic (HPLC, gel filtration etc.), high field NMR and mass spectroscopic techniques coupled with force field calculations and X-ray crystallography have markedly enhanced the isolation and structure elucidation of more complex and diverse natural products and along with advanced bioassay methods has opened new possibilities in discovery of drugs and agrochemicals from natural sources.

Acknowledgements

This study was supported by DST (Govt. of India), Young Scientist Scheme, grant no. 96282.

REFERENCES

Amsden G.W. (2004). Pneumococcal resistance in perspective: how well are we combating it? *Pediatr Infect Dis.* 23(Suppl): S125–S128.

Arthur, M., Reynolds, P.E. and Depardieu, F. (1996). Mechanisms of glycopeptide resistance in enterococci. *Jour. Infection* 32:11-16.

Aucken, H.M., Ganner, M., Murchan, S., Cookson. B.D. and Johnson, A.P. (2002). A new UK strain of epidemic methicillin-resistant *Staphylococcus aureus* (EMRSA-17) resistant to multiple antibiotics. *J Antimicrob. Chemother.* 50:171–175.

Barry, A.L., Fuchs, P.C. and Brown, S.D. (2001). *In vitro* activities of daptomycin against 2,789 clinical isolates from 11 North American medical centers. *Antimicrob. Agents Chemother.* 45(6): 1919-1922.

Biavasco, F., Vignaroli, C., Lupidi, R., Manso, E., Facinelli, B. and Varaldo, P.E. (1997). In vitro antibacterial activity of LY333328, a new semisynthetic glycopeptide. *Antimicrob. Agents Chemother.* 41(10): 2165-2172.

Blumberg, H.M., Rimland, D., Carroll, D.J., Terry, P. and Wachsmuth, I.K. (1991). Rapid development of ciprofloxacin resistance in methicillin-susceptible and -resistant *Staphylococcus aureus. J Infect Dis.* 163:1279–1285.

Bordoloi, G.N., Kumari, B., Guha, A., Bordoloi, M.J., Roy, M. K. and Bora, T.C. (2001). Isolation, and Structure Elucidation of a New Antifungal and Antibacterial antibiotic produced by *Streptomyces* sp. 201. *Biosci. Biotech. Biochem.* 65(8):1856-1858.

Bordoloi, G.N., Kumari, B., Guha, A., Thakur, D., Bordoloi, M., Roy, M.K. and Bora, T.C., (2002). Potential of a novel antibiotic, as a biocontrol agent against fusarial wilt of crucifers. *Pest Manag Sci.* 58:1-6.

Boruwa, J., Kalita, B., Barua, N.C., Borah, J.C., Majumder, S., Thakur, D., Gogoi, D.K. and Bora, T.C. (2004). Synthesis, absolute stereochemistry and molecular design of the new antifungal and antibacterial antibiotic produced by *Streptomyces* sp. 201. *Bioorganic Medicinal Chemistry Letters.* 14:3571-3574.

Boucher, H.W., Wennersten, C.B. and Eliopoulos, G.M. (2000). *In vitro* activities of the glycylcycline GAR-936 against grampositive bacteria. *Antimicrob. Agents Chemother.* 44(8):2225-2229.

Brickner, S.J., Hutchinson, D.K., Barbachyn, M.R., et al. (1996). Synthesis and antibacterial activity of U-100592 and U-100766, two oxazolidinone antibacterial agents for the potential treatment of multidrug-resistant gram-positive bacterial infections. *J. Med. Chem.* 39(3):673-679.

Cammarata, S.K., Timm, J.A., Hempsall, K.A., Todd, W.M., Oliphant, T.H. and Hafkin, B. (2000). Efficacy of linezolid in pneumonia due to penicillin intermediate and resistant *Streptococcus pneumoniae.* [abstract SO.OlO]. In: Programs and abstracts of the 9th International Congress on Infectious Diseases; Buenos Aires, Argentina, April 10-13 Boston: International Society of Infectious Diseases.

Cercenado, E., Garcia-Garrote, F. and Bouza, E. (2001). *In vitro* activity of linezolid against multiply resistant Gram-positive clinical isolates. *J. Antimicrob. Chemother.* 47(1): 77-81.

Chambers, H.F. (2001). The changing epidemiology of *Staphylococcus aureus*? Emerg Infect Dis. 7: 178 -182.

Cizman, M., Pokorn, M., Seme, K., Orazem, A. and Paragi, M. (2001). The relationship between trends in macrolide use and resistance to macrolides of common respiratory pathogens. *J. Antimicrob. Chemother.* 47: 475–477.

Cui, L., Ma, X. and Sato, K. (2003). Cell wall thickening is a common feature of vancomycin resistance in *Staphylococcus aureus*. J Clin Microbiol. 41:5–14.

Deresinski, S. (2005). Methicillin-resistant *Staphylococcus aureus*: an evolutionary, epidemiologic, and therapeutic odyssey. *Clin. Infect. Dis.* 40: 562–573.

Diekema, D.J., Pfaller, M.A., Schmitz, F.J., et al. (2001). Survey of infections due to Staphylococcus species: frequency of occurrence and antimicrobial susceptibility of isolates collected in the United States, Canada, Latin America, Europe, and the Western Pacific region for the SENTRY Antimicrobial Surveillance Program, 1997–1999. *Clin Infect Dis.* 32(suppl 2): S114 -S132.

Drlica, K. and Zhao X. (1997). DNA gyrase, topoisomerase IV, and the 4-quinolones. *Microbiol Mol Biol Rev.* 61: 377–392.

Edmond, M.B., Wallace, S.E., McClish, D.K., Pfaller, M.A., Jones, R.N. and Wenzel, R.P. (1999). Nosocomial bloodstream infections in United States hospitals: a 3-year analysis. *Clin. Infect. Dis.* 29:239 -244.

Enright, M.C. and Spratt, B.G. (1999). Multilocus sequence typing. *Trends Microbiol.* 7: 482-487.

Enright, M.C., Day, N.P., Davies, C.E., Peacock, S.J and Spratt, B.G. (2000). Multilocus sequence typing for characterization of methicillin-resistant and methicillin-susceptible clones of *Staphylococcus aureus*. *J. Clin. Microbiol.* 38: 1008-1015.

Enright, M.C., Robinson, D.A., Randle, G., Feil, E.J., Grundmann, H. and Spratt, B.G. (2002). The evolutionary history of methicillin-resistant *Staphylococcus aureus* (MRSA): *Proc Natl Acad Sci.* U S A. 99:7687–7692.

Espinal, M.A., Laszlo, A., Simonsen, L., Boulahbal, F., Kim, S.J., Reniero, A., Hoffner, S., Rieder, H., Binkin, N., Dye, C., Williams, R. and Raviglione, M. (2001). Global trends in resistance to antituberculosis drugs. *N. Engl. J. Med.* 344:1294–1303.

Falagas, M.E., Kasiakou, S.K., Kofteridis, D.P., Roditakis, G. and Samonis, G. (2006). Effectiveness and nephrotoxicity of intravenous colistin for the treatment of patients with infections due to polymyxin-only-susceptible (POS) Gram-negative bacteria. *Eur. J. Clin. Microbiol. Infect. Dis.* 25: 596–599.

Fekety, R. (2000). Vancomycin teicoplanin, and the streptogramins: quinupristin and dalfopristin. In: Mandell CL, Bennett JE, Dolin R, eds. Mandell, Douglas, and Bennett's Principles and practice of infectious diseases, 5th ed. pp. 382-392. London: Churchill Livingstone.

Fey, P.D., Safranek, T.J., Rupp, M.E., Dunne, E.F., Ribot, E., Iwen, P.C., Bradford, P.A., Angulo, F.J. and Hinrichs, S.H. (2000). Ceftriaxone-resistant Salmonella infection acquired by a child from cattle. *N. Engl. J. Med.* 342:1242–1249.

File, T.M. (2004). *Streptococcus pneumoniae* and community-acquired pneumonia: a cause for concern. *Am. J. Med.* 117(Suppl 3A): 39S–50S.

Foster, T.J. (2004). The *Staphylococcus aureus* "superbug". *J. Clin. Invest.* 114:1693–1696.

Fraise, A.P., Andrews, J. and Wise, R. (1997). Activity of a new glycopeptide antibiotic (LY333328) against enterococci and other resistant gram-positive organisms. *J. Antimicrob. Chemother.* 40(3): 423-425.

Gales, A.C. and Jones, R.N. (2000). Antimicrobial activity and spectrum of the new glycylcycline, GAR-936 tested against 1,203 recent clinical bacterial isolates. Diagn. *Microbial Infect. Dis.* 36(1): 19-36.

Garcia-Garrote, F., Cercenado, E., Alcala, L. and Bouza, E. (1998). *In vitro* activity of the new glycopeptide LY333328 against multiply resistant gram-positive clinical isolates. *Antimicrob. Agents Chemother.* 42(9): 2452-2455.

Garnacho-Montero, J., Ortiz-Leyba, C., Jimenez-Jimenez, F.J., et al. (2003). Treatment of multidrug-resistant *Acinetobacter baumannii* ventilatorassociated pneumonia (VAP) with intravenous colistin: a comparison with imipenem-susceptible VAP. *Clin. Infect. Dis.* 36:1111–1118.

Garrison, M.W., Rotschafer, J.C. and Crossley, K.B. (1989). Suboptimal effect of daptomycin in the treatment of bacteremias. *South Med. J.* 82(11): 1414-1415.

Gemmell, C.G. (2001). Susceptibility of a variety of clinical isolates to linezolid: a European inter-country comparison. *J. Antimicrob. Chemother.* 48(1): 47-52.

Gupta, M. (2003). WHO rings TB alarm, Telegraph, India, 11 August, pp. 5

Hanlon, G.W. (2005). The emergence of multidrug resistant *Acinetobacter* species: a major concern in the hospital setting. *Lett. Appl. Microbiol.* 41:375–378.

Helfand, M.S. and Bonomo, R.A. (2005). Current challenges in antimicrobial chemotherapy: the impact of extended-spectrum beta-lactamases and metallo-beta-lactamases on the treatment of resistant Gram-negative pathogens. Curr. Opin. *Pharmacol.* 5:452–458.

Hiramatsu, K., Cui, L., Kuroda, M., and Ito, T. (2001). The emergence and evolution of methicillin-resistant *Staphylococcus aureus*. *Trends Microbiol* 9:486-493.

Hiramatsu, K., Hanaki, H., Ino, T., Yabuta, K., Oguri, T. and Tenover, F.C. (1997). Methicillin-resistant *Staphylococcus aureus* clinical strain with reduced vancomycin susceptibility. *J Antimicrob Chemother.* 40:135–136.

Hofmann, J., Cetron, M.S., Farley, M.M., Baughman, W.S., Facklam, R.R., Elliott, J.A., Deaver, K.A. and Breiman, R.F. (1995). The prevalence of drug-resistant *Streptococcus pneumoniae* in Atlanta. *N. Engl. J. Med.* 333:481–486.

Hooper, D.C. (2005). Efflux pumps and nosocomial antibiotic resistance: a primer for hospital epidemiologists. *Clin. Infect. Dis.* 40:1811-1817.

Huovinen, P. (2002). Macrolide-resistant Group A Streptococcus – now in the United States. *N. Engl. J. Med.* 346:1243-1244.

Hussain, F.M., Boyle-Vavra, S., Bethel, C.D. and Daum, R.S. (2000). Current trends in community-acquired methicillin-resistant *Staphylococcus aureus*: at a tertiary cares pediatric facility. *Pediatr. Infect. Dis. J.* 19:1163-1166.

Jacobs, M.R. (2004). Streptococcus pneumoniae: epidemiology and patterns of resistance. *Am. J. Med.* 117 (Suppl 3A) :3S-15S.

Jacoby, G.A., Munoz-Price, L.S. (2005). The new β-lactamases. *N. Engl. J. Med.* 352:380-391.

Jaffe, H.W., Biddle, J.W., Johnson, S.R. and Wiesner, P.J. (1981). Infections due to penicillinase-producing *Neisseria gonorrhoeae* in the United States: 1976-1980. *J. Infect. Dis.* 144:191-197.

Jevons, M.P. (1961). "Celbenin"-resistant staphylococci. *Br. Med. J.* 1:124-125.

Jones, R.N. and Barry, A.L. (1987). Antimicrobial activity and spectrum of LY146032, a lipopeptide antibiotic, including susceptibility testing recommendations. *Antimicrob. Agents Chemother.* 31(4): 625-629.

Jones, R.N., Pfaller, M.A., Erwin, M.E. *et al.* (1999) Antimicrobial activity of linezolid tested against 3,808 strains of Grampositive organisms having resistance to various drugs [abstract]. In: Program and abstract of the 37th Annual Meeting of the Infectious Disease Society of America; November l&21, Philadelphia, PA.

Jorgensen, J.H. (1993).Update on mechanisms and prevalence of antimicrobial resistance in *Haemophilus influenzae*. *Clin. Infect. Dis.* 17: 1078-1079.

Kaplan, S.L., Hulten, K.G., Gonzalez, B.E., Hammerman, W.A., Lamberth, L., Versalovic, J. and Mason, E.O. Jr. (2005). Three-year surveillance of communityacquired *Staphylococcus aureus* infections in children. *Clin. Infect. Dis.* 40:1785-1791.

Kaplan, S.L., Hulten, K.G., Gonzalez, B.E., Hammerman, W.A., Lamberth, L., Versalovic, J. and Mason, E.O. Jr. (2005). Three-year surveillance of communityacquired *Staphylococcus aureus* infections in children. *Clin. Infect. Dis.* 40: 1785-1791.

Kenny, G.E. and Cartwright, F.D. (2001). Susceptibilities of *Mycoplasma hominis*, *M. pneumoniae*, and *Ureaplasma urealyticum* to GAR-936, dalfopristin, dirithromycin, evernimicin, gatifloxacin, linezolid, moxifloxacin, quinupristin-dalfopristin, and telithromycin compared to their susceptibilities to reference macrolides, tetracyclines, and quinolones. *Antimicrob. Agents Chemother.* 45(9): 2604-2608.

King, A. and Phillips, I. (2001). The in vitro activity of daptomycin against 514 Gram-positive aerobic clinical isolates. *J. Antimicrob. Chemother.* 48(2): 219-223.

Kreiswirth, B., Kornblum, J., Arbeit, R.D., Eisner, W., Maslow, J.N., McGeer, A., Low, D.E. and Novick, R.P. (1993). Evidence for a clonal origin of methicillin resistance in *Staphylococcus aureus*. *Science*. 259:227-230.

Kuo, L.C., Teng, L.J., Yu, C.J., Ho, S.W. and Hsueh P.R. (2004). Dissemination of a clone of unusual phenotype of pandrug-resistant Acinetobacter baumannii at a university hospital in Taiwan. *J. Clin. Microbiol.* 42:1759–63.

Leclercq, R. (2002). Mechanisms of resistance to macrolides and lincosamides:nature of the resistance elements and their clinical implications. *Clin. Infect. Dis.* 34: 482–492.

Levin, A.S., Barone, A.A., Penco, J., et al. (1999). Intravenous colistin as therapy for nosocomial infections caused by multidrug-resistant *Pseudomonas aeruginosa* and *Acinetobacter baumannii. Clin. Infect. Dis.* 28:1008–1011.

Levy, S.B. (2002). From tragedy the antibiotic era is born. In: Levy SB, ed. The Antibiotic Paradox: How the Misuse of Antibiotics Destroys Their Curative Powers, 2nd ed. Cambridge, MA: Perseus Publishing; pp. 1–14.

Levy, S.B. and Marshall, B. (2004). Antibacterial resistance worldwide: causes, challenges and responses. *Nat. Med.* 10: S122–S129.

Lind, I. (1990). Epidemiology of antibiotic resistant *Neisseria gonorrhoeae* in industrialized and developing countries. *Scand. J. Infect. Dis.* 69: 77–82.

Lowy FD. (2003). Antimicrobial resistance: the example of *Staphylococcus aureus. J. Clin. Invest.* 111:1265–1273.

Lowy, F.D. (1998). Medical progress: *Staphylococcus aureus* infections. *N. Engl. J. Med.* 339:520–532.

Marais, E., de Jong, G., Ferraz, V., Maloba, B. and Duse, A.G. (2004). Interhospital transfer of pan-resistant Acinetobacter strains in Johannesburg, South Africa. *Am. J. Infect. Control.* 32:278–81.

Martin, J.M., Green, M., Barbadora, K.A. and Wald, E.R. (2002). Erythromycin resistant Group A Streptococci in schoolchildren in Pittsburgh. *N. Engl. J. Med.* 346:1200–1206.

Martone, W.J. (1998). Spread of vancomycin-resistant enterococci: Why did it happen in the United States? Infect. Control. Hosp. *Epidemiol.* 19:539–545.

McGowan Jr, J.E. (2006). Resistance in nonfermenting gram-negative bacteria: multidrug resistance to the maximum. *Am. J. Med.* 119: 29–36.

McManus, M.C. (1997) Mechanisms of bacterial resistance to antimicrobial agents. *Am J Health Syst Pharm.* 54: 1420 –1433.

McManus, M.C. (1997). Mechanisms of bacterial resistance to antimicrobial agents. *Am. J. Health. Syst. Pharm.* 54:1420 –1433.

Michalopoulos, A.S., Tsiodras, S., Rellos, K., Mentzelopoulos, S. and Falagas, M.E. (2005). Colistin treatment in patients with ICU-acquired infections caused by multiresistant Gram-negative bacteria: the renaissance of an old antibiotic. *Clin. Microbiol. Infect.* 11: 115–121.

Musser, J.M. and Kapur, V. (1992) Clonal analysis of methicillin-resistant *Staphylococcus aureus* strains from intercontinental sources: association of the mec gene with divergent phylogenetic lineages implies dissemination by horizontal transfer and recombination. *J. Clin. Microbiol.* 30: 2058–2063.

National Institute of Allergy and Infectious Diseases (NIAID), Fact sheet on tuberculosis. www.niaid.nih.go.factsheets/tb.htm.

Neu, H.C. (1992). The crisis in antibiotic resistance. *Science.* 257:1064– 1073.

Nicas, T.I., Mullen, D.L., Flokowitsch, J.E. *et al.* (1996). Semisynthetic glycopeptide antibiotics derived from LY264826 active against vancomycin-resistant enterococci. Antimicrob. Agents. *Chemother.* 40(9): 2194-2199.

Noskin, G.A., Siddiqui, F., Stosor, V., Hacek, D. and Peterson, L.R. (1999). *In vitro* activities of linezolid against important grampositive bacterial pathogens including vancomycinresistant enterococci. Antimicrob. Agents. *Chemother.* 43(8):2059-2062.

Nuorti, J.P., Butler, J.C., Crutcher, J.M., Guevara, R., Welch, D., Holder, P. and Elliott, J.A. (1998). An outbreak of mutidrug-resistant pneumococcal pneumonia and bacteremia among unvaccinated nursing home residents. *N. Engl. J. Med.* 338:1861–1868.

Pablos-Mendez, A., Raviglione, M., Laszlo, A., Binkin, N., Rieder, H., Bustreo, F., Cohn, D.L., Lambregts-van Weezenbeek, C.S.B., Kim, S.J., Chaulet, P. and Nunn, P. (1998). Global surveillance for antituberculosis-drug resistance, 1994–1997. *N. Engl. J. Med.* 338: 1641–1649.

Palavecino, E. (2004).Community-acquired methicillin-resistant *Staphylococcus aureus* infections. *Clin. Lab. Med.* 24: 403– 418.

Panlilio, A.L., Culver, D.H., Gaynes, R.P., et al. (1992). Methicillin-resistant *Staphylococcus aureus* in US hospitals, 1975–1991. *Infect. Control. Hosp. Epidemiol.* 13:582–586.

Petersen, P.J., Jacobus, N.V., Weiss, W.J., Sum, P.E. and Testa, R.T. (1999). *In vitro* and in vivo antibacterial activities of a novel glycylcycline, the 9-t-butylglycylamido derivative of minocycline (GAR-936). Antimicrob. Agents. *Chemother.* 43(4): 738-744.

Petri, W.A.J. (2006). Antimicrobial agents: sulfonamides, trimethoprim-sulfamethoxazole, quinolones, and agents for urinary tract infections. In: Brunton LL, Lazo JS, Parker KL, eds. Goodman & Gilman's The Pharmacological Basis of Therapeutics, 11th ed. New York: McGraw-Hill. 1111–1126.

Rammelkamp, M. (1942). Resistances of *Staphylococcus aureus* to the action of penicillin. *Proc. R. Soc. Exp. Biol. Med.* 51:386–389.

Ramphal, R. and Ambrose, P.G. (2006). Extended-spectrum beta-lactamases and clinical outcomes: current data. *Clin. Infect. Dis.* 42:164–172.

Rice, L.B. (2006). Antimicrobial Resistance in Gram-Positive Bacteria. *Am. J. of Medicine.* 119 (6A): S11– S19.

Roberts, M.C. (1996). Tetracycline resistance determinants: mechanism of action, regulation of expression, genetic mobility, and distribution. FEMS. *Microbiol. Rev.* 19: 1–24.

Robinson, D.A., Kearns, A.M., Holmes, A., Morrison, D., Grundmann, H., Edwards, G., O'Brien, F.G., Tenover, F.C., McDougal, L.K., Monk, A.B. and Enright, M.C. (2005). Re-emergence of early pandemic *Staphylococcus aureus* as a community acquired methicillin-resistant clone. *Lancet.* 365: 1256–1258.

Roblin, P.M., Hammerschlag, M.R. (2000). In vitro activity of GAR-936 against *Chlamydia pneumoniae* and *Chlamydia trachomatis. Int. J. Antimicrob. Agents.* 16(1): 61-63.

Rossolini, G.M. and Mantengoli, E. (2005). Treatment and control of severe infections caused by multiresistant *Pseudomonas aeruginosa. Clin. Microbiol. Infect.* 11:17–32.

Rupp, M.E. and Fey, P.D. (2003). Extended spectrum β lactamase (ESBL)-producing Enterobacteriacea: considerations for diagnosis, prevention and drug treatment. Drugs. 63: 353–365.

Rybak, M.J., Hershberger, E., Moldovan, T. and Grucz, R.G. (2000). *In vitro* activities of daptomycin, vancomycin, linezolid and quinupristin dalfopristin against *Staphylococci* and *Enterococci*, including vancomycinintermediate and resistant strains. Antimicrob. Agents *Chemother.* 44:1062-1066.

Sahm, D.F., Marsilio, M.K. and Piazza, G. (1999). Antimicrobial resistance in key bloodstream bacterial isolates: electronic surveillance with the Surveillance Network Database–USA. *Clin Infect Dis.* 29: 259 –263.

Saravolatz, L.D., Markowitz, N., Arking, L., Pohlod, D. and Fisher, E. (1982). Methicillin-resistant *Staphylococcus aureus*. Epidemiologic observations during a community-acquired outbreak. *Ann. Intern. Med.* 96:11–16.

Sefton, A.M. (2002). Mechanisms of antimicrobial resistance. Drugs. 62:557–566.

Seppala, H., Klaukka, T., Lehtonen, R., Nenonen, E. and Huovinen, P. (1995). Outpatient use of erythromycin· link to increased erythromycin resistance in Group A streptococci. *Clin. Infect. Dis.* 21:1378–1385.

Sieradzki, K., Roberts, R., Haber, S.W. and Tomasz, A. (1999). The development of vancomycin resistance in a patient with methicillin-resistant *Staphylococcus aureus* infection. *N. Engl. J. Med.* 340: 517–523.

Smith, K.E., Besser, M.S., Hedberg, C.W., Leanon, F.T., Bender, J.B.. Wicklund, J.H., Johnson, B.P., Moore, K.A. and Osterholm, M.T. (1999). Quinoloneresistant Campylobacter jejuni infections in Minnesota, 1992–1998. *N. Engl. J. Med.* 340:1525–1532.

Smith, T.L., Pearson, M.L., Wilcox, K.R., Cruz, C., Lancaster, M.V., Robinson-Dunn, B., Tenover, F.C., Zervos, M.J., Band, J.D., White, E. and Jarvis, W.R. (1999). Emergence of vancomycin resistance in *Staphylococcus aureus. N. Engl. J. Med.* 340:493–501.

Snydman, D.R., Jacobus, N.V., McDermott, L.A., Lonks, J.R. and Boyce, J.M. (2000). Comparative in vitro activities of daptomycin and vancomycin against resistant grampositive pathogens. Antimicrob. Agents *Chemother.* 44(12): 3447-3450.

Storm, D.R., Rosenthal, K.S. and Swanson, P.E. (1977). Polymyxin and related peptide antibiotics. *Annu Rev Biochem.* 46: 723–763.

Sum, P.E. and Petersen, P. (1999). Synthesis and structure-activity relationship of novel glycylcycline derivatives leading to the discovery of GAR-936. *Bioorg. Med. Chem. Lett.* 9(10):1459-1462.

Tomasz, A. (1995). The pneumococcus at the gates. *N. Engl. J. Med.* 333:514–515.

Utsui, Y. and Yokota, T. (1985). Role of an altered penicillin-binding protein in methicillin- and cephem-resistant *Staphylococcus aureus.* Antimicrob. Agents *Chemother.* 28: 397–403.

van Ogtrop, M.L., Andes, D., Stamstad, T.J. et al. (2000). *In vivo* pharmacodynamic activities of two glycylcyclines (GAR-936 and WAY 152,288) against various gram-positive and gram-negative bacteria. Antimicrob. Agents *Chemother.* 44(4): 943-949.

Vanderkooi, O.G., Low, D.E., Green, K., Powis, J.E. and McGeer, A. (2005). Predicting antimicrobial resistance in invasive pneumococcal infections. *Clin. Infect. Dis.* 40:1288–1297.

Wadsworth, S.J., Kim Ki-Ho, Satishchandran, V., Axelrod, P., Truant, A. L. and Suh, B. (1992). Development of new antibiotic resistance in methicillin-resistant but not methicillin-susceptible *Staphylococcus aureus.* J. Antimicrob. *Chemother.* 30:821-826.

Waldvogel, F.A. (1999). New resistance in *Staphylococcus aureus. N. Engl. J. Med.* 340:556–557.

Waterer, G.W. and Wunderink, R.G. (2001). Increasing threat of gram-negative bacteria. *Crit. Care. Med.* 29: 75–81.

Wegener, H.C. (1999). The consequences for food safety of the use of fluoroquinolones in food animals. *N. Engl. J. Med.* 340:1581–1582.

White, D.G., Zhao, S., Sudler, R., Ayers, S., Friedman, S., Chen, S., McDermott, P.F., McDermott, S., Wagner, D.D. and Meng, J. (2001). The isolation of antibiotic resistant Salmonella from retail ground meats. *N. Engl. J. Med.* 345:1147–1154.

Whitney, C.G., Farley, M.M., Hadler, J., Harrison, L.H., Lexau, C., Reingold, A., Lefkowitz, L., Cieslak, P.R., Cetron, M., Zell, E.R., Jorgensen, J.H. and Schuchat, A. (2000). Increasing prevalence of multidrug-resistance *Streptococcus pneumoniae* in the United States. *N. Engl. J. Med.* 343:1917–1924.

Whittington, W.L., Roberts, M.C., Hale, J. and Holmes, K.K. (1995).Susceptibilities of *Neisseria gonorrhoeae* to the glycylcyclines. Antimicrob. Agents *Chemother.* 39:1864-1865.

Williams, J.D. and Moosdeen, F. (1986). Antibiotic resistance in *Haemophilus influenzae*: epidemiology, mechanisms, and therapeutic possibilities. *Rev. Infect. Dis.* 8:555–561.

Wisplinghoff, H., Bischoff, T., Tallent, S.M., Seifert, H., Wenzel, R.P. and Edmond, M.B. (2004). Nosocomial bloodstream infections in US hospitals: analysis of 24,179 cases from a prospective nationwide surveillance study. *Clin. Infect. Dis.* 39:309 -317.

Woodford, N., Morrison, D., Johnson, A.P. and George, R.C. (1993) Antimicrobial resistance amongst enterococci isolated in the United Kingdom: a reference laboratory perspective. J. Antimicrob. *Chemother.* 30:821-846.

Yao, J. and Moellering, R.J. (2003). Antibacterial agents. In: Murray PR, Baron EJ, Jorgensen JH, Pfaller MA, Yolken RH, eds. Manual of Clinical Microbiology, 8th ed. Washington, DC: ASM Press. 1039 -1073.

Yoon, J., Urban, C., Terzian, C., Mariano, N. and Rahal, J.J. (2004). In vitro double and triple synergistic activities of polymyxin B, imipenem, and rifampin against multidrug-resistant *Acinetobacter baumannii*. Antimicrob. Agents *Chemother.* 48:753–757.

Zeckel, M.L., Preston, D.A. and Allen, B.S. (2000). *In vitro* activities of LY333328 and comparative agents against nosocomial gram-positive pathogens collected in a 1997 global surveillance study. Antimicrob. Agents *Chemother.* 44:1370-1374.

□□□

CHAPTER **15**

Molecular Systematics of *Frankia*-Actinorhizal Symbiosis

Balwinder Bajwa, Manprit Gill, Saubashya Sur, Asim Bothra[1] and Arnab Sen*

Molecular Genetics Laboratory, Department of Botany, University of North Bengal, Siliguri 734013, India,[1]Department of Chemistry, Raiganj University College, Raiganj, India

ABSTRACT

Frankia is a gram-positive, microaerophilic, actinomycete that enters into symbiotic associations with a number of woody dicotyledonous plants called actinorhizal plants. *Frankia* can fix atmospheric nitrogen through such associations. The actinorhizal-*Frankia* association is beneficial for the environment, since it increases the fertility of soils. Extensive studies on the biochemical and molecular aspects helped in the classification of *Frankia*. PCR-RFLP studies of the 16S rRNA and the nif regions of *Frankia* revealed considerable information regarding the diversity of *Frankia* strains, isolated from different parts of the world. The completion of three whole genomes of Frankia has bolstered the study of the molecular organization of the genome. Comparative genomics study of the three complete genomes revealed that, the strains reflected the biogeographical history of the host plants they infected. Striking similarity in the codon usage patterns highlights the fact that they may have evolved as a unit. In the present review we highlighted the taxonomy of *Frankia* and the host plants, the *Frankia*-actinorhizal association, regulation of infectivity, biochemical and molecular characterization, phylogeny and genomics of *Frankia* and stressed the need for metabolomic study of the Frankia-actinorhizal association.

Key words : Frankia, symbiosis, molecular diversity, phylogeny, genome

1. INTRODUCTION

Nitrogen is the mineral nutrient, essential to all living organisms, which most often limits plant production. Though gaseous atmosphere has 80% dinitrogen (N_2), it cannot be utilized as such. Nitrogen from the atmosphere

*Email: senarnab_nbu@hotmail.com

can be fixed abiologically (industrial synthesis, ultraviolet irradiation, lightening) as well as biologically. Industrial synthesis contributes about 60 million tons of nitrogen per year for fertilizer uses whereas biological nitrogen fixation contributes about 100-175 million tons per year. Biological nitrogen fixation (BNF) accounts for 65% of nitrogen currently utilized in agriculture, and will be increasingly important in future crop productivity, especially for sustainable systems.

Biologically the ability to fix molecular nitrogen, occur in diverse groups of microorganisms that exist either as free-living or in symbiosis with plants (Young, 1992). Symbiotic nitrogen fixing bacteria include the cyanobacteria, the genera *Rhizobium, Bradyrhizobium, Azorhizobium, Sinorhizobium, Mesorhizobium* and *Frankia* (Paul and Clark, 1996; Brock, 2000). Unfortunately, the study of root nodules of non-leguminous plants was shrouded in the mist of misconceptions for a long period. Between 1886 and 1888, Hellrigel and Wilfarth published two papers, which brought an end to a 10-year-old controversy about sources of nitrogen for growth of plants and opened up new frontiers of plant microbial sciences. They showed that legume nodules fixed atmospheric nitrogen. They also proposed the difference between nitrogen users and nitrogen accumulators. Alders (Alnus nepalensis) were initially pushed into the nitrogen-accumulating group. However, Hiltner (1904) proved that actinorhizal plants could utilize atmospheric nitrogen. In all cases of rhizobia-legume or *Frankia*-actinorhizal symbioses, a new plant organ is developed in which the bacteria differentiate, express the enzyme nitrogenase, and fix nitrogen into ammonia. These compounds are then assimilated and transported to the rest of the plant. In some cases, so-called shoot nodules are formed (Prin *et al.* 1992), but these aerial nodules are modified adventitious roots, having the same structure as root nodules but are located above the soil or the stems.

Although both *Frankia*-actinorhizal and *Rhizobiu*m-legume symbiosis have been known for many years to benefit soil fertility, the nature of these nitrogen-fixing symbioses and the microsymbionts involved were discovered at the end of 19th century (Quispel, 1990). Various anatomical and cytological studies have been done to compare leguminous and actinorhizal microsymbionts. Actinorhizal and legume nodules can be easily distinguished at the anatomical level (Pawlowski and Bisseling, 1996). Legume nodules have a central infected tissue surrounded by nodule parenchyma and peripheral vascular bundles, whereas actinorhizal nodules are characterized by a central vascular bundle and peripheral infected tissue surrounded by cortical nodule parenchyma. Legume nodules have been proposed to have a shoot like anatomy, whereas actinorhizal nodules are ontogenically related to roots. In another study Beijerinck (1888) inoculated non-leguminous plants with his own culture of legume root nodule bacteria and observed that the bacteria failed to infect the non-leguminous plants which led to the conclusion that the two microsymbionts are two different microorganisms.

Study of the actinomycetes induced nodule symbiosis has lagged behind research on legumes largely because of the repeated failure of attempts to isolate and grow the bacterial endosymbiont in pure culture by many workers. The first cultured *Frankia*, isolated from *Alnus* root nodules was reported by Pommer (1959), but unfortunately the culture was lost. In 1978, the first successful isolation of *Frankia* was reported from *Comptonia peregrina* root nodules (Callaham *et al.*, 1978), beginning a new era in actinorhizal symbiosis research (Quispel, 1990). To date, *Frankia* strains from a number of other actinorhizal species are available in pure culture (Lechevalier, 1986; Benson and Silvester, 1993) and their growth requirements have been optimized (Akkermans *et al.*, 1983; Blom, 1982; Meesters *et al.*, 1985 and Tjepkema *et al.*, 1980).

2. *FRANKIA*- THE MICROSYMBIONT

The genus name *Frankia* was proposed in 1886 by J. Brunchorst to honor his mentor, A.B Frank, a Swiss microbiologist. Later, Krebber classified it as actinomycetes in 1932 (Quispel 1990). Brunchorst and Frank considered the microorganism *Frankia* to be a fungus but the name was not generally used until Becking, in 1970, revived it and redefined the genus as consisting of member of the order Actinomycetales in a new family, Frankiaceae (Becking, 1970). The genus *Frankia* is comprised of gram-positive and gram-variable actinomycetes (Lechevalier and Lechevalier, 1990). *In vitro* studies of *Frankia* were possible only after the isolation and subsequent axenic culture on yeast extract medium. But improvements in the composition of culture media now permitted *Frankia* to be grown *in vitro* much faster than earlier used yeast-extract medium (Lalonde and Calvert, 1979). In their classical reviews, Newcomb and Wood (1987) and Benson and Silvester (1993) gave vivid descriptions of structure and ultra structure of *Frankia* cells. Benson and Silvester (1993) also discussed the difficulties of its study with electron microscope because of the delicate nature of some components of cells.

In pure culture *Frankia* behave as microaerophilic and mesophilic microorganism (Burggraaf and Shipton, 1982). It forms dense mat of anastomosing hyphae with sporangia developing terminally and at intercalary positions when grows *in vitro* (Newcomb and Wood, 1987). Unlike other symbiotic nitrogen-fixing bacteria, *Frankia* is multicellular and differentiated. During the life cycle, vegetative hyphae, vesicles and sporangia develop in response to partially defined conditions (nitrogen limited). *Frankia* strains are characterized by the presence of vesicles and hyphae bearing large sporangia when cultivated under proper conditions. These spherical structures have not yet been found in other kind of actinomycetes and can be utilized as the feature of the genus *Frankia*. However, spore formation and induction of vesicles are largely dependent on the culture conditions. This makes the identification of the strain difficult. In many strains the production of vesicle is suppressed by the presence of combined nitrogen in the medium, i.e. conditions which also repress the nitrogenase. The most important physiological function of vesicles is to shelter nitrogenase, the key enzyme involved in nitrogen fixation. As in all prokaryotes the enzyme nitrogenase in *Frankia* is oxygen labile (Benson *et al.*, 1979). Within the vesicle, nitrogenase is protected from deleterious effects of oxygen. Special protection of the enzyme is provided in a variety of ways. One hypothesis is that the multi-laminate vesicle envelope developed in response to increase in ambient pO_2 provides a physiçal barrier to inward diffusion of oxygen, thereby protecting nitrogenase within the vesicle from denaturation by O_2 (Table 1).

TABLE 1. List of nitrogen fixing organisms and their strategy for escaping O_2 damage

Strategy	Mechanism	Organism
Avoid (Escape)	Move to lower O_2 pressure	*Azospirillum*
O_2 scavenging	High Oxygen uptake by respiration	*Azotobacter, Derxia*
	Hemoglobin combines efficiently with O_2	In nodules of rhizobia-legume symbiosis, *Casuarina-Frankia* symbiosis
Protect	Some proteins protect nitrogenase under O_2 exposure	*Azotobacter*
Separation in space	Nitrogenase in heterocysts (Heterocysts do not have photosynthetic ability, do not evolve O_2)	*Cyanobacteria*
	in Vesicles in *Frankia*-nodules symbiosis	*Frankia*
	In colony, differentiate into photosynthetic filamants and nitrogen fixing filaments	Marine Scenedesmium
Separation in time	Photosynthesis daytime, and nitrogen fixation at night.	Unicellular Cyanobacteria

Note: Generally, Nitrogen Fixation is active under low (<0.1% v/v) O_2 pressure

3. TAXONOMY OF THE GENUS *FRANKIA*

Becking (1970) has proposed ten species (*Frankia alni, F. elaeagni, F. brunchorstii, F. discariae, F. casuarinae, F. ceanothi, F. coriariae, F. dryadis, F. purshiae, and F. cercocarpi*) based on the source plant, and described the members as "obligate symbiotic organisms". In Bergey's Manual of Systematic Bacteriology, Vol.4 (1989), *Frankia* strains are included among "actinomycetes with multilocular sporangia" (Lechevalier and Lechevalier, 1989). Other genera in this group include *Geodermatophilus* and *Dermatophilus.* The initial classification proposed by Becking (1970) has been replaced with a system of registry numbers. The availability of numerous strains (Baker, 1982) helped in the clarification of the relationships among various members in this unique family Frankiaceae. It should be noted that only one genus and no species were recognized until 1988. This absence of recognition of species within the genus *Frankia* was mainly due to vast variations in the various morphological and physiological characters normally used for taxonomic purposes.

A vast number of *Frankia* strains have been obtained in pure cultures since its first successful isolation in 1978, and attempts have been made to classify the strains. So far all the strains have been classified within the genus *Frankia.* This taxon is characterized by - a) the ability to nodulate plants; b) the ability to fix nitrogen; c) its unique morphological properties, vesicles and sporangia; d) the presence of sugar, 2-O-methyl-D-mannose (Mort *et al.*, 1983); e) the presence of cell wall type III and phospholipid content of type I of actinomycetes (Lechevalier and Lechevalier, 1990); f) a high G+C% in the range of 68-72%.

The characters employed in actinomycete taxonomy in the past have been the size and shape of sporangia or spores, and production of cellular or soluble pigment. For *Frankia* these characters are not considered sufficiently stable to be useful (Lechevalier, 1984). According to Weber *et al.* (1988) the morphological characters are stable but too diverse for identification use. A combination of different observations (like Sporangial size, frequency and shape; pigment production; Urease activity and acetate decarboxylation) can be used to separate some of the strains, but do not allow general identification of individual *Frankia* strains. It should be noted that, although the available *Frankia* isolates can belong to various host specificity groups, this character is not used in the actual because some *Frankia* strains can infect two host specificity groups, such as *Alnus* and *Elaeagnus* groups and because some non infective strains of *Frankia* are known to exist. As mentioned above one stable character is the presence of a unique sugar, identified as 2-O- methyl D-mannose, which is a unique trait of Frankiaceae family (Mort *et al.* 1983).

Related strains can also be distinguished on the basis of electrophoretic separation of isozymes. Gardes *et al.* (1987) characterized *Frankia* strains by electrophoretic separation of isozyme of eight different enzymes. They analyzed diaphorase, leucine aminopeptidases, phosphoglucose isomerase, esterase, maltate dehydrogenase, phosphoglucomatase, and superoxide dismutase. The results obtained from this analysis were used to classify strains by numerical analysis based on pair-wise similarity coefficients, and further to propose the delineation of groups of strains.

On the basis of physiology two suprageneric or sub generic groups have been proposed (Lechevalier and Lechevalier, 1989; Lechevalier and Lechevalier, 1979). Main differences between the two groups are shown in Table 2.

TABLE 2. Two types of Frankial strains on the basis of physiological difference

Type A	Type B
They are physiologically active.	They are physiologically inactive
They grow relatively rapidly.	They grow very slowly
They are aerobic and can be maintained on slants.	They are strictly microaerophilic and cannot be maintained on slants
They do not reinfect host plant from which they are isolated, at least in the experimental conditions.	They can infect the host plant from which they are taken
They utilize a variety of monosaccharides and disaccharides with or without acid production.	They cannot utilize carbohydrates, proteins, or starch. The preferred carbon sources are organic acids or between compounds.
Many of them produce hydrolytic enzymes, such as pectinases, cellulases, amylases, and proteases	No such reports

Compiled from Lechevalier (1994).

The 'constant' (relatively invariable for all taxa) regions of the 16S rRNA nucleotide sequences of actinomycetes are valuable for determination of the proper placement of genera within the order Actinomycetales and for discerning the position of the order Actinomycetales in the prokaryote, while it is widely believed that the "hyper variable" regions may help in phylogenetic classification at the genus or species level (Harry *et al.*, 1991 and Mirza *et al.*, 1992). Recent 16S rRNA sequence analyses of representatives of these genera have revealed a relatively close phylogenetic relationship between *Frankia* and *Geodermatophilus* but not *Dermatophilus;* an invalidly described isolate from the Black Sea called "*Blastococcus*" also clustered with *Frankia* (Hahn *et al.*, 1989). Thus, future descriptions of the family Frankiaceae should include members of these genera. The family Dermatophilaceae will consequently contain only the genus *Dermatophilusm* (Hahn *et al.*, 1989). In a different approach by Simonet *et al.* (1991) two different sets of primers were designed. One of which was a universal set designed for nine nitrogen fixing genera and the other set was specific for the *nif*H-*nif*D regions of *Frankia* sp. Augmentation of different strains of *Frankia* with these two sets of primers resulted in fragments of the length specific for the genus. Similarly universal and specific primers targeted to 16S rRNA gene region also generated the fragments of the size characteristic of *Frankia* spp. and also of the closely related genus *Geodermatophilus* and were therefore presumably specific for members of the family Frankiaceae. Members of the genus *Geodermatophilus* and *Frankia* were further separated on the basis of intergenic sequence [IGS] the 16S and 23S rRNA genes by designing two additional primers. The amplification results showed that the 561bp fragment generated by the *Frankia* differed from the size of fragments generated from *Geodermatophilus* strains and the other microorganisms tested. Although this technique is quite promising, more strains have to be analyzed before general conclusion can be drawn. In an attempt to analyze molecular phylogeny among the nodules collected from different parts of Sub-Himalayan West Bengal, the authors group has successfully amplified the *nif* IGS region and Distal, middle and proximal part of 16S rRNA gene. The amplicons when further used for RFLP study revealed a high level of polymorphism in alder based *Frankia* of Darjeeling and adjoining areas (Bajwa, 2004).

3.1 Taxonomy of the Species

The most critical part of *Frankia* taxonomy is its speciation. Main difficulties include (1) individual strains often nodulate plants from different plant orders (Benson and Silvester, 1993), (2) strains from same genomic species have been isolated from members of different plant families, (3) some strains failed to re-infect their source plants (Mirza *et al.*, 1991) and (4) unpredictable nodulating abilities have been observed among the *Elaeagnus* strains (An *et al.*, 1985 and Fernandez *et al.*, 1989). Nevertheless, a considerable degree of host plant specificity does occur, with the result that the host plant origin and the ability to nodulate within certain host specificity groups are relevant but not determinative characteristics for strain identification or classification.

Arranging *Frankia* strains into phenotypically related groups have proven to be a difficult task. Classical physiological testing is of little use in grouping *Frankia* strains since they all grow slowly (doubling times of 15 to 48 h or more) and the results often vary depending on how long the strains are allowed to grow. Although the rate of Frankial growth differs from medium to medium, yet it is not a constant parameter for all the strains. In our laboratory, it is observed that some Frankial strains grow fast when grown on Q-mod medium (Lalonde and Calvert, 1979) in comparison to the widely used DPM medium (Baker and O Keefe, 1984). Also the growth rate has been observed to be faster when certain Frankial isolates from *Casuarina* growing on DPM medium were further subcultured to Q-mod. But when *Frankia* isolates from *Alnus* nepalensis were subcultured from DPM to Q-mod medium, no such enhancement was observed (unpublished data). Early attempts to correlate serological groups with host range or other phenotypic parameters were largely abandoned when such correlations proved more complex than was initially thought (Baker *et al.*, 1981 and Lechevalier *et al.*, 1983).

Techniques applied to resolve these problems are morphological differences, *in planta* sporangia formation (Normand and Lalonde, 1982), phylogeny of carbohydrate uptake (Lechevalier *et al.*, 1983 and Ganesh *et al.*, 1994), utilization of various substrates and enzyme production (Shipton and Burggraaf, 1982), serology (Baker *et al.*,1981 and Lechevalier *et al.*, 1983), host specificity (Bosco *et al.*, 1992) isozyme patterns (Maggia *et al.*,1990), whole cell difference in the ribosomal sequences of genomic species of isolates of Elaeagnaceae etc.

Lalonde *et al.* proposed the acceptance of the species *F. alni* and *F. elaeagni* and created *F. alni* subsp. *pommerii* and *F. alni* subsp. *vandijkii* (Lalonde *et al.*,1988). This proposal was based on the results of biochemical and physiological studies done on strains isolated mainly from members of the families Betulaceae, Myricaceae, Elaeagnaceae, and Casuarinaceae.

Fernandez *et al.* (1989) identified nine genomic species among 43 isolates examined. Three genomic species were found among the strains that infect *Alnus* species, five were found among strains that infect members of the family Elaeagnaceae, and one contained 11 strains isolated from members of the Casuarinaceae. Nine strains (four isolated from *Alnus* species, two isolated from members of the Elaeagnaceae, and three isolated from members of the Casuarinaceae) were not classified and may represent additional genomic species.

The largest group of *Alnus* infecting strains, genomic species 1, was first established as genogroup I by An *et al.* in their early DNA homology work

(An *et al.*, 1985). They established genogroup I with seven isolates from *Alnus* species plus one from *Comptonia peregrina* and one from *Myrica pensylvanica*, with DNA from ArI49 (from *A. rubra*) as a reference. The coherence of this group has since been confirmed by numerous phenotypic and genetic analyses and thus has a valid claim to the species epithet *F. alni* as described by Lalonde *et al.* (1988) with strain HFPCpI1 as the type strain.

The proposal that the subspecies *F. alni* subsp. *vandijkii* and *F. alni* subsp. *pommerii* be established has generated some confusion in the naming of strain. As Beyazova and Lechevalier (1992) have pointed out, a subspecies of *F. alni* should be *alni*. DNA-DNA re-association study has confirmed the diversity in the physiology, morphology, and biochemistry of *Elaeagnus* strains (An *et al.*, 1985 and Fernandez *et al.*, 1989).

An *et al.*, (1985) found no genogroup among 10 isolates (5 from *Elaeagnus* species, two from *Casuarina equisetifolia* and one each from *Ceanothus americanus, Purshia tridentate*, and *Alnus incana* subsp. *rugosa*) tested against DNA from EuI1a from *Elaeagnus umbellate*) or ArI4. Fernandez *et al.* (1989) found that out of the 11 strains tested 8 constituted genomic species 9, with percent reassociations ranging from 69 to 100%. The three strains that remained unclassified were "atypical", since they failed to reinfect members of the Elaeagnaceae.

Nazarat *et al.* (1991) developed a protocol based on the PCR amplification and sequencing of 16S ribosomal DNA sequence to measure phylogenetic relationships. They examined nine genomic species created by Fernandez *et al.* earlier in 1989 and found that with one exception, strains belonging to same genomic species had identical sequences and that they differed from other genomic species.

Another important method for understanding phylogeny of *Frankia* is restriction fragment length polymorphism (RFLP) of *nif* complexes (*nif*AB, *nif*K and *nif*H). The *nif* D-K DNA IGS region in *Frankia* spp. is larger and more variable than the *nif* H-D IGS region; thus the PCR-RFLP study of *nif* D-K IGS has proven to be more useful for distinguishing among closely related strains. Nazarat *et al.* (1989) found highly homologous results for infective isolates obtained from *Casuarina* spp. regardless of restriction enzyme used, whereas non-infective strains differed from first group and among themselves. Akimov and Dobritsa (1992) observed high level of DNA-DNA relatedness among both infective and non-infective isolates. Simonet *et al.* (1989) did RFLP analysis of 100 isolates from single alder stand using *nif*-HDK probes and they found similarity. When they used *nif*AB probes, five groups were found.

Jamann *et al.* (1993) employed PCR-RFLP technique. They amplified intergenic spacer (IGS) and a part of genes in *nif* cluster through PCR using two primers composed of sequences from conserved *nif*-D and *nif*-K regions flanking *nif* D-K IGS region. They cleaved the amplicons with four different restriction enzymes and patterns obtained were used as fingerprints for typing *Frankia* strains. They also revealed that the DNA of *Frankia* isolates obtained from members of Elaeagnaceae and Casuarinaceae could be amplified to yield a product of about 1380bp, whereas the DNA of isolates obtained from *Alnus* spp. were amplified only weakly or were not amplified. When these products were digested with various restriction enzymes like *Hinf-I*, a characteristic pattern is obtained i.e. products digested with *Hinf-I* grouped together the entire test isolates *Mip-I* and *Scfr-I* resolved these organisms into the genomic spp. previously determined by DNA-DNA pairing.

Marry P. Lechevalier and her colleague K. Beyazova used a novel and simple method to tackle species problem. They used restriction enzymes with less number of restriction sites (rare cutters) to cut DNA into fragments of bigger molecular weights. Fragments were separated with pulse field electrophoresis. This method is called 'low frequency restriction fragment analysis' (LFRFA). Findings of Beyazova and Lechevalier (1992) through LFRFA confirmed results of Nazarat *et al.* (1991) that type strain of *Frankia alni* subsp. *pomme* clusters at some distance from other *Frankia* genomic species isolated from *Alnus* spp. Each of these approaches has yielded phenotypic and in some cases genotypic subgroups within host specificity groups.

Not all approaches have been taken with classification in mind, some have been used primarily to assess the diversity of strains isolated from an individual plant species or of strains obtained from a limited geographical area (Benson *et al.*, 1984; Bloom *et al.*, 1989; Gardes and Lalonde, 1987). Of the techniques used, only DNA-DNA homology and perhaps the most recently developed restriction fragment length polymorphism can unequivocally place an unknown strain within a genospecies. The other techniques yield well-defined groups independent of phylogenetic relationships, but the sensitivity of the methods remains unknown. The available biochemical, physiological, and phylogenetic information on certain well-studied groups of strains has rekindled the desire to establish *Frankia* species. To do so, it is desirable to satisfy the consensus that new genotype species be defined in phylogenetic terms on the basis of DNA sequence data (Wayne *et al.*, 1987).

3.2 Actinorhizal Plants-the Host

The plants nodulated by *Frankia* are known as actinorhizal plants (Torrey, J.G. and J.D. Tjepkema 1979). According to Benson and Silvester (1993), there are 194 species of 24 genera, but there is no detailed account given. Till date about 288 species of 24 genera belonging to 8 families of 7 orders have been reported to have actinorhizal associations (Baker and Schwintzer, 1990). Table 3 shows some important symbiotic associations between *Frankia* and actinorhizal genera.

TABLE 3. Frankial isolates in symbiotic association with Actinorhizal plant general

Order	Family	Genus	Number of species	Isolates
Casuarinales	Casuarinaceae	Allocasuarina	54	+, I, E
		Casuarina	16	+, I, E
		Ceuthostoma	2	-
		Gymnostoma	18	+, I, E
Fagales	Betulaceae	Alnus	47	+, I, E
Myricales	Myricaceae	Comptonia	1	+, I, E
		Myrica	28	+, I, E
Protales	Elaeagnaceae	Elaeagnus	38	+, I, E
		Hippophae	2	+, I, E
		Shepherdia	2	+, I, E
Ranunculales	Coriariaceae	Coriaria	16	+, N
Rhamnales	Rhamnaceae	Ceanothus	31	+, N
		Colletia	4	+, N
		Discaria	5	+, N
		Kentrothamnus	1	-
		Retanilla	2	+, N
		Taalguenea	1	+, ?
		Trevoa	2	+, N
Rosales	Rosaceae	Cercocarpus	4	+, N
		Chamaebatia	1	-
		Cowania	1	+, ?
		Drayas	3	-
		Purshia	2	+, N
Violales	Datiscaceae	Datisca	2	+, N

Symbols: -, isolates not reported; +, isolates obtained; I, infective; N, non-infective; E, effective in fixing nitrogen;?, ineffectiveness unknown or unre-ported. From: Bond (1983), Torrey and Berg (1988), Newcomb and Wood (1987), Baker and Schwintzer (1990) and Benson and Silvester (1993).

Actinorhizal genera like to grow in marginally fertile soil and can be found in most of the climatic zones inhabiting a variety of ecosystems including arctic tundra (*Dryas* species), coastal dunes *(Casuarina, Hippophae, Myrica,* and *Elaeagnus* species), riparian (*Alnus* and *Myrica* species), glacial till *(Alnus* and *Dryas* species), forest (*Alnus, Casuarina, Coriaria,* and *Shepherdia* species), Chapparal and xeric (*Casuarina, Purshia, Ceanothus, Cercocarpus, Comptonia, and Cowania* species), and alpine (*Alnus* species) (Benson and Silvester, 1993).

Actinorhizal plants are distributed worldwide and occupy a wide range of habitat. They are native to Pakistan, eastern Nepal, Bhutan, northern India, southwestern China, upper Myanmar and parts of Indochina. They are introduced to various countries in Africa, Central America and South-East Asia. Table 4 shows the distribution of native actinorhizal families and genera.

3.3 Taxonomy of Host Plant

All the actinorhizal plants are trees or shrubs, except for the genus *Datisca,* which is herbaceous. Besides, all the members of a certain family may not be actinorhizal, while all the genera of the family Elaeagnaceae are actinorhizal. They cover a range of woody dicotyledons and the taxonomic relation is very less. Some species are very well adapted to flooded lands; warm arid and semiarid regions and areas of devastation (for example, rock slides). The taxonomic position of actinorhizal genera in the Magnoliopsida according to Cronquist (1988) is quite different from the position in Hutchinson's (1973) classification. Particularly, in Cronquist's classification, Rosales are not ascribed to a position anterior to the Hamamelidales and the Elaeagnaceae and Rhamnaceae are found in different orders and are shown to have diverged from the Rosales. Datiscaceae are found in the Violales (Dilleniidae and Casuarinales are grouped together with Fagales and Myricales in the Hamamelidae. Pollen records show, members of the orders Fagales and Myricales were the first among the actinorhizal plants to visit the earth. Rosales, Proteales and Rhamnales are relatively advanced.

4. ESTABLISHMENT OF SYMBIOSIS

Actinorhizal root nodules are established through a series of interactions between *Frankia* and host root cells. The nodule is system of structurally and developmentally modified lateral roots. The structural changes observed in the development of nodules reflect the molecular events in which normal patterns of plant development are altered. Despite the large genetic complexity of the host plant, analyses have ascertained several genes of eukaryotic origin necessary for nodulation and nitrogen fixation. Plant genes are defined either

TABLE 4. Continent wise distributions of native Actinorhizal families and their genera

Continent	Family	Genus
North America	Betulaceae	*Alnus*
	Coriariaceae	*Coriaria*
	Datiscaceae	*Datisca*
	Elaeagnaceae	*Elaeagnus*
		Shepherdia
	Myricaceae	*Myrica*
		Comptonia
	Rhamnaceae	*Ceanothus*
	Rosaceae	*Cercocarpus*
		Chamaebatia
		Cowania
		Dryas
		Purshia
South America	Betulaceae	*Alnus*
	Coriariaceae	*Coriaria*
	Myricaceae	*Myrica*
	Rhamnaceae	*Adolphia*
		Colletia
		Discaria
		Kentrothamnus
		Retanilla
		Talguenea
		Trevoa
Africa	Myricaceae	*Myrica*
Eurasia	Betulaceae	*Alnus*
	Coriariaceae	*Coriaria*
	Datiscaceae	*Datisca*
	Elaeagnaceae	*Elaeagnus*
		Hippophae
	Myricaceae	*Myrica*
	Rosaceae	*Dryas*
Australia and Oceania	Casuarinaceae	*Allocasuarina*
		Casuarina
		Ceuthostoma

Adapted from Baker (1988) and Baker and Schwintzer (1990) and updated based on Cruz-Cisneros and Valdes (1991).

by mutagenesis or molecular probes indicating specific nodule expression or amplification (Gresshoff and Delves, 1986; Nap and Besseling, 1989). The infection process begins in the rhizosphere. Conditions necessary for colonization and initial binding of *Frankia* to the host root have not been entirely elucidated. *Frankia* hyphae are embedded within a mucilage layer at the root hair or epidermal surface. This mucilage layer is evidently secreted by the root hairs (Berry and Torrey, 1983), although the root cap also produces mucilage. Elements of the host-derived matrix may participate in attachment phenomenon. As long as the conditions favorable for root growth are present, environmental factors such as pH or nutrient availability do not appear to limit infection and nodulation, at least for *Frankia* strains isolated to date. "Helper" microorganisms for example, *Pseudomonas* sp., improve nodulation of both *Alnus* and *Casuarina* under controlled growth conditions (Knowlton *et al.*, 1980).

Frankia infects its host plant in two different ways. In case of *Alnus, Casuarina Comptonia, Myrica* etc., infection takes place through root hair deformation (Berry and Torrey, 1983; Callaham *et al.*, 1979 and Torrey, 1976), while for other hosts like *Ceanothus, Elaeagnus, Shepherdia,* etc; early colonization of root tissues by *Frankia* is entirely intercellular (Miller and Baker, 1986; Racette and Torrey, 1989). Despite these differences in early stages of cellular interactions between host and microsymbiont, eventually *Frankia* penetrate host cell walls in developing nodule lobes and differentiate within the cells.

At the site of the wall penetrate; infection hyphae are continuous with *Frankia* in the rhizosphere. Characteristic changes occur in patterns of host wall deposition, as observed in *Alnus rubra* (Inner wall layers at the site of infection resemble transfer cell walls (Berry *et al.*, 1986). Within the infected hair, the microsymbiont may branch early in infection. Whether branched or unbranched, the microsymbiont become surrounded with host plasmalemma and by host derived wall material that is continuous with the wall accumulation at the site of penetration.

After *Frankia* infection, root hairs give the appearance of high metabolic activity, while uninfected hairs degenerate, infected hairs retain cytoplasmic contents, more so at the base, with numerous ribosome's and the organelles, including a nucleus, golgi bodies, endoplasmic reticulum (ER), and large plastids. Vacuoles are relatively small. Mitochondria are often numerous, particularly in the vicinity of *Frankia* hyphae within the cell. Several structural observations suggest that new wall deposition be closely linked to successful infection. *Frankia* infects host cells during cell expansion, when primary wall is deposited, both in tip growing root hairs and in pro-meristimatic derivatives in the nodule lobe. Wall encapsulation material is deposited around the invading and proliferating hyphae.

In cases reported to date, *Frankia* appears to transit from the root cortex into the developing nodule lobe primordium through cortical cells near the base of the lobe primordium. Further progress of the hyphae within the nodule lobe occurs either intercellularly or intracellularly depending on the mode of infection. In *Ceanothus,* for the most part, hyphae do not pass from cell to cell directly (Strand and Laetsch, 1977), while in *Alnus, Casuarina,* and *Myrica,* hyphae are clearly continuous with several cells in a file or group.

The initial intercellular colonization by *Frankia* occurs in mature root cortical tissue, while at the same time nodule lobe primodium is initiated in the pericycle. Neither cell division nor hypertrophy of root cortical cells characteristic of typical prenodules is observed in *Elaeagnus*. Ultimately hyphae penetrate and invade host cells in the developing nodule lobe.

5. REGULATION OF INFECTION PROCESS

Although the processes of infection and nodule formation are complex, there are certain key molecular and cellular phenomena that are likely to be pivotal for regulation. We presently know very little about the molecular levels of regulation. Cellular aspects of *Frankia* symbiosis that may determine developmental patterns includes the initial contact between *Frankia* and the host root cells, cell division in the host cortex, expansion of the host cell and deposition of the modified cell walls.

Root exudates are one category of signaling molecule generated by the host that could regulate *Frankia* genes concerned with nodulation. Such an interaction has been demonstrated conclusively to occur in legume-*Rhizobium* symbiosis where specific phenolic compounds secreted in host root exudates regulate the promoters of bacterial nodulation genes. Specific phenolics, or some other regulatory signal, might also mediate compatibility between *Frankia* and actinorhizal hosts. A host exudate factor is probably involved in regulating aspects of early nodulation. Extra cellular matrices such as mucilage, or the host derived encapsulation layer, may be important in regulating infection. Mucilage secreted by host, both in epidermal surfaces and within intercellular spaces, may facilitate bacterial colonization; *Frankia* and other microorganisms have been shown to adhere to mucilage at the root hair surface (Berry and Torrey, 1983). Binding of *Frankia* to the host cell exterior could directly trigger host responses, initially at the surface and secondarily within the host cell. Additionally, the polyanionic nature of these matrices may provide a medium for flux of cations, for example, calcium, which can affect a variety of cellular phenomena. Extra cellular polysaccharides of host origin may provide a substrate for *Frankia* enzyme activity. Seguin and Lalonde (1989) observed pectinolytic activity in culture and reported positive hybridization between *Frankia* DNA and genes coding for pectate lyase. These results suggest that *Frankia* might degrade pectic substances within nodule tissues. Resulting oligosaccharides might mediate compatibility in actinorhizal associations.

6. ROLE OF HOST IN SYMBIOSIS

The actinomyces symbiosis with the host plant suggests that the host cells and tissues exert a remarkable degree of control over the expression of the invading microorganism. This host control is part of the symbiotic relationship which keeps the invader under control and allows the symbiosis to work to the benefit of the host while providing the endophytic symbiont a site and substrates to function.

The nitrogen fixing effectiveness of host *Frankia* species have been studied primarily with *Casuarina* and *Alnus* species. Over three fold differences in growth and nitrogen fixation have been found for various *Casuarina-Frankia* combinations (Redell and Bowen 1985; Fleming *et al.*, 1988; Redell *et al.*, 1988; Sanginga *et al.*, 1990), although Rosbrook and Bowen (1987) found no significant *Frankia* strain effect. Differences in host provenance may be greater determinants of growth and nitrogen fixing effectiveness than variations among *Frankia* strains (Sougoufara *et al.*, 1992). Similar differences in growth and nitrogen fixation have been found for *Alnus-Frankia* combinations (Normand and Lalonde, 1982; Wheeler *et al.*, 1986; Hooker and Wheeler, 1987; Weber *et al.*, 1989). In the nodule *Frankia* sometimes produces sporangia. Depending upon theprescence or absence of sporangia Frankia strains can be divided into spore positive (sp^+) and spore negative (sp^-) strains (Sen, 1996). In *Alnus-Frankia* symbiosis, greater nitrogen fixing effectiveness has been reported for sp^- compared to sp^+ symbioses (Normand and Lalonde, 1982; Wheeler *et al.*, 1986), although this relationship is not always observed (Kurdali *et al.*, 1990). Despite the large variation in nitrogen fixing capacity of various host-*Frankia* combinations, no clear superior host genotype or *Frankia* strain have been recognized and any positive effect of inoculation with superior *Frankia* strain may soon be overshadowed in the field by indigenous *Frankia* (Hooker and Wheeler, 1987). In *Alnus-Frankia* symbiosis, it has been shown (Hall *et al.*, 1979) that an important improvement in the rate of nitrogen fixation can be gained by the selection of host plant, which is more active in photosynthesis, the key element for nitrogen fixation activity. Furthermore the susceptibility of the host plant to a *Frankia* strain is variable and consequently selectable (Tremblay *et al.*, 1984).

There is remarkable difference in the rate of symbiosis in terms of ultra-structure of *Frankia* in pure culture. The host genotype is shown to control the nodule morphology. The symbiosis in *Myrica gale* and *Comptonia* may represent an intermediate situation between *Alnus* and *Casuarina.* The terminal vesicle within the root nodule may vary in shape considerably from overtly enlarged club shaped terminal endings to only slightly enlarged elongate terminal filaments. The state of vesicle envelope within these nodules has not been determined. As in *Casuarina,* the cell walls of infected *Myrica* nodules show specialization (Vanden-Bosch and Torrey, 1985) and intermediate level of haemoglobin like compound has been reported (Tjepkema, 1983). Atmospheric levels of O_2 are available within the nodule to the infected cell (Tjepkema, 1983) and even in water logged sites, aeration of the nodules through air passages within the nodule roots allows the nodules to undergo aerobic respiration and would necessitate some special modifications of the infected cells to achieve O_2 protection of the nitrogenase within the *Frankia* filaments and vesicles. The interesting and peculiar orientation of the elongate vesicles in *Datisca* and *Coriaria* suggests yet another modification, which may have evolved to achieve the end of O_2 protection of the endophyte nitrogenase.

7. ALNUS-*FRANKIA* VERSUS LEGUME-RHIZOBIUM SYMBIOSIS

Although rhizobial symbiosis is by far the best studied of the nitrogen-fixing symbiotic associations with higher plants, many others exist. Symbiosis between woody dicotyledonous plants and the actinomycete, *Frankia* is believed to have a role in nitrogen-fixation comparable to that of rhizobial symbiosis.

Many similarities exist between rhizobial and the Frankial symbiosis. The mode of symbiosis and root nodule induction procedure of *Frankia* and *Rhizobium* are more or less similar, both lead to the formation of nodules on the roots of the host plant in which nitrogen fixation occurs. In both cases there is a complementarily with some of the *nif* gene sequences of *K. pneumoniae* and a number of associated proteins are related. However, *Frankia* doesn't invade the host cell and is not therefore surrounded by a poribacterial membrane. Instead it is encapsulated in carbohydrate material derived from the host plant. This may not be distinctive since *Andria* sp. (legume) has shown that similar encapsulation occurs. Other evidences suggest that the *Frankia* symbiosis involves hemoglobin similar to that found in legumes and that the mechanism regulating expression of nitrogenase in presence of oxygen and ammonium are very similar in both *Rhizobium* and *Frankia.* They differ markedly in development in morphology and development. Infected zone of legume nodule is central and is normally contending within an endodermis and an inner cortical layer of tightly packed cells. Actinorhizal nodules normally have a central stele that has infected tissue adjacent to it or around it (Benson and Silvester, 1993). The hydrogen metabolism of *Frankia* seems to be an interesting physiological characteristic. All nitrogen-fixing organisms evolve hydrogen from the nitrogen-fixing enzyme, nitrogenase, when nitrogen is reduced to ammonia. This means that some of the energy used in the nitrogenase reaction is wasted on nitrogen evolution. Root nodules of many legumes (*Rhizobium* symbioses) evolve considerable amounts of hydrogen with as much as 30%-60% of the electron flow through lost as hydrogen (Schubert and Evans, 1976). In contrast, *Frankia* symbioses were claimed to have little or no net hydrogen evolution (Moore, 1964; Schubert and Evans, 1976), Roelofsen and Akkermans, 1979). In some strains of *Rhizobium* and *Frankia* hydrogenase is present (Roelofsen and Akkermans, 1979; Benson *et al.*, 1980) and recycles a part of the hydrogen evolved from nitrogenase. Hereby some energy is regained and no or little net hydrogen evolution is measured.

8. PHYLOGENETIC STUDIES ON ACTINORHIZAL PLANTS AND *FRANKIA*

Extensive studies on symbiotic nitrogen fixing associations have shown that unlike leguminous plants which form symbiotic association with rhizobia, the actinorhizal plants are considered to be polyphyletic in their origin. On the other hand, *Frankia*-the microsymbionts forming symbiotic association with these actinorhizal plants have been assumed to be phylogenetically homologous group on the basis of a variety of their morphological, biochemical, phenotypical and nitrogen fixing capacity (Wall, 2000) (Fig. 1).

A comparison of sequences and catalogs of certain regions of 16S rRNAs by Stackebrandt and his associates showed that two type-B *Frankiae* could be shown to belong to one of the major sub-lines of descent to members of actinomycete of genus *Geodermatophilus*. The 16S rRNA sequences of type-A strains have yet to be determined. Sequencing of the rRNA genes of a single *Frankia* strain showed that the 5′ to 3′ order is 16S-23S-5S, which is typical of bacterial in general.

A study of the phylogeny of seed plants, based on *rbcL* gene sequences, revealed that all nitrogen fixing and nodulated plants cluster in the Rosid I lineage of the angiosperms. This result suggests that the predisposition to develop nitrogen-fixing nodules of any type arose only once during the evolution of the angiosperm (Doyle, 1998). Nodulated plants within the Rosid I clade can be grouped into four major lineages: three of them include actinorhizal plants (Soltis *et al.*, 1995). One includes the Hamamelid families, Myricaceae, Betulaceae, and Casuarinaceae, whereas a second includes the

Rosid families Elaeagnaceae, Rhamnaceae, and Rosaceae, as well as the *Bradyrhizobium*-infected *Parasponia* (Ulmaceae). Coriariaceae and Datiscaceae (Swensen and Mullin, 1997) define the third line of actinorhizal plants. The fourth line of nodulated plants includes the rhizobia-infected legumes of the Fabaceae. Morphologic and anatomic features of actinorhizal nodules correlate with a more detailed analysis of *rbcL* grouping (Swensen and Mullin, 1997). To date, four clades of actinorhizal plants have been defined. One of the above-mentioned groups of Rosid families is divided into two sub-clades: one including Elaeagnaceae and Rhamnaceae, and the second defined by the Rosaceae (Fig. 1). Fossil records and the geographical distribution of

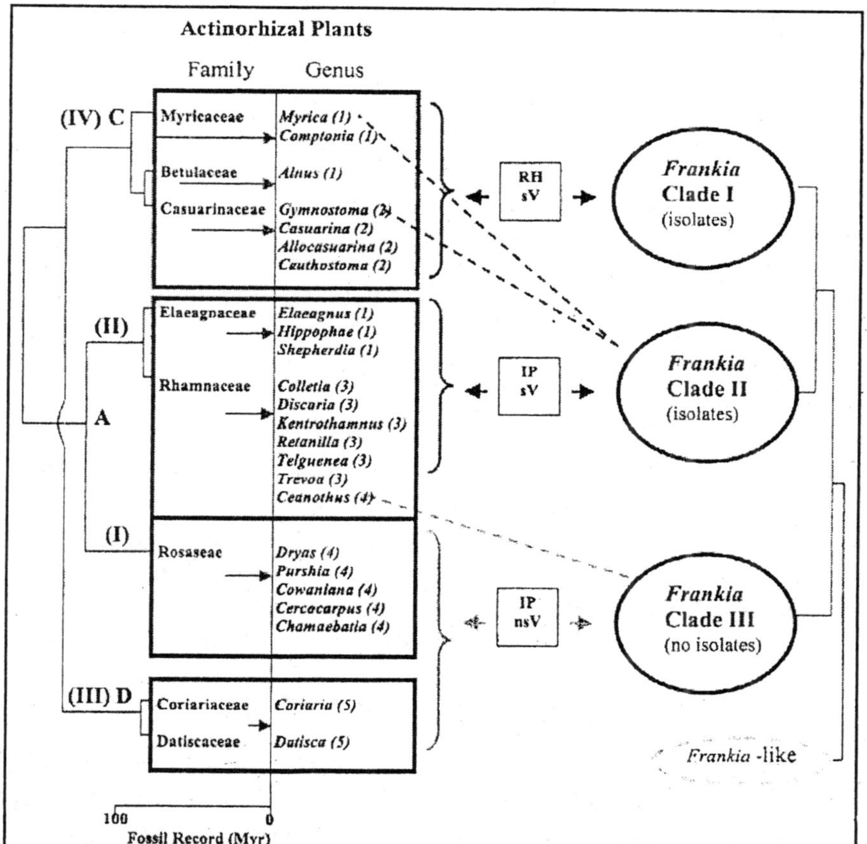

FIG. 1. Phylogenetic grouping of actinorhizal plants and *Frankia*. Number between brackets of plant genus indicates native geographical distribution *(1)* to most continents, *(2)* to Australia and western Pacific, *(3)* to South America and southern New Zealand, *(4)* western North America, *(5)* disjunct distribution in northern and southern temperate zones. *RH*, root hair infection; *IP*, intercellular penetration; *sV*, septated vesicles in nodule; *ns* V, nonseptated vesicles in nodule (after Wall , 2000).

actinorhizal species give extra support to these groupings (Benson and Clawson, 2000). Phylogenetic studies on *Frankia* have focused mainly on 16S RNA gene sequences (Benson *et al.*, 1996; Jeong *et al.*, 1999; Normand *et al.*, 1996; Ritchie and Myrold, 1999). Similar results have been obtained with *nifD* gene sequences (Normand *et al.*, 1992), and recently confirmed using *recA* and *glnII* sequences (Cournoyer and Lavire, 1999). In all these studies, there has been difficulty with isolating *Frankia* from root nodules. This problem has been partially overcome by direct amplification of total nodule DNA, using specifically designed primers for *Frankia.* Consensus phylogenetic trees generated from 16S rDNA sequences consistently yield three major groups of *Frankia,* and a fourth "*Frankia*-like" clade of Nod. / Fix actinomycetes (Fig. 1). Subgroups can be found, although these are not statistically well supported. There are many well-known isolates included in groups I and II, whereas no one isolate has been obtained from group III, which is defined only on the basis of analysis of nodule-extracted DNA. Physiologically, at least, the absence of a septum in vesicles of nodules of host plants infected with group III *Frankia* agrees with the proposed division. Although the phylogenies of the microsymbiont are not congruent with the four host clades, a close relationship exists between the plant and bacterial groups. Further analysis shows that the plant clades diverged earlier than the *Frankia* clades, suggesting that the *Frankia*-actinorhizal symbiosis evolved independently, at least three or four times, rather than co-evolving from an ancestral symbioses. Nevertheless, once the symbiosis was established, the plants or *Frankia* were retained within certain taxonomic groups, with limited lateral transfer and probable co-evolution from that point onwards (Simonet *et al.*, 1998). These analyses as a whole reinforce a model for host preferences of specificity in actinorhizal symbioses.

9. GENOMICS OF *FRANKIA*

In recent years our understanding of biological nitrogen fixation has been bolstered by a diverse array of scientific techniques (Raymond *et al.*, 2003). Genomics in its broadest sense includes research in genome structure, bioinformatics, and genome expression and engineering. Modern methodologies for DNA sequencing have brought about a scientific revolution in the understanding of the genome as a whole. Bioinformatics contends with the challenge of decoding and manipulating the vast number of genomic sequences available in gene banks. This knowledge is then used as a basis to state questions relevant to the life sciences. The knowledge of how the genome is expressed under different conditions will provide an intimate understanding of cell physiology and design whole metabolic pathways.

The complete genome sequences of three *Frankia* strains are now available in the public domain. Their sizes varies from 5.43 Mbp for a narrow host range strain (*Frankia* sp. strain CcI3) to 7.50 Mbp for a medium host range strain (*Frankia alni* strain ACN14a) to 9.04 Mbp for a broad host range strain (*Frankia* sp. strain EANIpec.) This size divergence is the largest yet reported for such closely related soil bacteria (97.8-98.9% identity of 16S rRNA genes) (Normand *et al.*, 2007). In order to get insight into the evolutionary pathways followed by these closely related, yet host range and geographically diverse *Frankia* strains a comparative study of the three genomes were carried out by an international group of scientists. The coding capacity of the three genomes is quite high, at 89% for ACNI14a, 84% for CcI3 and 86% for EANIpec. From the comparative study of the three genomes, Normand *et al.* (2007) found that the unusual size divergence displayed by the *Frankia* genomes arose by the processes of deletion, duplication and retention. These processes might have driven them in various directions reducing CcI3, expanding EANIpec and keeping ACNI14a relatively stable. They concluded that the *Frankia* strains reflected the biogeographic history of the host plants they infect and the differential genome contraction and expansion might be linked to the evolution of their hosts on a global scale.

The authors group studied the synonymous codon usage patterns of the protein coding genes with special reference to the nitrogen fixing genes in *Frankia* sp. strain CcI3 and EANIpec. It was observed that the genes are highly biased and have moderate expression levels. Mutational bias is the major source of variation in codon usage in *Frankia* and GC3 composition influences codon usage variation among most of the genes in this organism. There is very little difference in codon bias among these strains. We inferred that they co-evolved as a unit. It is probable that, due to the association of *Frankia* with actinorhizal plants in nitrogen poor soils, nitrogen fixation may have played a vital role in its early expansion (Normand *et al.*, 2007).

Although these analyses provided some significant information regarding the molecular nature of the *Frankia* genome, an extensive study of the host genomes, especially the metabolomics of *Frankia*-actinorhizal association may provide an excellent picture of the association.

10. IMPORTANCE OF *FRANKIA*-ACTINORHIZAL SYMBIOSIS

Study of the *Frankia* actinorhizal symbiosis is important for several reasons, as a pioneer, nitrogen-fixing species actinorhizal plants are suitable for soil improvement and rehabilitation of degraded lands, including the use of actinorhizal shrubs and trees for reforestation and reclamation of depauperate, nitrogen limiting soils. Seeds have been broadcast to stabilize landslides (Gordon and Wheeler, 1983). In Burma, alder has been used with success to reforest abandoned Taungya areas. In agro- forestry systems, it can be inter-planted with a number of crops. The wood of many actinorhizal plants are important source of firewood and charcoal. As more and more land, especially in the tropics, becomes deforested, the utilization of fast growing, nitrogen fixing trees for reforestation purposes becomes critical. According to the Office of the Technology Assessment (OTA) (1984), about 11.3 million hectares of what remains of tropical forests are destroyed annually. In addition many non-leguminous, nitrogen-fixing plants are colonizers of poor soils. After the Pleistocene glaciation's, plants such *as Alnus, Dryas, Elaeagnus, Hippophae,* and *Shepherdia* played an important role in soil reconstruction (Lawrence *et al.*, 1967). Currently their role in nitrogen fixation is recognized as being at least comparable to that *Rhizobium*-legume symbiosis and attempts are being made to utilize actinorhizal trees in reforestation and soil reclamation in developing countries.

In exchange for a water and nutrient supply, the bacteria enable alders to fix much needed atmospheric nitrogen, which in turn enriches the surrounding soils. This increases the fertility of the soil for many species and makes alders as important species in ecological successions, land reclamation and remediation. Thus, *Frankia* have a direct and important impact on the environment and it makes the actinorhizal plants as an important and unique group in the world.

11. CONCLUSION

Symbiotic association is a perfect example of division of labour between two organisms, the outcome of which may lead to endeavors like nitrogen fixation. Reduction of atmospheric nitrogen to ammonia and its further assimilation into amino-acids and other biomolecules enables gaseous nitrogen to incorporate life processes. As all organisms need nitrogen to survive, nitrogen fixation is probably the second most important biochemical pathway after carbon dioxide fixation. However the ability to fix nitrogen is found only in one biologic kingdom, the Prokaryota (Sprent and Sprent, 1990). Thus, other organisms have exploited the ability of prokaryotes to fix nitrogen by establishing various types of interactions (Werner, 1992). Frankia is a prominent member among the symbiotic nitrogen fixers. It is seen that *Frankia* - actinorhizal symbioses play an important role in reforestation, soil rehabilitation and reclamation. Nitrogen fixation achieved by this symbiosis increases the fertility of soil. All these have a positive impact on the environment. With the completion of three genome sequences of *Frankia*, much of the studies are now focused on the molecular organization of the genomes. Efforts have already been started in the authors' laboratory along with the laboratory of Prof. Louis Tisa of University of New Hampshire, USA to sequence more Frankia strains. Analysis of already sequenced genomes through various bioinformatics approaches and molecular modeling of major nitrogen fixing genes of Frankia are being done. This is significant because genome characteristics of *Frankia* may be linked to the evolutionary history of their hosts on a global scale.

REFERENCES

Akimov, V. N., and Dobritsa, S.Y. (1992). Grouping of *Frankia* strains on the basis of DNA relatedness. *System. Appl. Microbiol.* 15:372-379.

Akkermans, A. D. L., Roelofsen, W., Blom, J., HussDanell, K. and Harkink R. (1983). Utilization of carbon and nitrogen compounds by *Frankia* in synthetic media and in root nodules of *Alnus glutinosa, Hippophae rhamnoides* and Datisca cannabina. *Ca. J. Bot.* 61:2793-2800.

An, C. S., Riggsby, W. S. and Mullin, B. C. (1985). Relationships of *Frankia* isolates based on deoxyribonucleic acid homology studies. *Int. J. Syst. Bacteriol.* 35:140-146.

Bajwa, B.S (2004) Molecular characterization of *Frankia* and Alder-*Frankia* symbiosis in Eastern India. Ph.D Thesis, N.B.U, Siliguri.

Baker, D. (1982). A cumulative listing of isolated Frankiae, the symbiotic, nitrogen-fixing actinomycetes. *Actinomycetes,* 17:35-42.

Baker, D., Pengelly, W. and Torrey, J. G. (1981). Immunochemical analysis of relationships among isolated frankiae (*Actinomycetales*). *Int. J. Syst. Bacteriol.* 31:148-151.

Baker, D.D. and Keefe, D.O. (1984). A modified sucrose fractionation procedure for the isolation of Frankiae from actinorhizal root nodules and soil samples. *Plant Soil.* 78: 23-28.

Baker, D.D. and Schwintzer, C.R. (1990). Introduction. pp. 1-13. *In* C.R. Schwintzer and J.D. Tjepkema (ed.) The biology of *Frankia* and actinorhizal plants. Academic Press, San Diego.

Becking, J. H. (1970). *Frankiaceae* fam. novo *(Actinomycetales)* with one new combination and six new species of the genus *Frankia* Brunchorst 1886, 174. *Int. J. Syst. Bacteriol.* 20:201-220.

Beijerinck, M.W. (1888). Die bakterien der papilionaceen knolichen. Bot. Zeit. 46:725-735 (original paper not seen).

Benson D.R. and Clawson, M.L. (2000). Evolution of the actinorhizal plant symbiosis. *In*: Triplett E.W, editor. Prokaryotic nitrogen fixation: A model system for analysis of a biological process. Wymondham, UK: Horizon Scientific Press. 207-224.

Benson D.R., Arp, D.J. and Burris, R.H. (1979). Cell free nitrogenase and hydrogenase from actinorhizal root nodules. *Science,* 205:688-689.

Benson D.R., Arp, D.J. and Burris, R.H. (1980). Hydrogenase in actinorhizal root nodules and root nodule homogenates. J. *Bacteriol.* 142: 138-144.

Benson, D. R., S. E. Buchholz, and D. G. Hanna. (1984). Identification of *Frankia* strains by two-dimensional polyacrylamide gel electrophoresis. *Appl. Environ. Microbiol.* 47:489494.

Benson, D.R. and Silvester, W.B. (1993). Biology of *Frankia* strains, actinomycete symbionts of actinorhizal plants. *Microbiol. Rev.* 57:293-319.

Berry A. M. and Torrey, J. G. (1983). Root hair deformation in the infection process of *Alnus rubra. Can. J. Microbiol.* 61:2863-2976.

Berry A. M., McIntyre, L. and McCully, M. E. (1986). Fine structure of root hair infection leading to nodulation in the *Frankia-Alnus* symbiosis. *Can. J. Bot.* 64:292-305.

Berry A.M., Kahn, R.K.S and Booth, M.C. (1989). Identification of indole compounds secreted by *Frankia* HFPArI3 in defined culture medium. *Plant Soil,* 118:205-209.

Beyazova, M. and Lechevalier, M. P. (1992). Low-frequency restriction fragment analysis of *Frankia* strains (Actinomycetales*). Int. J. Syst. Bacteriol.* 42:422-433.

Blom, J. (1982). Carbon and nitrogen source requirements of *Frankia* strains. *FEMS Microbiol. Lett.* 13:51-55.

Bloom R. A., Mullin, B. C. and Tate, R. L. (1989). DNA restriction patterns and DNA-DNA solution hybridization studies of *Frankia* isolates from *Myrica pensylvanica* (bayberry). *Appl. Environ. Microbiol.* 55:2155-2160.

Bosco, M., Fernandez, M. P., Simonet, P., Materassi, R. and Normand, P. (1992). Evidence that some *Frankia* sp. strains are able to cross boundaries between Alnus and *Elaeagnus* host specificity groups. *Appl. Env.. Microbiol.* 58:1569-1576.

Brock. (2000). Microbiol ecology. In M.T. Madigan, J.M. Martinko and J. Parker (eds.), Biology of microorganisms. Upper Saddle River, New Jersey. Pp. 642-720.

Burggraaf, A.J.P. and Shipton, W.A. (1982). Estimates of *Frankia* growth under various pH and temperature regimes. *Plant Soil.* 69:135-147.

Callaham, D., DelTredici, P. and Torrey, J. G. (1978). Isolation and cultivation *in vitro* of the actinomycete causing root nodulation in *Comptonia. Science,* 199:899-902.

Callaham, D., Newcomb, W. Torrey, J. G. and Peterson, R. L. (1979). Root hair infection in actinomycete-induced root nodule initiation in *Casuarina, Myrica and Comptonia. Bot. Gaz.* (Suppl.)140: SI-S9.

Cournoyer, B. and Lavire, C. (1999). Analysis of *Frankia* evoliutionary radiation using glnII sequences. *Microbiol. Lett* 177:29-34.

Cronquist, A. (1988). The evolution and classification of Flowering plants. Allen, Lawrence and Kansas.

Doyle, J.J. (1998). Phylogenetic perspectives on nodulatiion: evolving views of plants and symbiotic bacteria. *Trends Plant Sci..* 3:473-478.

Fernandez, M. P., Meugnier, H., Grimont, P. A. D. and Bardin, R. (1989). Deoxyribonucleic acid relatedness among members of the genus *Frankia. Int. J. Syst. Bacteriol.* 39:424-429.

Fleming, E.I., Williams, E.R and Trunbull, J. W. (1988). Growth and nodulation and nitrogen fixation of provenances of *Casuarina cunninghamiana* inoculated with a range of *Frankia* sources. *Austral. J. Bot..* 36:171-181.

Ganesh, G., Misra, A.K., Chapleon, C. and Normand, P. (1994). Morphological and molecular characterization of *Frankia* sp. Isolates from nodules of *Alnus nepalensis* Don. *Arch. Microbiol.* 161:152-155.

Gardes, M. and M. Lalonde. (1987). Identification and sub-grouping of *Frankia* strains using sodium dodecyl sulfatepolyacrylamide gel electrophoresis. *Physiol. Plant,* 70:237244.

Gardes, M., Bousquet, J. and Lalonde, M. (1987). Isozyme variation among 40 *Frankia* strains. *Appl. Env.. Microbiol.* 53:1596-1603.

Gordon, J.C. and Wheeler, C.T. (1983). Biological Nitrogen Fixation in Forest Ecosystem: Foundations and Applications. Martinus Nijhoff/junk, The Hague.

Gresshoff, P.M. and Delves, A.C. (1986). Plant genetic approaches to symbiotic nodulation and nitrogen fixation in legumes. Pp. 159-206. In: A.D. Blonstein and P.J. King (eds.). Plant gene research III. A genetical approach to plant biochemistry. Springer Verlag, Wien.

Hahn, D., Lechevalier, M.P., Fischer, A. and Stackebrandt, E. (1989). Evidence for a close phylogenetic relationship- between members of the genera *Frankia, Geodermatophilus*and "*Blastococcus*" an emendation of the family *Frankiaceae. Syst. Appl. Microbiol.* 11:236-242.

Hall, R.B. and Maynard, C.A (1979). Considerations in the genetic improvement of alder. In Symbiotic Nitrogen Fixation in the Management of Temperate Forests. Eds. JC Gordon, CT Wheeler and DA Perry. Oregon State University, Corvallis. Pp. 322-344.

Harry, D.E., Yang, D.C. and Dawson, J.O. (1991). Nucleotide sequence and diversity in 16S ribosomal RNA from *Frankia. Plant Soil,* 131:143-146.

Hiltner, L. (1904). Über neuere Erfahrungen und Probleme auf dem Gebiet der Bodenbackteriologie und unter besonderer Berücksichtigung der Gründüngung und Brache. *Arbeiten Deutscher Landwirtschafts Gesellschaft* 98:59–78 (original paper not seen).

Hooker, J.E., and Wheeler, C.T. (1987). The effectivity of *Frankia* for nodulation and nitrogen fixation in *Alnus rubra* and *A. glutinosa*. *Physiol. Plant.* 70:333-341.

Huss-Danell, K. (1997). Actinorhizal symbioses and their N_2 fixation. *New Phytol.* 136:375-405.

Hutchinson, J. (1973). The families of flowering plants arranged according to a new system of their s from *Alnus incana* subsp. *rugosa*. *Can. J. Bot.* 61:28262833.

Lechevalier, M.P. (1984). The taxonomy of the genus *Frankia*. *Plant Soil* 78: 1-6.

Maggia, L., Prin, Y., Picard, B. and Goullet, P. (1990). Esterase patterns of 60 isolates of *Frankia* from Casuarina equisetifolia grown in Senegambia. In M.H.EI-Lakany, J.W. Turnbull, and J.L. Brewbaker (ed.), Advances in *Casuarina* research. Desert Development Center, American University, Cairo. pp. 141-148.

Meesters, T. M., vanGenesan, S. T., and Akkermans A.D.L.(1985). Growth, acetylene reduction activiy and localization of nitrogenase in relation to vesicle formation in *Frankia* strains Cc1.1 and Cp1.2. *Arch. Microbiol.* 143:137-142.

Miller, I. M. and Baker, D. D. (1986). Nodulation of actinorhizal plants by *Frankia* strains capable of both root hair infection and intercellular penetration. *Protoplasma,* 131:82-91.

Mirza, M.S., Hahn, D. and Akkermans, A.D.L. (1992). Isolation and characterization of *Frankia* strains from *Coriaria nepalensis*. *Syst. Appl. Microbiol.* 15:289-295.

Mirza, M.S., Janse, J.D., Hahn, D. and Akkermans, A.D.L (1991). Identification of atypical *Frankia* strains by fatty acid analysis. *FEMS. Microbiol. Lett.* 83:91-98.

Moore, A. W. (1964). Note on non-leguminous nitrogen-fixing plants in Alberta. *Can. J. Bot.* 42:952-955.

Mort, A., Normand, P. and Lalonde, M. (1983). 2-O-methyl-D-mannose, a key sugar in the taxonomy of *Frankia*. *Can. J. Microbiol.* 29:993-1002.

Nap, J.P. and T. Besseling. (1989). Nodulin function and nodulin gene regulation in root nodule development. In: P.M. Gresshoff (ed.). The molecular biology of symbiotic nitrogen fixation, CRC Press, Boca Raton, Pp. 181-230.

Nazaret, S., Cournoyer, B., Normand, P. and Simonet, P. (1991). Phylogenetic relationships among *Frankw* genomic species, determined by use of amplified 16S rDNA sequences. *J. Bacteriol.* 173:4072-4078.

Nazaret, S., P. Simonet, P. Normand, and R. Bardin. (1989). Genetic diversity among *Frankia* isolated from *Casuarina* nodules. *Plant Soil,* 118:241-247.

Newcomb, W. and Wood, S. M. (1987). Morphogenesis and fine structure of *Frankia (Actinomycetales):* the micro symbiont of nitrogen-fixing actinorhizal root nodules. *Int. Rev. Cytol.* 109:1-88.

Normand, P. and Lalonde, M. (1982). Evaluation of *Frankia* strains isolated from rovenances of *twoAlnus* species. *Can. J. Microbiol.* 28:1133-1142.

Normand, P., Cournoyer, B., Simonet, P. and Nazaret, S. (1992). Analysis of a ribosomal RNA operon in the actinomycete *Frankia. Gene,* 111:119-124.

Normand, P., Lapierre, P., Tisa, L. S., Gogarten, J. P., Alloisio, N., Bagnarol, E., Bassi, C. A., Berry, A.M., Bickhart, D. M., Choisne, N., Couloux, A., Cournoyer, B., Cruveiller, S., Daubin, V., Demange, N.,. Francino, M. P., Goltsman, E., Huang, Y., Kopp, O. R , Labarre, L. , Lapidus, A Lavire, C., Marechal, J., Martinez, M., Mastronunzio, J. E., Mullin, B.C., Niemann, J., Tania, P.P., Zoe Rouy, R., Schenowitz, C.,. Sellstedt, A., Tavares, F., Tomkins, J. P., Vallenet, D., Valverde, C., Wall, L. G., Wang, Y., Medigue, S.C. and Benson, D. R. (2007). Genome characteristics of facultative symbiotic *Frankia* sp. strains reflect host range and host plant biogeography. *GenomeResearch* 7:7-15.

Normand, P., Orso, S., Cournoyer, B ,. Jeannin, P., Chapelon, C., Dawson, J., Evtushenko, L. and Misra, A.K. (1996). Molecular phylogeny of the genus *Frankia* and related genera and emendation of the family Frankiaceae. *Int. J. Syst. Bacteriol.* 46: 1-9.

Paul, E.A. and Clark, F.E. (1996). Soil microbiology and biochemistry. E.A. Paul and F.E. Clark (eds.). Academic Press, San Diego, California. pp. 340.

Pommer, E.H. (1959). Uber die Isolierung des Endophyten ausden Wurzelknollehen *Alnus* glutinosa Gaertn. Und uber erfolgreiche reinfections versuche. *Ber. Dtsch. Bot. Ges.* 72:138-150.

Prin, Y., Kodja, H., Duhoux, E., Diem, H.G., Roederer, Y., Dommergues, Y.R. (1992). Aerial nodulation in Casuarina spp: field survey and preliminary .experimental data. *Acta. Ecol.* 13: 479-486.

Quispel, A. (1990). Discoveries, discussions, and trends in research on actinorhizal root nodule symbioses before 1978, Pp. 15-33. *In* C. R. Schwintzer and J. D. Tjepkema, (ed.), The biology of *Frankia* and actinorhizal plants. Academic Press, Inc., New York.

Racette, S. and Torrey, J. G. (1989). The isolation, culture and infectivity of a *Frankia* strain from *Gyffinostoma papuanum* (Casuarinaceae). *Plant Soil* 118:165-170.

Raymond, J., Siefort, J. L., Staples, C.R. and Blankenship, R. E. (2004). The Natural History of nitrogen fixation *Mol. Biol. Evol* 21:541-554

Reddell, P. and Bowen, G.D. (1985). *Frankia* source affects growth, nodulation and nitrogen .fixation in *Casuarina* species. *New Phytol.* 100:115-122.

Reddell, P., Rosbrook, P.A., Bowen, G.D and Gwaze, D. (1988). Growth responses in *Casuarina cunninghamiana* plantings to inoculation with *Frankia. Plant Soil,* 108:79-86.

Ritchie, N.J. and Myrold, D.D (1999). Phylogenetic placement of uncultured *Ceanothus* microsymbionts using 16S rRNA gene sequences. *Can. J. Bot.* 77:1208-1213.

Roelofsen, W. and. Akkermans, A. D. L (1979). Uptake and evolution of H_2 and reduction of C_2H_2 by root nodules and homogenates of *Alnus glutinosa. Plant Soil*, 52:571-578.

Rosbrook, P.A. and Bowen, G.D. (1987). The ability of three *Frankia* isolates to nodulate and fix nitrogen with four species of *Casuarina. Physiol. Plant.* 70:373-377.

Sanginga, N., Bowen, G.D. and Danso, S.K.A. (1990). Genetic variability in symbiotic nitrogen fixation within and between provenances of two *Casuarina* species using the 15N labeling methods. *Soil Biol. Biochem.* 22:539-547.

Schubert, K. R. and Evans, H. J. (1976). Hydrogen evolution: a major factor affecting the efficiency of nitrogen fixation in nodulated symbionts. *Proc. Natl. Acad. Sci.* USA 73:12071211.

Seguin, A. and Lalonde, M. (1989). Detection of pectolytic activity and *pel* homologous sequences in *Frankia. Plant Soil,* 118:221-229.

Sellstedt, A. and Huss-Danell, K. (1984). Growth, nitrogen fixation and relative efficiency ofnitrogenase in *Alnus incana* grown in different cultivation systems. *Plant Soil* 78:147-158.

Sen, A. (1996). Electron microscopy and molecular biology of Frankia. PhD Thesis, NEHU, Shillong.

Shipton, W.A. and Burggraaf, A.J.P. (1982). A comparison of the requirements of various carbon and nitrogen sources and vitamins in some *Frankia* isolates. *Plant Soil,* 69:149-161.

Simonet, P. Navarro, E., Rouvier, C., Reddell, P., Zimpfer, J., Dawson, J., Dommergues, Y., Bardin, R., Combarro, P., Domenach, A.M., Hamelin, J., Gourbiere, F., Prin, Y. and Normand, P. (1998). 11th International and actinorhizal plant conference, University of Illinois. Book of abstracts.

Simonet, P., Grosjean, M.C., Misra, A.K., Nazaret, S., Cournoyer, B. and Normand, P. (1991). *Frankia* genus-specific characterization by polymerase chain reaction. *Appl. Env. Microbiol.* 57:3278-3286.

Simonet, P., Thi Le, N., Moiroud, A. and Birdin, R. (1989) Diversity of *Frankia* strains isolated from single Alder stand. *Plant and Soil,* 188:13-22.

Soltis, D.E., Soltis, P.S., Morgan, D.R, Swensen, S.M., Mullin, B.C., Dowd, J.M. and Martin, P.G. (1995). Chloroplast gene sequence data suggest a single origin of the predisposition for symbiotic nitrogen fixation in angiosperms. *Proc Natl Acid Sci,* USA, 92:2647-2651.

Sougoufara, B., Maggia, L., Duhoux, E. and Domrnergues, Y.R. (1992). Nodulation and nitrogen fixation in nine *Casuarina* clone-*Frankia* strain combinations. *Acta Ecol.*13:497-503.

Sprent, J.I and Sprent, P. (1990). Nitrogen fixing microorganisms. London, Newyork, Tokyo, Melbourne, Madras: Chapman and Hall.

Strand, R. and Laetsch, W. (1977). Cell and endophyte structure of the nitrogen-fixing root nodules of *Ceanothus integerrimus* H. and A. II. Progress of the endophyte into young cells of the growing nodule. *Protoplasma* 93:179-190.

Swensen, S.M. (1996). The evolution of actinorhizal symbioses: Evidence for multiple origins of the symbiotic association. *Amer J Bot* 83:1503-1512.

Swensen, S.M. and Mullin, B.C. (1997). Phylogenetic relationships among actinorhizal plants. The impact of molecular systematics and implications for the evolution of actinorhizal symbioses. *Physiol Plant,* 99:565-573.

Tjepkema, J. D. (1983). Hemoglobins in the nitrogen fixing root nodules of actinorhizal plants. *Can. J. Bot.* 61:2924-2929.

Torrey J. G. (1976). Initiation and development of root nodules of *Casuarina* (Casuarinaceae). Am. J. Bot. 63:335-344.

Torrey, J. G. and Tjepkema, J. D. (1979). Symbiotic nitrogen fixation in actinomycete-nodulateCl plants. *Bot. Gaz.* 140(Suppl.):i-ii.

Trembly, F.M., Nesme, X and. Lalonde, M. (1984). Selection and micropropagation of nodulating and non-nodulating clones of Alnus crispa (Ait.) Pursh. *Plant and soil.* 78:171-179.

Vanden-Bosch, K.A. and Torrey, J.G. (1985). Development of endophytic *Frankia* sporangia in field and laboratory gown nodules of *Comptonia peregrina* and *Myrica gale. Am. J. Bot.* 72:99-108.

Wall, L.G. (2000). The Actinorhizal Symbiosis. *J. Plant Growth Regul.,* 19:167-182.

Wayne, L.G., Brenner, D. J., Colwell, R. R., Grimont, P. A. D., Kandler, O., Krichevsky, M. I., Moore, L. H., Moore, W. E. C., Murray, R. G. E., Stackebrandt, E., Starr, M. P. and Triiper, H. G. (1987). Report of the ad hoc committee on reconciliation of approaches to bacterial systematics. *Int. J. Syst. Bacteriol.* 37:463-464.

Weber, A., Sarsa, M.L. and Sundman, V. (1989). *Frankia* Alnus incana symbiosis: Effect of endophyte on nitrogen fixation and biomass production. *Plant Soil.* 120:291-297.

Weber, A.., Smolander, A., Nurmiaho-Lassila, E.L. and Sundman, V. (1988).Isolation and characterization of *Frankia* strains from *Alnus incana* and *Alnus glutinosa* in Finland. *Symbiosis,* 6:97-116.

Werner, D. (1992). Symbiosis of plants and microbes. London, UK: Chapman and Hall.

Wheeler, C.T., Hooker, J.E., Crowe, A. and Berrie, A.M.M (1986). The improvement and utilization in forestry of nitrogen fixation by actinorhizal plants with special reference to *Alnus* in Scotland. *Plant Soil,* 90:393-406.

Young, J.P.W (1992). Phylogenetic classification of nitrogen fixing organisms. *In* Biological Nitrogen Fixation. Eds. GS Stacey, RH Burris and HJ Evans. Chapman and Hall, New York. pp. 43-85.

□□□

Index

P

Q

R

S

T

□□□